T0074530

Security of Internet of Things Nodes

Chapman & Hall/CRC Internet of Things: Data-Centric Intelligent Computing, Informatics, and Communication

The role of adaptation, machine learning, computational Intelligence, and data analytics in the field of IoT Systems is becoming increasingly essential and intertwined. The capability of an intelligent system is growing depending upon various self-decision-making algorithms in IoT Devices. IoT based smart systems generate a large amount of data that cannot be processed by traditional data processing algorithms and applications. Hence, this book series involves different computational methods incorporated within the system with the help of Analytics Reasoning, learning methods, Artificial intelligence, and Sense-making in Big Data, which is most concerned in IoT-enabled environment.

This series focuses to attract researchers and practitioners who are working in Information Technology and Computer Science in the field of intelligent computing paradigm, Big Data, machine learning, Sensor data, Internet of Things, and data sciences. The main aim of the series is to make available a range of books on all aspects of learning, analytics and advanced intelligent systems and related technologies. This series will cover the theory, research, development, and applications of learning, computational analytics, data processing, machine learning algorithms, as embedded in the fields of engineering, computer science, and Information Technology.

Series Editors
Dac-Nhuong Le, Souvik Pal

Security of Internet of Things Nodes: Challenges, Attacks, and Countermeasures
Chinmay   Chakraborty,   Sree   Ranjani   Rajendran   and   Muhammad   Habibur   Rehman

Security of Internet of Things Nodes

Challenges, Attacks, and Countermeasures

Edited by
Chinmay Chakraborty, Sree Ranjani Rajendran,
and Muhammad Habibur Rehman

CRC Press
Taylor & Francis Group
Boca Raton London New York

CRC Press is an imprint of the
Taylor & Francis Group, an **informa** business

A CHAPMAN & HALL BOOK

First edition published 2021
by CRC Press
6000 Broken Sound Parkway NW, Suite 300, Boca Raton, FL 33487-2742

and by CRC Press
2 Park Square, Milton Park, Abingdon, Oxon, OX14 4RN

Library of Congress Cataloging-in-Publication Data
Names: Chakraborty, Chinmay, 1984- editor. | Rajendran, Sree Ranjani, editor. | Rehman, Muhammad Habibur, editor.
Title: Security of internet of things nodes : challenges, attacks, and countermeasures / edited by Dr. Chinmay Chakraborty, Dr. Sree Ranjani Rajendran ' Dr. Muhammad Habibur Rehman.
Description: Boca Raton, FL : CRC Press, 2021. | Series: Chapman ' Hall CRC internet of things | Includes bibliographical references and index. |
Summary: "The book Security of Internet of Things Nodes: Challenges, Attacks, and Countermeasures covers a wide range of research topics on the security of the Internet of Things nodes along with the latest research development in the domain of Internet of Things. It also covers various algorithms, techniques, and schemes in the field of computer science with state-of-the-art tools and technologies. This book mainly focuses on the security challenges of the Internet of Things devices and the countermeasures to overcome security vulnerabilities. Also, it highlights trust management issues on the Internet of Things nodes to build secured Internet of Things systems. The book also covers the necessity of a system model for the Internet of Things devices to ensure security at the hardware level"-- Provided by publisher.
Identifiers: LCCN 2021010097 (print) | LCCN 2021010098 (ebook) | ISBN 9780367650490 (hbk) | ISBN 9780367650513 (pbk) | ISBN 9781003127598 (ebk)
Subjects: LCSH: Internet of things--Security measures.
Classification: LCC TK5105.8857 .S45 2021 (print) | LCC TK5105.8857 (ebook) | DDC 005.8--dc23
LC record available at https://lccn.loc.gov/2021010097
LC ebook record available at https://lccn.loc.gov/2021010098

ISBN: 978-0-367-65049-0 (hbk)
ISBN: 978-0-367-65051-3 (pbk)
ISBN: 978-1-003-12759-8 (ebk)

Typeset in Palatino
by MPS Limited, Dehradun

Contents

Preface

This book covers a wide range of research topics on the security of Internet-of-Things (IoT) nodes. It also covers challenges and countermeasures to mitigate relevant security issues in multiple perspectives. The wide-range coverage of security issues differentiates this book from other relevant publications. The book aims systematically to collect and present quality research and give a wide benefit to a huge community of researchers, educators, practitioners, and industries. IoT is the interconnection of a large number of resource-constrained devices such as sensors, actuators, and nodes that generate large volumes of data. This is then processed into useful actions in areas such as home and building automation, intelligent transportation and connected vehicles, industrial automation, smart healthcare, smart cities, and others. Connected devices are data collectors and data processors. Personal information collected and stored by these devices, such as name, age, health data, location, and more, can help criminals committ identity theft. At the same time, IoT is a growing trend, with a stream of new products hitting the market. Here's the problem: when one is connected to everything, there are more ways to steal information. That makes users attractive targets for people who want to profit from stolen data. Important challenges remain to fulfil the IoT vision, including data provenance and integrity, trust management, identity management, and privacy. This book aims to describe how software, embedded and hardware security approaches address these security challenges. The devices connected to IoT should be secured from vulnerabilities, like software threats and hardware threats, which can cause the loss of several billions of dollars to semiconductor industries. This book focuses on the security challenges of IoT devices and countermeasures to overcome those vulnerabilities. Also, it highlights the issue of managing trust on IoT nodes to build secured IoT systems. IoT devices should authenticate and identify the correct users; otherwise, unauthorised users may attack the devices. Thus there is a necessity of a system model for the IoT devices to ensure security at the hardware level.

The book content is structured into 12 chapters. Chapter 1 discusses robust hologram-based obfuscation technique to enable hardware-level security in IoT nodes. In this approach, two Digital Signal Processing (DSP) designs are merged, such that the functionality of one design is camouflaged in another. This camouflaging is analogous to a security-image hologram, which is the reason this obfuscation technique has been called "hologram-based obfuscation". The hologram-based obfuscation methodology accepts the scheduled CDFG of two DSP kernel applications as primary inputs and generates an obfuscated common data-path of both DSP kernels. Hologram-based obfuscation affects a larger number of gates compared to other structural obfuscation techniques. Moreover, it achieves a higher area efficiency. However, hologram-based obfuscation is only applicable if two applications have partial similarity in their structure.

Chapter 2 focuses on a random number generator that could be useful everywhere, especially in the IoT environment. The authors design a simple but effective CMOS-Memristor-based random number generator, which shows good statistical results in simulations and has low hardware and energy requirements. The proposed architecture harvested the randomness of memristor along with the traditional randomness caused due to manufacturing process variations. In their design, the authors use two identical ring oscillators and compare the delay between them to generate random bits. Such design is more useful where space and energy both are primary concerns. The primary advantage of their design is that they may produce multiple bits in a single cycle, which significantly enhances the throughput of the random number generator. This can be done easily by tapping numerous nodes of both ring oscillators instead of tapping the nodes at last, as in the case of a single-bit ring oscillator. The authors also modified their architecture and simulated the same design after the removal of the memristor components from the architecture to evidence the difference in statistical randomness in both designs.

Chapter 3 discusses cryptographic algorithms used for security purposes in IoT devices. Various side-channel attacks on AES cryptographic ICs are reported from the literature. The chapter also discusses existing countermeasures for securing the scan chain which is typically inserted in the AES cryptographic ICs. It presents design and simulation results of the scan-inserted AES crypto-module. Also, the chapter discusses proposed methods used to enhance the security of the scan chain of the AES crypto-module. It then presents the results and analysis of the implemented crypto-module. The chapter concludes with highlights of the application of security towards the testing architecture of crypto-chips.

Chapter 4 discusses biometric-based secure authentication approaches for IoT-enabled devices and applications. The chapter considers the IoT system from the perspectives of a consumer, vendor, and researcher to figure out the present scenario and give future direction to the authentication-related security issues in IoT subsystems.

Chapter 5 presents an improved verification scheme based on user biometrics. The purpose of this study is to introduce a novel and well-structured threat-modeling approach which is specifically tailored for IoT devices.

Chapter 6 presents countermeasures for hardware security vulnerabilities in IoT devices. The authors discuss the origin of hardware security and highlight the types of security attacks on IoT devices. Also, the authors elaborate on the consequences and challenges of security attacks on IoT nodes. Moreover, the authors present the discussion of various artificial intelligence and machine-learning techniques to countermeasure hardware attacks in IoT devices. Finally, the chapter deals with the implementation of hardware obfuscation for DSP through suitable signal- processing transformations like folding, parallel processing, pipelining, and retiming, to mitigate vulnerability in the computing nodes of IoT systems.

Chapter 7 presents lightweight security solutions for IoT using physical-layer-key-generation methods. In this chapter, a secure key- generation scheme from physical layer characteristics is introduced as a possible lightweight security alternative to traditional upper-layer security approaches. The proposed scheme is based on the wiretap channel model; security is achieved by generating keys at both the communicating ends independently based on inherent common channel characteristics like randomness. It is based on the principle of spatial decorrelation and channel reciprocity for identical carrier frequency.

Chapter 8 presents threats and attack models in IoT devices. The authors discuss the need for security in IoT devices and present a detailed discussion on IoT architecture. Further, a taxonomy of security threats and attack models is presented for different layers, such as physical layer, data link layer, network layer, transport layer, application layer, and multilayer attacks. Also, various malware attacks and their impact on security objectives is presented.

Chapter 9 presents a review of hardware attacks and security challenges in IoT edge nodes. The authors discuss hardware attacks, protection algorithms, secure hardware levels, attacks during the manufacturing process, attacks during design, and sophisticated attacks. Counterfeiting and debug security in IoT edge nodes are mostly interlinked with the perception layer in IoT. Challenges in hardware-based IoT designs need a new security architecture, especially in the edge nodes in IoT. The next-generation system-on-chip devices' security features have the solution for intrinsic hardware security. The open problems such as trade-off between security and power are also discussed broadly. These problems affect the design of secured devices in IoT hardware.

Chapter 10 presents a study of hardware attacks on the smart system design lab. This chapter focuses on the simulation of the smart system design lab using an IoT builder, and studies recent attacks related to the physical structure with a future direction towards countermeasures.

Chapter 11 presents a novel threat-modeling and analysis approach for IoT applications. The proposed approach consists of a seven-step threat- modeling process. This methodology also addresses the aspects of IoT devices which directly affect the user, which are Privacy threats, Safety threats, and Malfunction threats (PSM). This method increases the performance and effectiveness of threat-modeling, which leads to increased mitigation of the identified threats. The final part of the study presents practical mitigation techniques to eliminate threats in IoT devices.

Chapter 12 discusses security-related trust management issues in IoT devices. First, the authors discuss trust management issues to understand the properties of trust and the goals of IoT trust management. The goal of this chapter is also to review various confidence models in the wireless sensor networks (WSN) and other network domains, along with the impact and need for IoT confidence management, trust computing, and IoT management issues.

We are sincerely thankful to all the contributors including editors, authors, reviewers, and the CRC staff for supporting and actively contributing to this project. We are also very grateful to the series editors for their feedback on this book.

About the Editors

Dr. Chinmay Chakraborty is an Assistant Professor (Sr.) in the Department of Electronics and Communication Engineering, BIT Mesra, India. His primary areas of research include Wireless body area network, Internet of Medical Things, point-of-care diagnosis, Wireless Networks, Telemedicine, m-Health/e-health, and Medical imaging. Dr. Chakraborty is co-editing eight books on Smart IoMT, Healthcare Technology, Sustainable Smart Cities, and Sensor Data Analytics with CRC Press, IET, Pan Stanford, and Springer.

Dr. Sree Ranjani Rajendran is a passionate and curious researcher, in pursuit of knowledge and expertise in the broader domain of hardware security, with specific interest in "Design for a secured hardware". She has more than ten years of professional experience, comprising research and university teaching. She is presently a post-doctoral researcher at the Indian Institute of Technology Madras, engaged in the "Detection of Hardware Trojans in AES module using a Formal Verification method".

Dr. Muhammad Habibur Rehman is currently working as a research scientist at Khalifa University of Science and Technology in UAE. Overall, he has authored or co-authored 42 international publications including journal articles, conference proceedings, book chapters, and peer-reviewed magazine articles. His research interests include blockchain technologies, artificial intelligence, cyber-physical systems, differential privacy, big data, edge computing, and industrial IoT.

1

Securing Dedicated DSP Co-processors (Hardware IP) using Structural Obfuscation for IoT-oriented Platforms

Anirban Sengupta[1] and Dipanjan Roy[2]

Indian Institute of Technology Indore
Institute for Development and Research in Banking
Technology, Hyderabad

1.1 Introduction

IoT is omnipresent in today's world. IoT applications find space in smart homes, smart cities, smart healthcare systems, wearable electronics, smart grids, connected cars, and in many other applications. The Internet of Things can exist with conventional micro-controllers and System on Chips (SoCs), but issues such as low power requirements and wireless support have accelerated the development of platforms designed for IoT applications. Thus, DSP coprocessors play a pivotal role for developing SoC-based IoT applications (Schneiderman, 2010). However, considering globalization in the design process, hardware security support for such devices is a must. Many industries use specialised coprocessors for IoT platforms such as 32-bit DSP for sensor fusion support, a 128-neuron pattern-matching accelerator, and other peripheral modules such as Bluetooth Low Energy (BLE) support and a six-axis accelerometer and gyroscope. In an SoC, these DSP coprocessors are mostly used as a reusable intellectual property (IP) core to cope with the time to market pressure (Castillo, Meyer-Base, Garcia, Parrilla, & Lloris, 2007). For all-inclusive hardware security, the protection of these DSP coprocessors is pivotal (Sengupta & Roy, 2019). Due to the globalization of the supply chain, the design process of a SoC is distributed among multiple countries in the world. Multiple third parties with their specific expertise contribute in various phases. Therefore, the possibility of the presence of an attacker is very high. These DSP IP cores are the primary targets for these kinds of attackers. The major attacks which target these DSP IP cores are Reverse Engineering (RE) (Torrance & James, 2009), IP counterfeiting and Trojan insertion. Though RE is permitted for analysis, evaluation and education purposes, it is prohibited to use that knowledge for any illegal purpose. In the RE attack, the objective of the attackers is to gain complete information about the design. They try to identify the device technology, extract the gate-level netlist, and infer the IP functionality. This knowledge helps them to identify a suitable place to insert a malicious logic, known as hardware

Trojan. Additionally, a successful RE attack also helps the attackers to perform IP counterfeiting. They can clone or copy then resell the design to other vendors. Therefore, the RE attack has a deep-rooted impact on the security of IoT devices. Researchers have incorporated multiple solutions using different security algorithms during the design process of these reusable DSP IP cores. However, such processes are segregated into multiple design abstraction levels: architectural, register-transfer, gate, physical, etc. Integrating the security algorithms in the early design phases has multiple benefits. Moreover, as most of the DSP applications have a complex function associated with them, starting the design process with the topmost design abstraction level (i.e. architectural level), is compulsory. The architectural level has the advantages of securing the design in the lower design phases and having the flexibility to perform a trade-off between multiple design parameters such as silicon area, execution latency and power consumption. One of the popular solutions against the RE attack is obscuring the design or making it unidentifiable to the attackers. This process is known as hardware obfuscation (Zhang, 2016). Hardware obfuscation prevents attacks by enhancing RE complexity. The hardware obfuscation process, which converts the DSP IP design into an unobvious form without affecting the functionality of the DSP application, is known as structural obfuscation. Compared to watermarking (Newbould, Carothers, & Rodriguez, 2002; Ni & Gao, 2005; Colombier & Bossuet, 2015; Sengupta & Roy, 2017; Le Gal & Bossuet, 2012; Ziener & Teich, 2008; Hong & Potkonjak, 1999; Koushanfar, Hong, & Potkonjak, 2005), fingerprinting (Roy & Sengupta, 2017), and forensic engineering-based detective approach, structural obfuscation (Li & Zhou, 2013; Sengupta, Roy, Mohanty, & Corcoran, 2017; Sengupta, Roy, Mohanty, & Corcoran, 2018; Lao & Parhi, 2015) is more robust as it is a preventive hardware security approach. It prevents the attackers from identifying the functionality of a DSP application from its structure. Thus, it hinders the RE process. Structural obfuscation also indirectly protects the design from hardware Trojan insertion, IP cloning, IP piracy, and other methods. Normally, the RE process requires lots of tools and techniques. It is also quite titime-consuming. Therefore, converting a standard DSP structure into an unobvious form makes it too difficult to launch an attack. The robustness of the secured design after incorporating structural obfuscation is measured using a metric known as Strength of Obfuscation (SoO) or Power of Obfuscation (PoO) (Sengupta et al., 2017). This metric measures the amount of dissimilarity between the non-obfuscation design (also known as baseline design) and the obfuscated design. Because it incorporates the structural obfuscation algorithm, it can also be measured by the number of gates affected (Sengupta & Rathor, 2020; Sengupta & Rathor, 2019). A higher percentage of dissimilarity or affected gate count indicates higher robustness. The typical attack and protection scenario for DSP coprocessors used in IoT devices is shown in Figure 1.1. This chapter discusses four state-of-the-art structural obfuscation approaches proposed by various researchers. All these approaches target DSP application for providing protection. The analysis of their corresponding results is shown in the subsequent section.

1.2 Discussion on Contemporary Structural Obfuscation Approaches used for Securing DSP Hardware/Coprocessor

Workflow of the methodology is shown in Figure 1.2.

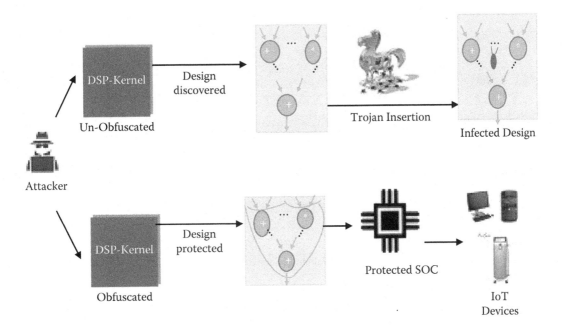

FIGURE 1.1
Typical attack and protection scenario for IoT devices.

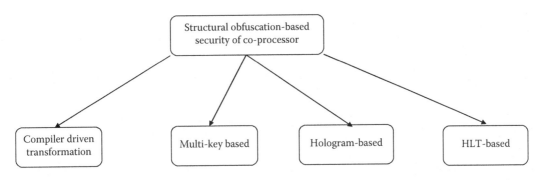

FIGURE 1.2
Workflow of structural obfuscation-based security methodology for co-processor design.

1.2.1 *Securing DSP Designs Using Compiler Driven Transformation Based Structural Obfuscation*

The methodology proposed by Sengupta et al. (2017) secures a DSP core through multiple compiler-driven transformation- based structural obfuscation techniques. It is capable of providing protection for complex loop-based DSP applications. To increase the reverse engineering complexity, five compiler driven transformation techniques were employed in succession. They are Redundant Operation Elimination (ROE), Logic Transformation (LT), Tree Height Transformation (THT), Loop Invariant Code Motion (LICM) and Loop Unrolling (LU). Each of these transformation techniques accepts a loop-based DSP application in the form of a Control Data Flow Graph (CDFG). Thereafter, it generates a structurally obscured yet functionally equivalent Register Transfer Level (RTL) design of

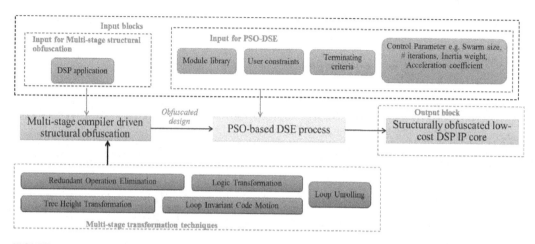

FIGURE 1.3
Overview of low-cost compiler driven structural obfuscation approach.

that application. A Design Space Exploration (DSE) framework in high level synthesis (HLS) is used on top of the transformed design to minimise the design overhead (Krishnan & Katkoori, 2006). It helps to meet the parametric constraints by generating a low-cost obfuscated DSP design. The overview is shown in Figure 1.3.

Redundant Operation Elimination: ROE is one of the compiler-driven transformation techniques used for generating a structurally obscured but functionally equivalent RTL DSP design. This technique removes redundant operations from the application. As mentioned earlier, the inputs are accepted as a CDFG; therefore, each operation of the application is represented as a node. Each node accepts input data from its parent nodes and after processing passes the output to its child node(s). A redundant operation can be identified if two nodes have the same parents as well as the same operator to process the data. A search algorithm based on the aforementioned rules runs on all the operations of the application based on their node number in ascending order to identify the list of redundant operations. Whenever a pair of redundant nodes is identified, the node with the higher node number is removed from the CDFG. Thereafter, the child of the removed node is adjusted as the child of the equivalent node. Thus, the correctness of the design is preserved. To verify the same, functional equivalence testing is performed before and after applying ROE. ROE can decrease the total node count of the application. It can also change the dependencies between nodes (Sengupta et al., 2017).

Logic Transformation: LT is another compiler-driven transformation technique used for generating a structurally obscured but functionally equivalent RTL DSP design. This technique breaks the application into multiple sub-functions and then generates logical equivalence of each sub-function. It then replaces each sub-function with the currently generated sub-function. In the CDFG representation of an application, LT may alter one node with multiple nodes or vice-versa with different input values or operators, or both. To identify the list of sub-functions where LT can be applied, a search algorithm is run on all the possible sub-graphs that can be generated from the CDFG. Each sub-graph is then transformed and replaced with its logical equivalence graph. The parents (primary inputs), child and the dependencies of each sub-graph are adjusted accordingly to preserve the correctness of the design. To verify the same, functional equivalence testing is performed before and after applying LT. Here, LT can increase or decrease the total node count of the application. It can also change the operator type, the number of primary inputs, and the dependencies between nodes (Sengupta et al., 2017). All these significantly contribute in structurally obfuscating (obscuring) when its equivalent RTL datapath is generated using DSE and HLS.

Tree Height Transformation: THT is another compiler-driven transformation technique used for generating a structurally obscured but functionally equivalent RTL DSP design. This technique converts the consecutive computations of serial operations into multiple parallel sub-computations. In other words, it introduces parallelism in subcomputations by modifying a long chain-like serial computation. In the CDFG representation of an application, THT replaces the serial dependencies with larger tree height into several parallel dependencies with lesser height. Therefore, THT not only decreases the height of the tree but increases the width of the tree (i.e., number of possible parallel executions). It may also reduce the length of the critical path of the CDFG. To identify the list of serial dependencies where THT can be applied, a search algorithm is run on all the paths starting from primary inputs to the final outputs. It identifies each associative operation that is executed serially; if the path length is three or more, THT can be applied to it. The parents, child and the dependencies of each sub-computation are adjusted accordingly to preserve the correctness of the design. To verify the same, functional equivalence testing is performed before and after applying THT. THT can increase or decrease the total height of the graph, the total number of possible parallel executions and can change the dependencies between nodes (Sengupta et al., 2017). All these significantly contribute in structurally obfuscating (obscuring) when its equivalent RTL datapath is generated using DSE and HLS.

Loop Unrolling: LU is another compiler driven transformation technique used for generating a structurally obscured but functionally equivalent RTL DSP design. It is only applicable for loop-based applications. This technique converts the loops into their unwound form. In the CDFG representation of an application, LU unrolls the graph into multiple copies and executes the same calculation several times in parallel. The number of copies is decided by what is known as the unrolling factor. It may increase the area consumption of the design but decreases the execution latency, as multiple iterations can be executed in parallel. LU is applicable to all the loop-based applications. The primary inputs and outputs of each unroll are adjusted accordingly to preserve the correctness of the design. To verify the same, functional equivalence testing is performed before and after applying LU. LU can increase the number of nodes of the application (Sengupta et al., 2017). The amount of obfuscation, also known as 'Strength of Obfuscation (SoO)' depends on the unrolling factor besides the other transformations explained earlier.

Loop Invariant Code Motion: LICM is another compiler driven transformation technique used for generating a structurally obfuscated but functionally equivalent RTL DSP design. It is also only applicable to loop-based applications. This technique splits up the loop's independent operations from the loop body. In the CDFG representation of an application, LICM moves out those nodes from the loop body; this doesn't make any difference if it executes iteratively inside the loop or for a single time outside the loop. Therefore, it decreases the execution latency and alters the structure of the graph, without affecting the area consumption. To identify the list of a loop's independent operations, a search algorithm is run on all the nodes of the application based on their node number in ascending order. The parents, child and the dependencies of the loop's independent nodes are adjusted accordingly to preserve the correctness of the design. To verify the same, functional equivalence testing is performed before and after applying LICM (Sengupta et al., 2017). LICM has capability to alter the structure of the CDFG, thus obfuscating its corresponding RTL datapath.

It may be possible that all the aforementioned compiler-driven transformation techniques are not applicable to all the DSP applications at once. However, at least one or two of them can always be used in conjunction for obfuscating the RTL design of the respective DSP application. Moreover, if multiple transformation techniques are applicable to a single CDFG, the SoO will depend on the order of applying each technique. The reason is that the transformed graph yielded from one technique is fed as an input to the next technique. For instance, applying ROE followed by THT may give higher robustness in terms of SoO compared to the reverse order. According to Sengupta et al. (2017), the order which may give the highest robustness is LICM, followed by LU, ROE, LT, and THT, respectively. All these significantly contribute to structurally obfuscating (obscuring) when its equivalent RTL coprocessor datapath is generated using DSE and HLS.

The same resource configuration, which was used before applying these transformation techniques, may incur higher design costs. Authors have adopted a Particle Swarm Optimization (PSO) (Chakraborty, 2017) based DSE process for generating a low-cost structurally obfuscated DSP design. It will accept the compiler transformed CDFG as the input along with the other design inputs (such as module library, user constraints, the weight of each design parameter, etc.) and will generate a resource configuration that is used to generate a low cost RTL datapath for the coprocessor.

A non-obfuscated DSP circuit for a sample application is shown in Figure 1.4 (Sengupta et al., 2017). The resource configuration used in this DSP datapath circuit design is three adders (shown in red) and four multipliers (shown in deep blue). The primary inputs (RegX, RegY, RegU[a], RegV[a+1], RegW[a] etc.) are shown in green and the final output, RegA, is shown in light blue. All the inputs and outputs are stored in the registers. The delay elements are shown in black. The switching elements are shown in red lines. The complete design employs nine 4:1 switches, six 2:1 switches, ten input registers, one output register and eight delay elements. The obfuscated design of the same DSP circuit is shown in Figure 1.5. The resource configuration used in this datapath circuit is three adders and one multiplier. As mentioned earlier, this resource configuration is explored via PSO- driven DSE process. Additionally, the number of switches changed from nine 4:1 and six 2:1 to twelve 8:1, the number of input registers changed from ten to eighteen, the number of output register changed from one to two, and the number of delay elements changed from eight to thirteen. As evident, these changes due to structural obfuscation techniques employed earlier are clearly reflected in the RTL or module level datapath and in the gate-level netlist. The analysis of results in terms of SoO and design costs of this approach will be discussed in the next section.

1.2.2 *Enhanced Security of DSP Circuits Using Multi-key Based Structural Obfuscation*

The methodology proposed by Sengupta and Rathor (2020) secures the hardware accelerators through a key-controlled transformation based structural obfuscation technique. It is capable of providing protection against RE and counterfeiting/cloning of DSP IP cores. Due to the current distributed model, an attacker may be present in the supply chain aiming either to insert a Trojan covertly after reverse engineering the design, or copying the design. Generating an obfuscated RTL datapath can hinder the RE process for an attacker. Five different key-controlled techniques are employed to generate a structurally obfuscated design of a DSP application. Each of these keys should be chosen by the designer based on his/her choice. A different key value will generate a differently obfuscated design, while always preserving the actual functionality of the application.

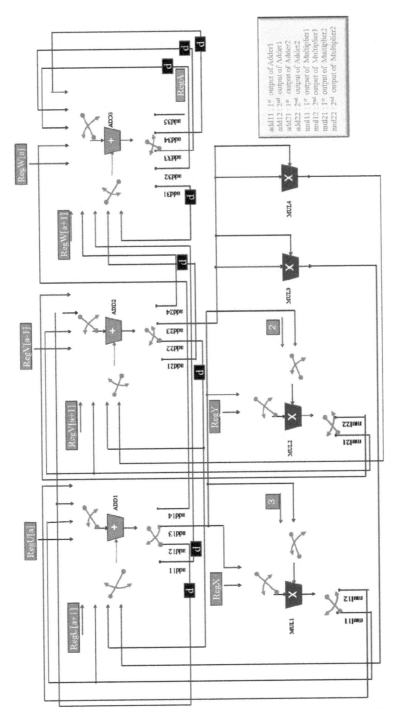

FIGURE 1.4
Sample non-obfuscated DSP circuit.

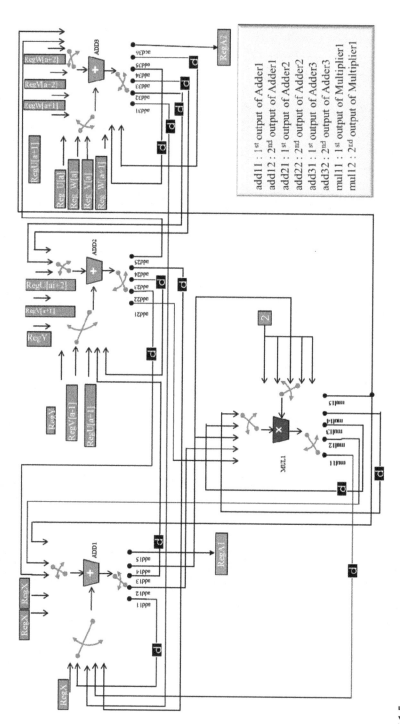

FIGURE 1.5
Obfuscated design of the sample DSP circuit.

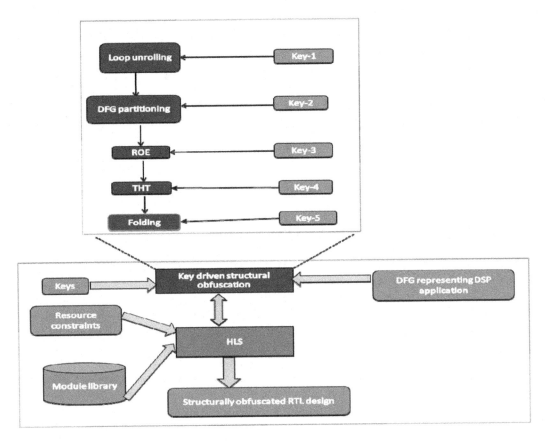

FIGURE 1.6
Overview of multi-key based obfuscation approach.

The five different key-controlled obfuscation techniques proposed by the authors to make the design unobvious (non-interpretable) to an attacker are: key-based loop unrolling, key-based partitioning, key-based redundant operation elimination and key-based tree height transformation. The overview of this approach is shown in Figure 1.6. As shown, the inputs accepted in this approach are the DSP application in the form of a C-code or corresponding DFG, designer-provided resource constraints, a module library and the five different obfuscation keys (K1 – K5). Thereafter, it generates a key-controlled structurally camouflaged yet functionally equivalent RTL DSP coprocessor design through behavioral synthesis. Each of these key-controlled obfuscation techniques is described below.

Loop-unrolling: The secret key that controls the loop unrolling is named K1. The value of secret key K1 defines the number of times the loop body should be unwound. In other words, K1 indicates the unrolling factor for a loop-based DSP application. The maximum possible bit size of K1 can be defined as $\lceil \log_2 (\text{UFmax}) \rceil$, where 'UFmax' is the maximum possible loop iteration count of the application. Along with the interconnectivity between different components/modules, the value of K1 can alter the size and number of muxes/demuxes, resource configurations, and registers; thus generating an unobvious/non-interpretable RTL datapath for an attacker's perspective. Figure 1.7 shows a sample DFG and its corresponding transformed structure, based on K1 = 00000011. The UFmax of the sample DFG is 160. Therefore, 8 bit is required to represent the UFmax. However, the

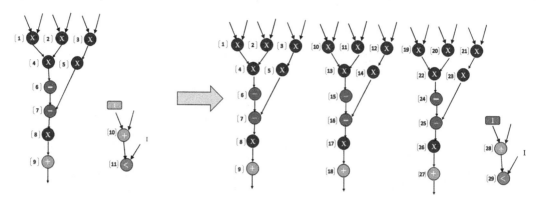

FIGURE 1.7
Loop unrolling based transformation of a sample DFG for K1 = 00000011.

DFG is unrolled three times as the decimal equivalent of the input K1 is three. Due to this transformation, the total node counts increases from 11 to 29.

Partitioning: The secret key that controls the partitioning of the loop unrolled DFG representing the DSP application, is termed 'K2'. The value of secret key K2 defines the number of cuts to be applied. It may be noted that 't' number of cuts will generate 't+1' partitions of the input DFG. A valid cut must generate a partition with at *'least two operations/nodes'*. The maximum possible bit size of secret key K2 can be defined as ⌈log2(Max. cut)⌉ , where 'Max cut' is the maximum valid cuts possible in the DFG. The value of K2 can alter the interconnectivity between resources and muxes/demuxes; thus generating an unobvious/non-interpretable RTL datapath from an attacker's perspective. The sample unrolled DFG for UF=3 shown in Figure 1.7, is taken as an input for demonstration. Its corresponding transformed structure based on secret key K2 = 0010 is shown in Figure 1.8.

The maximum possible cut of the sample DFG is eleven. Therefore, 4 bit is required to represent the Max. cut. However, two cuts are applied on the sample DFG as the decimal equivalent of the input K2 is two. Due to this transformation, the DFG is

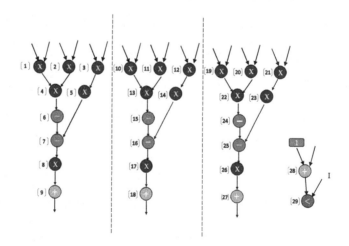

FIGURE 1.8
Partitioning based transformation of a sample DFG for K2 = 0010.

partitioned into three parts. The impact of partitioning-based transformation is also explained here using IIR filter DSP benchmark. The generic equation of an IIR filter is mentioned below:

$$Y[n] = b0 * X[n] + b1 * X[n-1] + b2 * X[n-2] + b3 * X[n-3] + b4 * X[n-4]$$
$$- a1 * Y[n-1] - a2 * Y[n-2] - a3 * Y[n-3] - a4 * Y[n-4] \qquad (1.1)$$

The respective DFG of the IIR filter and its corresponding transformed structure based on K2 = 010 is shown in Figure 1.9. The maximum possible cut of the sample DFG is 7. Therefore, 3 bit is required to represent the Max. cut. However, two cuts are applied to the sample DFG as the decimal equivalent of the input K2 is two. Due to this transformation, the DFG is partitioned into three parts.

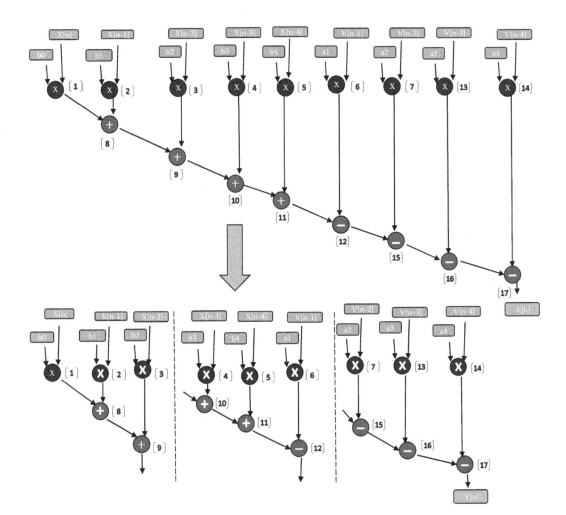

FIGURE 1.9
Partitioning based transformation of IIR for K2 = 010.

Redundant Operation Elimination: The secret key that controls the ROE process of each partitioned DFG is named K3. The value of secret key K3 defines the number of redundant operations to be removed among all the available redundant operations. The maximum possible bit size of K3 can be defined as $\lceil log2(Max. RO) \rceil$, where 'Max. RO' is the maximum redundant operations available across all the partitions. The value of K3 can also alter the interconnectivity and sizes of muxes/demuxes. Thus, an unobvious/non-interpretable RTL datapath is generated, from an attacker's perspective. Figure 1.10 shows the partitioned DFG of the ARF filter for key value = 3 and its corresponding transformed structure based on K3 = 100. The maximum number of redundant nodes present in all the partitioned DFG is four. Therefore, 3 bit is required to represent the Max. RO. However, four redundant nodes are removed from the sample DFG as the decimal equivalent of the input K3 is four. Due to this transformation, nodes number 11, 12, 21 and 22 are removed from the DFG.

Tree Height Transformation: The secret key that controls the THT of the each partitioned DFG is named K4. The value of K4 defines the number of partitions where THT has to be applied among all the available partitions. The maximum possible bit size of K4 can be defined as $\lceil log2(Max. THT) \rceil$, where 'Max. THT' is the maximum partitions possible where THT can be applied. THT introduces parallel sub-computation in the DFG. The value of secret key K4 can alter the interconnectivity and size of muxes/demuxes and functional units. Thus, it generates an unobvious/non-interpretable RTL datapath from an attacker's perspective. Figure 1.11(a) shows the DFG of the FIR filter before applying any transformation technique. Figure 1.11(b) shows the partitioned DFG of the FIR filter for cut value = 4. Its corresponding transformed structure based on K4 = 101 is shown in Figure 1.12. The maximum number of partitions where THT can be applied is five. Therefore, 3 bit is required to represent the Max. THT. Here, THT is applied to all the five partitions of the FIR DFG as the decimal equivalent of the input K4 is five. Due to this, interconnectivity between nodes has changed in the DFG.

Folding Knob: After performing all four aforementioned structural transformation techniques, each partition needs to be scheduled before applying folding knob based transformation. It indicates the number of FUs needed to be shared. For instance, if the folding value (also known as folding factor) is two, two same FU needs to be shared by a single FU of the same type. In that case, this transformation can be applied on each scheduled partition if two same FUs present in direct dependencies need to share one FU. The secret key that controls the folding of each scheduled partition is named K5. The maximum possible bit size of K5 can be defined as $\lceil log2(Max. folding) \rceil$, where 'Max folding' is the maximum possible folding in all the partitions. The value of K5 can alter interconnectivity, decrease the number of FUs due to sharing, and decrease the size of muxes/demuxes and storage elements. Thus it will generate an unobvious/non-interpretable RTL datapath from an attacker's perspective.

A non-obfuscated DFG of FIR filter and its corresponding datapath are shown in Figure 1.13(a). The resource configuration used in this datapath is one adder, one multiplier and one comparator. The primary inputs are a, b, i, 160, 1 etc. All the inputs and outputs are stored in the registers. The complete design employs three 2:1 muxes, one 1:2 demuxes, six input registers, and one output register. The obfuscated datapath design of the same DSP circuit is shown in Figure 1.13(b). The resource configuration used in this circuit is changed to one adder, four multipliers and one comparator. The primary inputs are a0 to a15, b0 to b15, i, 16, and so on. Additionally, the number of muxes changed from three 2:1 muxes to eighteen 4:1 and two 8:1, the number of demuxes changed from one 1:2

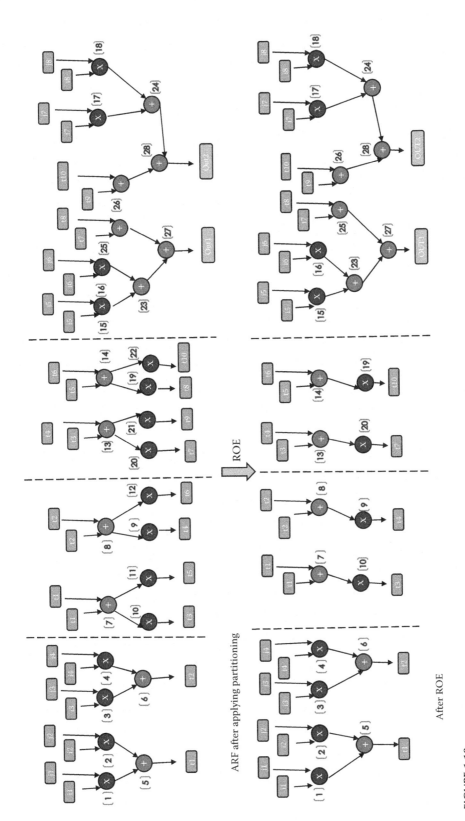

ROE

ARF after applying partitioning

After ROE

FIGURE 1.10
ROE based transformation of ARF for K3 = 100.

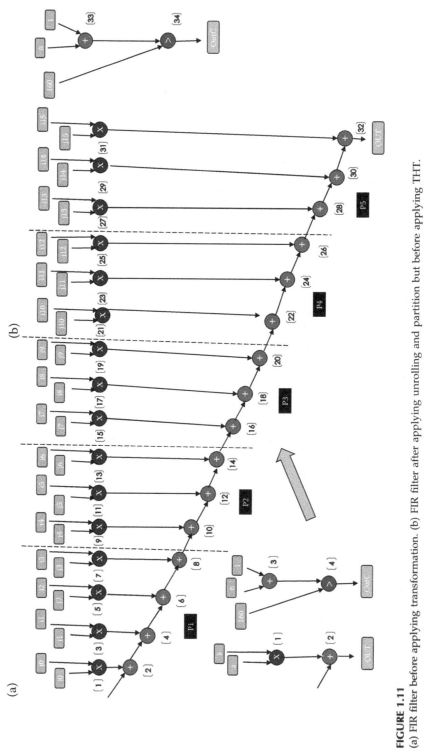

FIGURE 1.11

(a) FIR filter before applying transformation. (b) FIR filter after applying unrolling and partition but before applying THT.

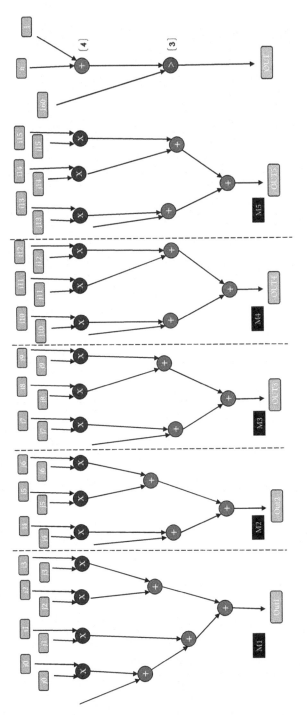

FIGURE 1.12
THT based transformation of FIR for K4 = 101.

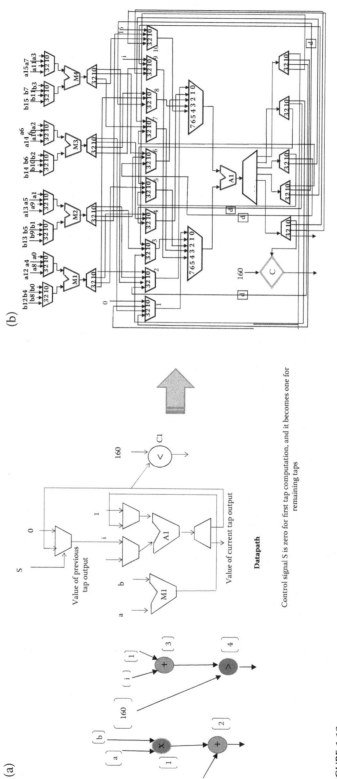

FIGURE 1.13

(a) Non-obfuscated DFG of FIR and its corresponding datapath. (b) Obfuscated datapath of FIR.

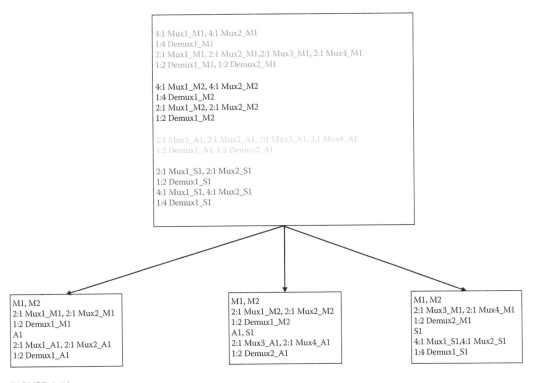

FIGURE 1.14
Component info of scheduled IIR.

demuxes to nine 4:1 and one 1:8. Similarly, the number of input and output registers is also changed accordingly. Four delay elements are also introduced in the obfuscated design. All these changes are reflected in the generated RTL of the coprocessor and its equivalent gate-level netlist of the design.

Further, we also show another example for IIR filter. First, the partitioning-based transformed IIR filter (shown in Figure 1.9) is scheduled. Two multipliers, one adder and one subtractor, are used to perform the scheduling. As mentioned earlier, the DFG of the IIR filter is partitioned into three parts using K2 = 010. The component info of the complete scheduled IIR and its transformed obfuscated structure of each part is shown in Figure 1.14. The analysis of results in terms of SoO and design overhead of this approach will be discussed in the next section.

1.2.3 *Securing DSP Kernels Using Robust Hologram Based Obfuscation*

The methodology proposed by Sengupta and Rathor (2019) secures the DSP IP through hologram-based structural obfuscation technique. It integrates two DSP kernel architectures in a camouflaged fashion without change of functionality of either, such that it becomes complex and un-interpretable to an adversary. In the security image hologram, two or more images are camouflaged using flip-flop (switch) elements. When images are moved in different viewing angles, different images evolve through the hologram (2016Matrix Technologies, Hologram Features, 2018). This feature of image holograms is exploited by the authors to perform a structural obfuscation of two DSP kernels.

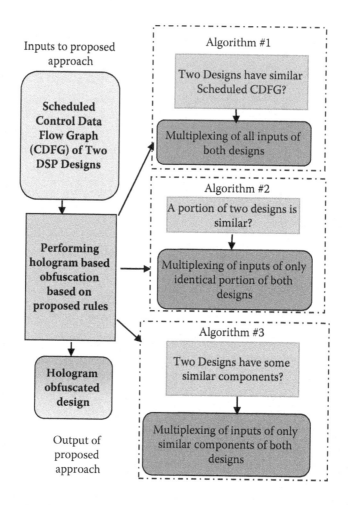

FIGURE 1.15
Overview of hologram-based structural obfuscation.

Overview

In this approach, two DSP designs are merged, such that the functionality of one design is camouflaged in another. This camouflaging is analogous to security image hologram, which is the reason this obfuscation technique has been called 'hologram-based obfuscation'. Figure 1.15 depicts its methodology, which accepts the scheduled CDFG of two DSP kernel applications as primary inputs and generates an obfuscated common datapath of both DSP kernels (Sengupta & Rathor, 2019).

Algorithms

Hologram based obfuscation is performed based on three algorithms described as follows:
   Algorithm #1: Two DSP designs having similar scheduled CDFGs are merged by multiplexing of all inputs and outputs of both designs.

Algorithm #2: Two DSP designs having a portion identical are camouflaged by multiplexing inputs of only the identical portion of both designs.

Algorithm #3: If two DSP designs have only some similar components, then both designs are camouflaged by multiplexing inputs of only those similar components of both designs.

Thus, a structurally obfuscated common register transfer level (RTL) datapath of two DSP 21designs is obtained based on these algorithms. Each design functions at a specific bit pattern and only one design remains active at a time. These algorithms have been applied on several DSP kernel designs to generate their hologram-based structurally obfuscated design, i.e., IIR - FIR filter designs, 8-point DCT-IDCT, 4-point DCT-IDCT, 8-point and 4-point DIT-FFT, and so on. This type of obfuscation is explained in the next sub-section by demonstratiting a hologram-based obfuscated design of 4-point DCT and 8-point DCT.

Demonstration

Generic equations of first output value of both designs are presented below:

$$4 - \textbf{Point DCT}: X'[0] = k'1 *x'[0] + k'2 *x'[1] + k'3 *x'[2] + k'4 *x'[3] \quad (1.2)$$

$$8 - \textbf{Point DCT}: X[0] = k1 *x[0] + k2 *x[1] + k3 *x[2] + k4 *x[3] + k5 *x[4]$$
$$+ k6 *x[5] + k7 *x[6] + k8 *x[7] \quad (1.3)$$

As explained in Figure 1.15, scheduled CDFG of 4-point DCT and 8-point DCT are the primary inputs to perform the hologram- based obfuscation. Scheduling of 4-point DCT and 8-point DCT based on one adder and one multiplier is shown in Figure 1.16(a) and Figure 1.16(b), respectively. The blue nodes indicate the multiplication operations and the green nodes the addition operations. The 4-point DCT is scheduled in 5 control steps and the 8-point DCT is scheduled in 9 control steps. Referring to Equations (1.2) and (1.3), the tabular representation of scheduled DFGs of both designs is shown in Table 1.1 and Table 1.2, respectively. Since a portion of 8-point DCT is identical to 4-point DCT, hologram-based obfuscation is applied to both DSP designs using algorithm #2. Thus, this methodology generates a common obfuscated datapath of both DSP kernels, as shown in Figure 1.17. The design has used sixteen total registers (R1-R16) shown in the blue box, six latches shown in the red box, fifteen 2:1 & four 8:1 normal muxes shown in the green shape, fifteen 1:2 normal demuxes shown in the orange shape, one adder (A1) shown in the green border, one multiplier (M1) shown in the blue border, and nine 2:1 additional muxes for controlling the inputs and output of the design shown in the red border. As shown in the figure, the inputs and outputs are controlled via these additional multiplexers, where 'S' is the common select line for all of them. The datapath will behave like an 8-point DCT function when the select line 'S'=0 and will behave like a 4-point DCT when 'S' is changed to 1.

Using this type of obfuscation, a common obfuscated datapath of two DSP designs can be obtained, which results in huge savings in the area (compared to individually obfuscating two DSP designs) and provides security to both DSP designs simultaneously by making designs uninterpretable to an adversary. The strength of this obfuscation technique can be analyzed in terms of the number of gates that are obscured (modified or affected). This process affects a larger number of gates compared to other structural

(a)   (b)

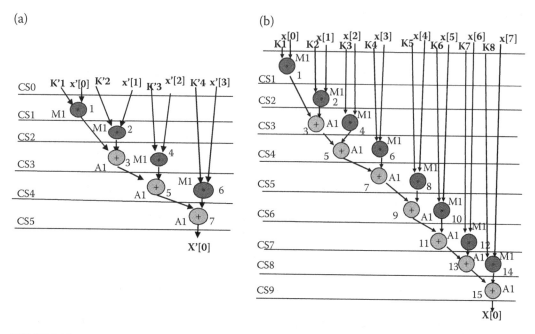

FIGURE 1.16

(a) Scheduling of 4-point DCT based on 1 adder and 1 multiplier. (b) Scheduling of 8-point DCT based on 1 adder and 1 multiplier.

TABLE 1.1

Scheduling of 4-point DCT based on 1 adder (A1) and 1 multiplier (M1)

Control Steps	Operation #, Type	Binding of Resources	Dependency Information	
			Input1	Input2
CS1	1, *	M1	k'1	x'[0]
CS2	2, *	M1	k'2	x'[1]
CS3	3, +	A1	1	2
CS3	4, *	M1	k'3	x'[2]
CS4	5, +	A1	3	4
CS4	6, *	M1	k'4	x'[3]
CS5	7, +	A1	5	6

obfuscation techniques. Hence, the methodology achieves more area efficiency and stronger obfuscation. Additionally, it can secure two DSP kernel designs simultaneously by thwarting reverse engineering. The analysis of results in terms of SoO and design costs of this approach will be discussed in the next section.

1.2.4 Securing DSP Designs Using HLT Based Structural Obfuscation

Another structural obfuscation approach for securing the DSP circuits was proposed in the literature by Lao and Parhi (2015). In that approach, high-level transformation (HLT) techniques are used to hinder the reverse engineering process for DSPs. Additionally, this approach incorporates a trade-off between design area/power consumption and execution

TABLE 1.2

Scheduling of 8-point DCT based on 1 adder (A1) and 1 multiplier (M1)

Control Steps	Operation #, Type	Binding of Resources	Dependency Information Input1	Input2
CS1	1, *	M1	k1	x[0]
CS2	2, *	M1	k2	x[1]
CS3	3, +	A1	1	2
CS3	4, *	M1	k3	x[2]
CS4	5, +	A1	3	4
CS4	6, *	M1	k4	x[3]
CS5	7, +	A1	5	6
CS5	8, *	M1	k5	x[4]
CS6	9, +	A1	7	8
CS6	10, *	M1	k6	x[5]
CS7	11, +	A1	9	10
CS7	12, *	M1	k7	x[6]
CS8	13, +	A1	11	12
CS8	14, *	M1	k8	x[7]
CS9	15, +	A1	13	14

speed of the obfuscated circuit. By considering the primary objective of structural obfuscation, this approach has exploited two scenarios. The first one is that multiple circuits having different structures can unveil the same functionality. The other one is that multiple circuits with the same structure can unveil different functionalities. To achieve the same, the authors have used folding-based structural obfuscation. In this type of obfuscation, multiple meaningful modes, known as folding variants, are generated. These folding variants are generated based on a value known as the folding factor. The folding factor is an integer value where the minimum value is one and the maximum value is the maximum number of resources that can be used to design the circuit. The maximum folding factor indicates the complete unrolled circuit, which will have the highest number of folding variants, but will consume the most area and power. Conversely, the minimum folding factor indicates the non-unrolled circuit, which will have a single folding variant, but will have the highest execution latency. Therefore, it is clear that the folding factor plays a crucial role in the design trade-off. The structural obfuscation part of this approach is built upon the assumption that the functionality of the circuit is not known to the attacker. Otherwise, this HLT based structural obfuscation will fail to secure the design against RE. Therefore, to enhance the robustness of the design, along with the HLT based structural obfuscation, the authors have also used a key-based obfuscation technique. A key-controlled finite state machine (FSM) with a reconfigurator is used to achieve that. The job of this tool is to reconfigure the input circuit (for example, changing the order of the finite impulse response filter) based on the keys provided to the FSM. The reconfigurator will be activated if a valid key is given as the input. In such a case, it will reconfigure the order of the input filter in multiple folding variants to achieve obfuscation. The primary objective of this type of obfuscation is to conceal the functionality of the DSP circuit. As the attacker will not be aware of the FSM key to activate the reconfigurator, the design of the DSP circuit will remain secured against piracy, cloning and overbuilding attacks. Only legitimate vendors with the actual FSM key and the correct folding variant can use it.

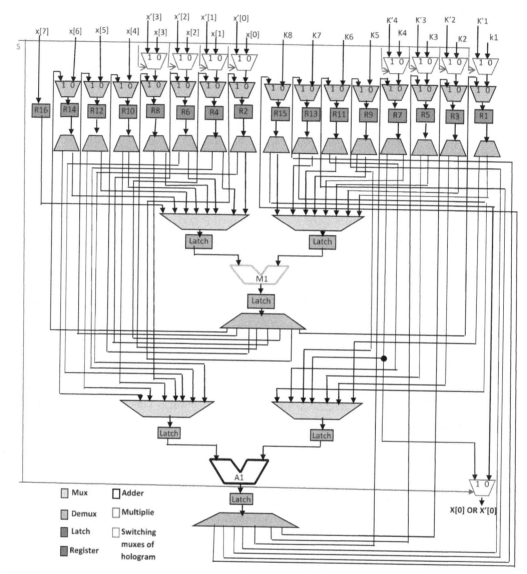

FIGURE 1.17

Hologram-based structurally obfuscated RTL datapath of 8 point DCT- 4 point DCT.

Based on this information, they can activate the reconfigurator, and generate the meaningful mode (out of all the available folding variants) to make the circuit functional. The key can be used either to activate the fabricated integrated circuit or shared with the user when selling the product. However, key-based obfuscation is vulnerable to removal attacks. In such a case, the attacker tries to identify the FSMs and the reconfigurator, which helps to generate key-based obfuscation and then remove those parts from the synthesised netlist of the design. Thus attackers can remove the key generating and mode controlling part of the design, which are directly involved in generating the key-based obfuscated design. Therefore, the unlocked but structurally obfuscated netlist of the design can be accessed. An attacker then tries to identify the functionality of the unlocked netlist through simulation-based RE. The overview of this approach is shown in Figure 1.18. As shown in

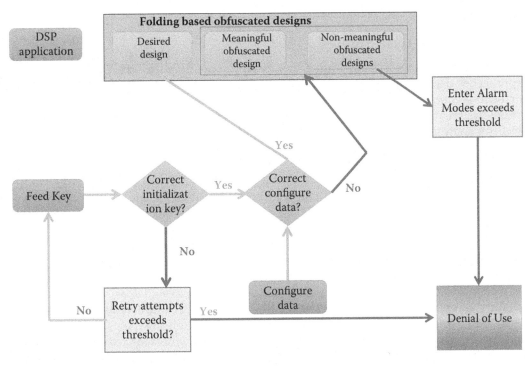

FIGURE 1.18
Overview of HLT based structural obfuscation.

the figure, upon applying the correct initialization key, the reconfigurator gets activated. However, entering a wrong key will ask again for a new key value if the total number of attempts doesn't reach the predefined threshold. In case the number of invalid attempts does reach the threshold, the FSM will go to the "Denial of Use" state. This will block the invalid attempts permanently. On the other hand, after activating the reconfigurator by applying the correct key, different functionality modes will be generated. Some modes may produce the desired structurally obfuscated design. Some may produce either a meaningful but functionally non-equivalent design or a completely non-meaningful design. However, in the case of a completely non-meaningful design, a new set of different functionality modes will be generated again if the total number of repetitions doesn't exceed the predefined threshold. Otherwise, it will enter the alarm mode and the FSM will go to the "Denial of Use" state. This will help to protect the design against a brute-force attack, where an attacker tries to identify the correct key by providing all possible combinations of keys. Incorporating the "Denial of Use" state will block an attacker from making multiple invalid attempts. It is possible to generate several equivalent modes/ structures for the same circuit using folding based HLT, which leads to structural obfuscation. Similarly, due to folding based HLT, structurally similar datapaths with a different controller can represent DSP circuits with different functionalities. However, this approach has some limitations, such as: (i) no explicit method of selecting an optimal value for folding factor is available, (ii) a poor selection of folding factor may lead to higher design costs, and (iii) using a single HLT technique to obfuscate the design structurally is not always robust enough (Lao & Parhi, 2015).

1.3 Analysis of Case Studies

This section presents a comparative analysis of contemporary structural obfuscation approaches used for DSP hardware. The analysis is explained in two sub-sections. The first sub-section discusses the results of different structurally obfuscated designs in terms of the design area, execution latency, and design cost. The next sub-section discusses the security analysis of the design using the strength of obfuscation metric.

1.3.1 *Design Analysis*

A structurally obfuscated DSP IP core can be designed using multiple resource configurations. However, the same DSP IP core, designed using different resource configurations, will consume different hardware area and vary in execution latency. Therefore, any random selection of resource configuration will not only incur heavy design costs but also may violate the designer- specified constraints. Therefore, while securing an IP core, the IP designer should develop a low overhead security algorithm as well as select an optimal resource configuration using design space exploration. To identify an optimal resource configuration, the cost of each candidate design solution is evaluated using some algorithm. The design cost doesn't indicate the price of the design but the fitness of the design. The metric for evaluating the design cost involves all the design parameters. It is a normalised weighted metric. The structural obfuscation approach explained in section 1.2.1 has considered both design area and execution latency for calculating the design cost.

The design cost (C (Xi)) metric is defined as follows (Sengupta et al., 2017):

$$F(X_i) = \Phi_1 \frac{H_T^{OBF} - H_{cons}}{H_{max}^{OBF}} + \Phi_2 \frac{L_E^{OBF} - L_{cons}}{L_{max}^{OBF}} \tag{1.4}$$

In Equation (1.4), $F(X_i)$ represents the design cost for the resource configuration X_i. The user defined constraints for hardware area and execution latency is represented through H_{cons} and L_{cons} respectively. The maximum possible hardware area and execution latency of the obfuscated design is represented through H_{max}^{OBF} and L_{max}^{OBF} respectively. The total hardware area and execution latency of the obfuscated design based on resource configuration X_i is represented through H_T^{OBF} and L_E^{OBF} respectively. The weighting of both the design parameters is controlled through $\emptyset_1$ and $\emptyset_2$. Their values are provided by the designer and lies between 0 to 1. To maintain an equal preference during design space exploration, both the weighting control parameters are kept as 0.5. The total hardware area of an obfuscated DSP IP core is calculated as follows (Sengupta et al., 2017):

$$H_T^{OBF} = \sum_{i=1}^{m} H(R_i)_*C(R_i) + H(mux)_*C(mux) + H(buffer)_*C(buffer) \tag{1.5}$$

In Equation (1.5) the total area is calculated as the sum of the areas of each individual component used in the design, such as function units, interconnecting units and storage units, which are then summed over all 'm' component types present in the design. A(Ri) represents the area of the 'Ri' th functional units such as adder, multiplier, comparator etc. A(mux) represents the area of the interconnecting units such as multiplexer, de-multiplexers etc. A(buffer), represents the area of the storage units such as latches, registers etc. N(Ri), N(mux) and N(buffer) represents the corresponding number of instances each component is present in the design. On the other hand, the execution latency of an obfuscated DSP core is calculated as follows (Sengupta et al., 2017):

$$L_E^{OBF} = \left(L_{body}^{OBF} * \left\lfloor \frac{I}{UF} \right\rfloor \right) + (I \quad Mod \quad UF) * L_{first}^{OBF} \tag{1.6}$$

In Equation (1.6) the total execution latency is calculated as the sum of latency to execute the loop part and the non-loop part of the application. For a loop-based DSP application, the execution latency of the loop part depends on the unrolling factor (UF). UF represents a magnitude defined as the number of titimes the loop portion is unrolled in the design, compared with the total number of loop iterations (I). Therefore, the number of instances to execute the loop part is calculated by dividing (I with UF). It is then multiplied with the execution delay of each loop represented as L_{body}^{OBF} to calculate the execution latency of the loop part. The execution latency of the non-loop part is calculated by multiplying the number of non-loop instances present in the design with the execution latency of a single non-loop portion, represented as L_{first}^{OBF} (Sengupta et al., 2017).

Based on all these equations, the design area, execution latency and the design cost of a structurally obfuscated DSP design is calculated. The results of the compiler-driven transformation based structural obfuscation approach (Sengupta et al., 2017) explained in section 1.2.1 are discussed first. Seven different loop based and non-loop based applications are used for that. They are 2D Autoregression Lattice Filter (ARF), 6-tap Finite Impulse Response (FIR), Auto-Correlation (Auto-cor), Differential Equation (Diff-Equ), DHMC, Adaptive Filter for noise cancellation (Ada-NC) and Adaptive Filter for least mean square (Ada-LMS). As mentioned earlier, the PSO driven DSE framework is used in this approach to explore the optimal resource configuration. During the exploration process, different swarm sizes are used. Figure 1.19 shows the comparison of hardware area consumption for different swarm sizes 3, 5 and 7. The area of all the aforementioned benchmarks is reported for swarm size 3, 5 and 7. It can be observed that except Auto-cor benchmark, the area of all other benchmarks is the same for different swarm sizes. The Auto-cor benchmark has consumed lesser design area for swarm size 3 compared to

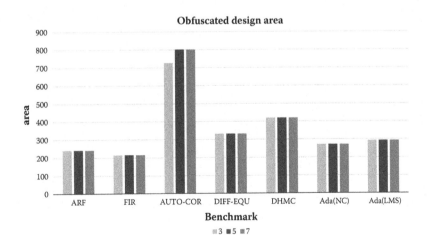

FIGURE 1.19
Comparison of design area for different swarm size.

5 and 7. Figure 1.20 shows the comparison of execution latency for different swarm size. The latency of all the benchmarks is reported for swarm size 3, 5 and 7. It can be observed that except for Auto-cor benchmark, the latency of all other benchmarks is the same for different swarm sizes. The Auto-cor benchmark has consumed higher design latency for swarm size 3 compared to 5 and 7. Figure 1.21 shows the comparison of design cost calculated as a normalised weighted sum of hardware area and execution latency for different swarm sizes. The cost of all the benchmarks is reported for swarm size 3, 5 and 7. The cost value in negative indicates it satisfies the designer provided area-delay constraints. It can be observed that except Auto-cor benchmark, the cost of all other benchmarks is the same for different swarm sizes. These optimal resource configurations for all the DSP case studies in this approach are used for design analysis.

Figure 1.22 shows the comparison between the non-obfuscated approach (baseline), the compiler-based structural obfuscation approach (Sengupta et al., 2017) discussed in section 1.2.1 and the HLT-based structural obfuscation approach (Lao & Parhi, 2015) discussed in section 1.2.4 in terms of hardware area consumption. It can be observed that the design area of baseline and (Sengupta et al., 2017) is the same; however, the design area of

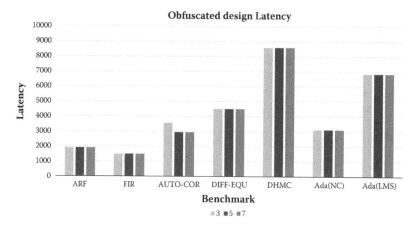

FIGURE 1.20
Comparison of design latency for different swarm size.

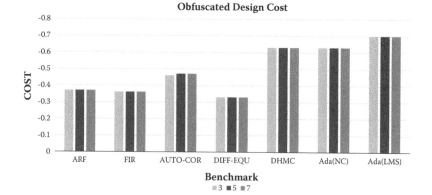

FIGURE 1.21
Comparison of design cost for different swarm size.

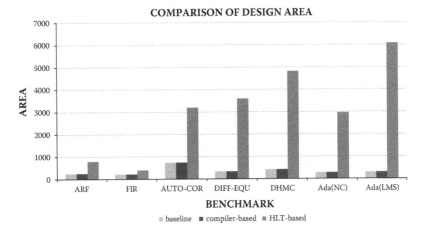

FIGURE 1.22
Comparison of design area of baseline, complier based obfuscation and HLT based obfuscation approach.

Lao and Parhi (2015) is too high because of a lack of exploring optimal folding factor. As mentioned earlier, the value of the folding factor has a direct control on the overall design parameters. Therefore, any arbitrary selection of folding factor consumes a higher design area. Figure 1.23 shows the comparison between the compiler-based structural obfuscation (Sengupta et al., 2017) discussed in section 1.2.1 and the HLT-based structural obfuscation approach (Lao and Parhi, 2015) discussed in section 1.2.4, in terms of execution latency. For some benchmarks, it can be observed that the design latency of the baseline is higher than the Sengupta et al. (2017). After applying multiple compiler-driven transformation techniques for obfuscation, the application may be further optimised. This be could reflected in the design's total execution time. Moreover, as the design area and design latency are inversely proportional, Lao and Parhi (2015) are able to achieve lower design latency by sacrificing higher design area. Figure 1.24 shows the comparison between the non-obfuscated approach (baseline), the compiler-based structural obfuscation

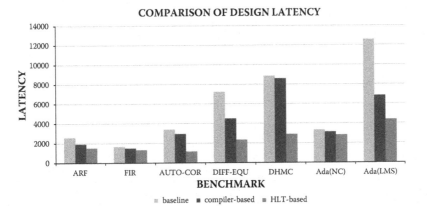

FIGURE 1.23
Comparison of design latency of baseline, complier based obfuscation and HLT based obfuscation approach.

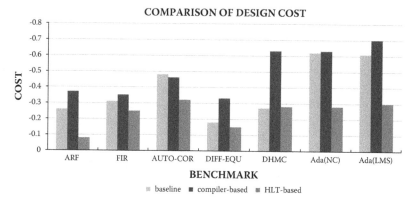
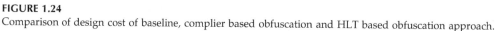

FIGURE 1.24
Comparison of design cost of baseline, complier based obfuscation and HLT based obfuscation approach.

approach (Sengupta et al., 2017) discussed in section 1.2.1, and the HLT-based structural obfuscation approach (Lao & Parhi, 2015) discussed in section 1.2.4, in terms of design cost. It can be observed that Sengupta et al. (2017) is able to achieve a lower design cost (higher negative value) for all the reported benchmarks compared to baseline and Lao and Parhi (2015). Additionally, Lao and Parhi (2015) incur higher design overhead in terms of design cost for all the reported DSP case studies.

The results of the multi-key based structural obfuscation approach (Sengupta & Rathor, 2020) explained in section 1.2.2 are discussed here. Five different loop based and non-loop based DSP applications are used for that. They are Autoregression Filter (ARF), Finite Impulse Response (FIR), Infinite Impulse Response (IIR), Discrete Cosine Transformation (DCT), and Differential Equation (Diff-Equ). Figure 1.25 shows the comparison of design cost between the non-obfuscated approach (baseline) design and the multi-key based structurally obfuscated design (Sengupta & Rathor, 2020), calculated as a normalised weighted sum of hardware area and execution latency. In this approach, the designer-provided area and latency constraints are not taken as input for calculating the design cost. It can be observed that the design of Sengupta and Rathor (2020) is able to achieve a

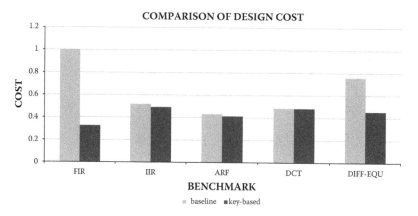

FIGURE 1.25
Comparison of design cost of baseline and multi-key based obfuscation approach.

lower design cost for all the reported DSP case studies compared to the baseline design. The reason is that the multi-key based structural obfuscation approach optimises execution latency by introducing parallelism into the process via a key-controlled THT technique. Similarly, it optimises the number of operations in the application by removing redundant nodes from the DFG of the DSP application via a key-controlled ROE technique. This optimises both hardware area and execution latency. The overall hardware area and execution latency are also impacted via a key-controlled unrolling factor. It helps to control the number of duplicate iterations. Due to all these key-controlled transformation techniques, the overall design cost of this approach (Sengupta & Rathor, 2020) is reduced compared to the baseline design for all the reported DSP case studies.

1.3.2 *Security Analysis*

The robustness of each structural obfuscation approach is determined by the Strength of Obfuscation (SoO) metric that represents the number of gates impacted or modified due to the structural transformation. The SoO metric may also calculate the impact due to obfuscation at operation level, i.e., the number of nodes impacted or modified in the DFG before and after applying structural obfuscation. The SoO metric is defined below as (Sengupta et al., 2017).

$$S^{obf} = \frac{\sum_{i=1}^{5} s_i^{obf}}{C(T)} \tag{1.7}$$

$$s_i^{obf} = \frac{n_i}{n_i^T} \tag{1.8}$$

In Equation (1.7), S^{obf} represents the SoO, which is an average for each individual transformation technique. $C(T)$ represents the total number of transformation techniques applied on the DSP application. In Equation (1.8), s_i^{obf} represents the SoO for an applied transformation technique 'i'. It is determined by dividing the number of transformed or modified nodes represented by n_i with the total number of nodes present in the DFG represented by n_i^T. As shown in Equation (1.6), the SoO will have a normalised value in the range of 0 to 1.

Figure 1.26 shows the comparison between the compiler-based structural obfuscation (Sengupta et al., 2017) discussed in section 1.2.1 and the HLT-based structural obfuscation approach (Lao & Parhi, 2015) discussed in section 1.2.4, in terms of robustness of the design via SoO. It can be observed that the Sengupta et al. (2017) is able to provide stronger security for all the reported DSP case studies compared to Lao and Parhi (2015).

Further, Figure 1.27 shows the comparison between the baseline design and the multi-key based structural obfuscation approach (Sengupta & Rathor, 2020) discussed in section 1.2.2, in terms of total number of gates modified (obfuscated) in the design. It can be observed that a large number of gates have been added, removed and/or modified (resulting in structural obfuscation) due to the multi-key based structural obfuscation approach. For IIR and ARF benchmarks, the total gate count decreases compared to the baseline, whereas for the rest of the applications, it increases. A gate is considered as a modified one if its interconnectivity information is changed without affecting the overall functionality of the DSP design.

The results of the hologram based structural obfuscation approach (Sengupta & Rathor, 2019) explained in section 1.2.3 are discussed here. Ten different non-loop based applications are used for this purpose. In this approach, two applications are represented

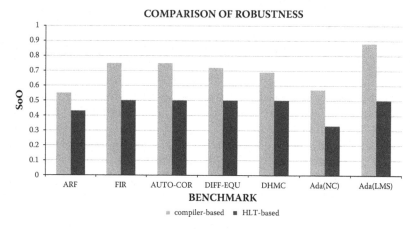

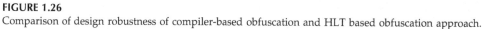

FIGURE 1.26
Comparison of design robustness of compiler-based obfuscation and HLT based obfuscation approach.

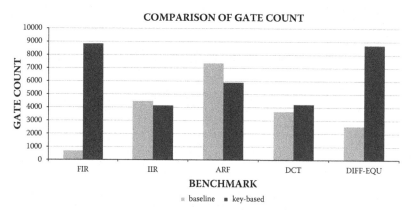

FIGURE 1.27
Comparison of number of gates of baseline and multi-key based obfuscation approach.

through one design. They are IIR + FIR, 8-pt DCT + 8-pt IDCT, 4-pt DCT + 4-pt IDCT, 8-pt DCT + 4-pt DCT, 8-pt DIT-FFT + 4-pt DIT-FFT. Figure 1.28 shows the comparison between the baseline design and the hologram based structural obfuscation approach (Sengupta & Rathor, 2019) discussed in section 1.2.3, in terms of the total number of gates modified due to structural obfuscation in the DSP design. It can be observed that a large number of gates have been added, removed and/or modified due hologram based structural obfuscation process. As two benchmarks are combined through one design, the total gate count also decreases compared to the baseline design of individual applications. The percentages of gates affected by hologram based structural obfuscation (Sengupta & Rathor, 2019) are 29.2%, 45.5%, 45.1%, 18.1%, and 29.7% for IIR + FIR, 8-pt DCT + 8-pt IDCT, 4-pt DCT + 4-pt IDCT, 8-pt DCT + 4-pt DCT, and 8-pt DIT-FFT + 4-pt DIT-FFT respectively. Further, Figure 1.29 shows the comparison between the compiler driven structural obfuscation approach (Sengupta et al., 2017) discussed in section 1.2.1 and the hologram based structural obfuscation approach (Sengupta & Rathor, 2019) discussed in section 1.2.3, in terms of the total number of affected gates modified/affected. It can be observed that more gates are affected due to hologram- based structural obfuscation

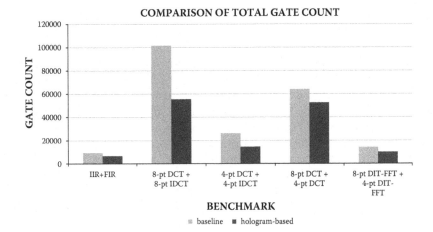

FIGURE 1.28
Comparison of number of gates of baseline and hologram based obfuscation approach.

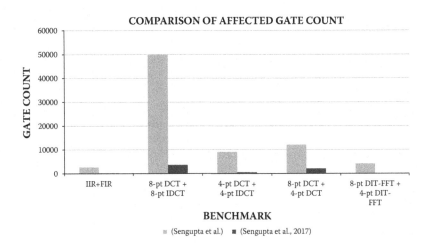

FIGURE 1.29
Comparison of number of gates of compiler-based and hologram-based obfuscation approach.

compared to compiler-driven structural obfuscation approach. The percentages of gates affected by compiler-driven structural obfuscation technique are 2.4%, 3.5%, 2.0%, 3.2%, and 0% for IIR + FIR, 8-pt DCT + 8-pt IDCT, 4-pt DCT + 4-pt IDCT, 8-pt DCT + 4-pt DCT, and 8-pt DIT-FFT + 4-pt DIT-FFT respectively. Thus the hologram-based structural obfuscation approach provides on average ~14 times higher security than the compiler-driven structural obfuscation approach for all the reported DSP case studies.

1.4 Conclusion

Structural obfuscation based hardware security at the behavioral level is an emerging area. There are only a few approaches available in the literature. In this chapter, we discussed four popular structural obfuscation approaches, their limitations and design costs. The HLT based structural obfuscation is the first approach that targets DSP circuits for security. It also uses key-based obfuscation to enhance security. As this approach uses only one transformation technique, i.e., folding for obfuscation, security may be compromised against a strong attack. To overcome this, a compiler-driver structural obfuscation approach is proposed. It uses five different transformation techniques: ROE, LT, THT, LICM, and LU. Moreover, to reduce design costs, PSO based DSE framework is also used. However, in this obfuscation approach, the design can be changed with much less designer control over the cost. The key-based structural obfuscation approach is used to control the degree of obfuscation by changing the key value. It gives the designer flexibility in the overall design cost. Moreover, applying different keys will generate a differently obfuscated design for the same input application. The hologram-based structural obfuscation approach generates a common obfuscated datapath of two DSP designs. Thus, it results in huge savings in area (compared to two individually obfuscated DSP designs) and provides protection to both DSP designs simultaneously. The robustness of all these approaches is analyzed in terms of the number of gates or nodes that are obscured (modified or affected). A higher the number of modified or affected gates/nodes, the more robust the system. The hologram-based obfuscation affects a larger number of gates compared to other structural obfuscation techniques. Moreover, it also achieves higher area efficiency. However, hologram-based obfuscation can only be used if two applications have partially similar structures.

Security through structural obfuscation can be strengthen by incorporating more algorithms which are capable of producing functionally similar but structurally different DSP applications. Moreover, implementing structural obfuscation based security at lower and multiple design levels can be a new research direction. The robustness of hardware security can be enhanced by combining both structural and functional obfuscation, where it can not only hide structure but also protect functionality.

References

Castillo, E., Meyer-Baese, U., Garcia, A., Parrilla, L., & Lloris, A. (May 2007). IPP@HDL: Efficient intellectual property protection scheme for IP cores. *IEEE Transactions on Very Large Scale Integration Systems*, 15(5), 578–591.

Chakraborty, C. (2017). Chronic wound image analysis by particle swarm optimization technique for tele-wound network. *Springer: International Journal of Wireless Personal Communications*, 96(3), 3655–3671. ISSN: 0929-6212, doi: 10.1007/s11277-017-4281-5

Colombier, B., & Bossuet, L. (2015). Survey of hardware protection of design data for integrated circuits and intellectual properties. *IET Computers & Digital Techniques*, 8(6), 274–287, doi: 10.1049/iet-cdt.2014.0028.

Hong, I., & Potkonjak, M. (1999). Behavioral synthesis techniques for intellectual property protection. *Proceedings of Design Automation Conference*, New Orleans, LA, U.S.A, pp. 849–854, doi: 10.1109/DAC.1999.782161.

Koushanfar, F., Hong, I., & Potkonjak, M. (2005). Behavioral synthesis techniques for intellectual property protection. *ACM Transactions on Design of Automation of Electronic Systems*, 10(3), 523–545, doi: 10.1145/1080334.1080338.

Krishnan, V., & Katkoori, S. (2006). A genetic algorithm for the design space exploration of datapaths during high-level synthesis. *IEEE Transactions of Evolutionary Computation, 10*(3), 213–229, doi: 10.1109/TEVC.2005.860764.

Lao, Y. & Parhi, K. K. (2015). Obfuscating DSP circuits via high-level transformations. *IEEE Transactions on Very Large Scale Integration (VLSI) Systems, 23*(5), 819–830, doi: 10.1109/TVLSI.2 014.2323976.

Le Gal, B., & Bossuet, L. (2012) Automatic low-cost IP watermarking technique based on output mark insertions. *Design Automation for Embedded Systems, 16*(2), 71–92, doi: 10.1007/s10617-012-9085-y.

Li, L., & Zhou, H. (2013). Structural transformation for best-possible obfuscation of sequential circuits. *Proceedings of IEEE International Symposium on Hardware-Oriented Security and Trust (HOST), 2013 IEEE International Symposium on Hardware-Oriented Security and Trust (HOST),* Austin, TX, pp. 55–60, doi: 10.1109/HST.2013.6581566.

2016Matrix Technologies, Hologram Features. [Online]. Available: http://www.matrixtechnologies.in/hologram-features.html, last accessed on August 2018.

Newbould, R. D., Carothers, J. D., & Rodriguez, J. J. (2002). Watermarking ICs for IP protectio. *Electronics Letters, 38*(6), 272–274, doi: 10.1049/el:20020143.

Ni, M., & Gao, Z. (2005). Detector-based watermarking technique for soft IP core protection in high synthesis design level. *Proceedings of International Conference on Communications, Circuits and Systems,* Hong Kong, pp. 1348–1352, doi: 10.1109/ICCCAS.2005.1495356.

Roy, D., & Sengupta, A. (2017). Low overhead symmetrical protection of reusable IP core using robust fingerprinting and watermarking during high level synthesi. *Future Generation Computer Systems, 71*, 89–101.

Schneiderman, R. (2010). DSPs evolving in consumer electronics applications. *IEEE Signal Processing Magazine, 27*(3), 6–10, doi: 10.1109/MSP.2010.936031.

Sengupta, A., & Roy, D. (April 2017). Antipiracy-aware IP chipset design for CE devices: A robust watermarking approach. *IEEE Consumer Electronics Magazine, 6*(2), 118–124.

Sengupta, A., & Roy, D. (2019). Low cost dual-phase watermark for protecting CE devices in IoT framework. In R. Chakraborty, J. Mathew, A. Vasilakos (Eds.), Security and fault tolerance in internet of things. Internet of things (technology, communications and computing). Cham: Springer.

Sengupta, A., & Rathor, M. (Feb 2019). Protecting DSP kernels using robust hologram based obfuscation. *IEEE Transactions on Consumer Electronics (TCE), 65*(1), 99–108.

Sengupta, A., & Rathor, M. (May 2020). Enhanced security of DSP circuits using multi-key based structural obfuscation and physical-level watermarking for consumer electronics systems. *IEEE Transactions on Consumer Electronics (TCE), 66*(2), 163–172

Sengupta, A., Roy, D., Mohanty, S.P., & Corcoran, P. (2017). DSP design security in CE through algorithmic transformation based structural obfuscation. *IEEE Transactions on Consumer Electronics, 63*(4), 467–476, doi: 10.1109/TCE.2017.015072.

Sengupta, A., Roy, D., Mohanty, S.P., & Corcoran, P. (2018). Low-cost obfuscated JPEG CODEC IP core for secure CE hardware. *IEEE Transactions on Consumer Electronics, 64*(3), 365–374, doi: 10.1109/TCE.2018.2852265.

Torrance, R. & James, D. (2009). The state-of-the-art in IC reverse engineering. *Proceedings of the CHES, 5747*, 363–381, https://doi.org/10.1007/978-3-642-04138-9_26.

Zhang, J. (2016). A practical logic obfuscation technique for hardware security. *IEEE Transactions on Very Large Scale Integration (VLSI) Systems, 24*(3), 1193–1197, doi: 10.1109/TVLSI.2015.2437996.

Ziener, D., & Teich, J. (2008). Power signature watermarking of IP cores for FPGAs. *Journal of Signal Processing Systems, 51*(1), 123–136, doi: 10.1007/s11265-007-0136-8.

2

Multi-bit True Random Number Generator for IoT Devices using Memristor

V. K. Rai, S. Tripathy, and J. Mathew
Indian Institute of Technology Patna

2.1 Introduction

Random Number Generator is one of the essential tools used in protocols to enhance security features. The primary uses of the random numbers are the encryption and decryption keys, one-time password, nonce, and initial vectors, among others. The random numbers' effectiveness plays a crucial role in maintaining the high quality of any security protocols and algorithms. In other words, we can say that the quality of random numbers determines the quality of the protocols that make use of them.

We can broadly categorise the random number generators into two categories. The first one is the pseudo random number generator. The second category is called the true random number generator. The true random number generator is generally referred as TRNG. The pseudo random number generator uses mathematical algorithms, hence becomes deterministic. So, they often are called deterministic random number generators. The design of pseudo random number generators make them simpler to implement. However, if an attacker knows the mathematical algorithms used in it and some previous output knowledge, the numbers may be predictable. Thus, it is not advised to use pseudo random number generators in a highly secure environment (Massey, 1969).

On the other hand, TRNGs use physical phenomena like clock jitter, and thermal, atmospheric , flicker, or shot noise to extract the randomness. Along with these noise sources, TRNGs also require a noise extraction and digitization mechanism. The quality of the random bit stream can be improved by post processing. The widely used high-level architecture of TRNGs is recommended by the AIS-31 standard (Killmann & Schindler, 2011). A generic architecture (Fischer, 2014) based on this recommendation is depicted in Figure 2.1.

With the rapid and dynamic growth in the internet of things technology, the devices which suit the constraints environment of the IoT paradigm are highly in demand among both the industries and academia. So here in our work, we focus on a random number generator that could be useful everywhere, especially in the IoT environment. IoT devices are deployed everywhere and rapidly gaining popularity (Mandal, Mondal, Banerjee, Chinmay, & Biswas, 2020; Gupta, Chakraborty, & Gupta, 2019). Most applications in IoT devices require a security protocol to protect data. Various security

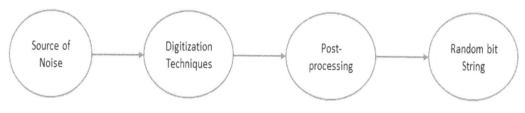

FIGURE 2.1
Generic architecture of a TRNG circuit used as a cryptographic primitive.

protocols which are implemented in IoT devices use random numbers as input to perform their functions. The primary reason that our design is useful in the IoT environment is the high throughput of the design. In the IoT paradigm, the devices have space and energy resource constraints. So, if we need to produce a large random number or generate it at high speed, our proposed design is better than any single-bit random number generator. We design a simple but effective CMOS-memristor based random number generator whose simulation shows good statistical results and proves low hardware and energy requirements. Our architecture harvested the randomness of memristor along with the traditionfal randomness caused due to manufacturing process variations. In our design, we use two identical ring oscillators and compare the delay between them to generate random bits. Such design is more useful where efficient space and energy use both are crucial. The primary advantage of our design is that we may produce multiple bits in a single cycle, which significantly enhances the throughput of the random number generator. This can be done easily by tapping the numerous nodes of both the ring oscillators instead of tapping the nodes at last, as in the case of a single-bit ring oscillator. We also modified our architecture and simulated the same design after the removal of the memristor components from the architecture to make evident the difference in the statistical randomness in both designs. Thus, the main contributions of this work are as following.

- We propose simple and resource constrained architecture for a random number generator.
- We modify the design and remove the memristor to test the statistical differences.
- We test the statistical randomness of the generated output by applying two widely used statistical testing techniques called the NIST and Dieharder test suites.
- We compare the proposed scheme with that of some existing true random number generators and find our proposed design to be better in terms of throughput and randomness.

The rest of the chapter is organised as follows. The relevant background regarding TRNGS, statistical testing techniques, memristor, memristor-based TRNGS and some recent works in the field of true random number generator are presented in section 2. The design, operation, and architecture of the proposed random number generator are presented in section 3. Section 4 discusses the simulation results. In section 5, the comparison between the proposed TRNG design with other TRNGS is presented. Finally, we conclude the chapter in section 6 with suggestions for future research.

2.2 Background and Related Work

In this section, we provide some relevant background information. We discuss true random number generators, memristors, along with memristor-based TRNGs. We also give some information regarding the statistical test suite for randomness like NIST and Die-harder tests.

2.2.1 TRNGs and Statistical Randomness Testing

Many random number generators have been proposed by different researchers from industries and academia. Few true random number generators (TRNGs) are presented in (Sunar, Martin, & Stinson, 2007; Wold & Tan, 2008; Lao, Tang, Kim, & Parhi, 2016; Haddad, Fischer, Bernard, & Nicolai, 2015). A TRNG, with self-repair capabilities, is discussed in (Martin, Natale, & Entrena, 2017). A TRNG design that is resistant to a side-channel attack is presented in (Chari et al., 2010).

Although evaluation of the unpredictability and the reliability of a random bit stream generated by TRNG is a difficult task, there are still some widely accepted techniques to test the statistical randomness of a bit stream. The NIST statistical test suite (Rukhin et al., 2010) is widely used and accepted to test the statistical randomness of a binary bit string. This test suite consists of 15 tests that can be further classified into those which are binomial based and others which are chi-square tests based. The Dieharder test suite, another popular choice, has a group of 26 different tests. It is an enhancement of the Diehard battery of the test presented in (Marsaglia, 1995). The results of the proposed true random number generator are verified through both these test suites.

The following are included in the NIST test suite.

    I. Frequency test: The main purpose of this test is to determine the proportion of 0s and 1s in a given random bit sequence. The proportion of 0s and 1s should be equal for an ideal random bit stream. So, it tests the closeness of 1s to the ½. In other words, we can say that the numbers of 1s and 0s should be equal for an ideal random bit stream. The passage or failure of this test determines whether further tests should be performed.

    II. Frequency test in a block: The idea behind this test is to find out the proportion of 1s in a fixed-length block instead of the entire string. If the length of the block is L, then the proportion of 1s should be L/2 for an ideal assumption of randomness. If L becomes 1 then it would be converted in a frequency mono bit test.

    III. RUNS test: A string of similar bits is called a run. For a run of length l, the string should contain exactly l number of similar bits and sandwiches with different bits at the start and the end. The primary motivation behind this test is to observe the continuous 1s or 0s, i.e., runs. If the number of runs of 1s and 0s does not match the required criteria, then the sequence is not random. In simpler terms, the oscillation between ones and zeroes is calculated, checking whether oscillation is slower or faster.

    IV. Longest RUN: Here, we test the maximum length run of 1s in a fixed-sized block. The main intention here is to find out if the span of the longest run in the input bit stream is in accordance with the expected length for a truly random bit sequence. So, here we can discover asymmetry in the span of longest runs of 1s. Also, there is no need to check the 0s as irregularities in 1s mirror those in 0s.

V. Binary Matrix Rank test: In this test, the whole bit stream is scanned and forms disjointed sub matrices. Then these matrices are ranked. This test assesses the linear dependency between the substrings of a certain length in the given random bit strings. This test is also performed in the Dieharder randomness test suite.

VI. Spectral test: In this test, we want to identify peaks in the Discrete Fourier Transform of the random bit stream. In this test, the purpose is to identify periodic patterns in mutually close proximity. If there are many repetitive periodic patterns, that means the bit streams diverge from the assumption of randomness. Ideally, we have to look out for peaks which exceed the limit of 95%, which is substantially different than 5%.

VII. Non-Overlapping Template Matching test: Here, we seek to know how many times a pre-determined string appears in the sequence. The primary aim of this test is to find out the generators, which are the reason behind the production of multiple appearances of a non-periodic pattern. This test and the next, which matches the overlapping patterns, use a window of size s to look for the s-bit pattern. The window is shifted 1 bit right if the pattern is not found; otherwise, the window starts from the right side of the entire pattern.

VIII. Overlapping Template Matching test: In this test also, we seek to know the number of appearances of pre-determined bit strings. Similar to previous tests, an s-bit window is employed to locate the s-bit pattern. In the previous test, in the absence of the pattern, the window would move one bit right but, in this test, the window shifts one bit right if the pattern is found before starting a new search.

IX. Universal Statistical test: This test looks for matching patterns to calculate the bits between them. This measurement is related to the span of the bit stream, which is compressed. The main purpose of this test is to identify if the span of a bit stream could be substantially compressed and whether the information is intact or not. For a sequence to be truly random, it should not be significantly compressible.

X. Linear Complexity test: In this test, a linear feedback shift register's length is evaluated. The aim of this test is to identify that a bit sequence is complex enough or not to consider as a random sequence. The length of an LFSR determines its randomness. If the LFSR is too long, the bit sequence is random; if too short, it is not random.

XI. Serial test: The identification of the frequency of every possible s-bit pattern in the whole bit sequence is the goal of this test. The main purpose of this test is to determine whether the number of appearances of the 2 s-bit pattern is almost equal to what is required for a truly random bit stream. Since the random sequence possesses uniformity, the chance of the appearance of two s-bit patterns is equal. If the s=1, the serial test becomes the frequency test discussed previously.

XII. Approximate Entropy test: In this test, the whole bit stream is scanned to look for every possibility of s-bit pattern (overlapping). Then the frequencies for all such patterns are calculated. Further, the blocks (overlapping) which differ in block length by 1 (for example, length s and s+1), have their frequencies compared against the required frequency for random bit streams.

XIII. Cumulative Sums test: This test's goal is to find the maximal excursion of the random walk, which is actually determined by the cumulative sum of the adjusted digits in the entire sequence. In this test, the partial bit sequence's cumulative sum is calculated. The cumulative sum of the partial sequence should be neither too large nor too small as compared to the value that is required for a truly random sequence. The excursion of a random walk is closer to zero for a bit stream (random). The excursion of a random walk is away from 0 for a non-random binary sequence.

XIV. Random Excursions test: The number of cycles with s visits in a cumulative sum random walk is the focus of this test. In this testing, the sequence 0s and 1s are converted into -1 and +1, respectively. Further, from the partial sum, the cumulative sum of a random walk is produced. The main purpose of this test is to identify whether the frequency of visits to a certain state within the limit of a cycle is as expected or not. This is in actuality a series of eight tests and eight conclusions for eight states, which are 1,2,3,4, and -1,-2,-3,-4.

XV. Random Excursion Variant test: The primary attention of this test is the frequency of visits to a particular state in a cumulative sum random walk. In this test, we check how much deviation there is from the required amount in the visits to different states in the random walk. Unlike the previous one, this is a combination of 18 tests and 18 conclusions, one for every state, which are 1,2, 3,..., 9 and -1,-2,-3,...,-9.

After the NIST testing suite, we use Dieharder randomness testing, which consists of 26 tests. Here are some of the tests:

I. Birthdays test: The idea behind this test is to choose k and l where k is the birthday, and l is the number of days in the year. Then we find out the r, which is asymptotically Poisson distributed having the mean of k3 ÷ (4l). Here we take a sample of approximately 500 r and use the chi-square goodness of fit test to provide a p-value. The bits 1–24 is used in the first test. Then the next 2–25 bits use in the second test, and thereafter 3–26 up to 9–32 bit are used to generate the p-value.

II. Overlapping 5-Permutation test: This test is also called the OPERM5 test. It operates on a string of 1 million random integers, each 32-bit. Every continuous combination of five integers could have a one-state among 120 states, which is a possible number of states for five numbers in different orders. So, here every number would contribute to a state. Here a lot of state transitions take place, so the focus is to count the cumulative sum of each state. This test considers one million integers two times to produce the test results.

III. 32X32 Binary Rank test: In this test, a 32X32 binary matrix is constructed to find out the rank of the matrix. Every row of this matrix is a random integer of 32-bit. Now, the rank of the matrix is calculated. It is well established that a rank of less than 29 is rare, so any rank less than 29 is kept with rank 29.

We calculate the rank on 40000 matrices, and then we perform chi-square test for the rank,<=29, 30, 31, and 32.

IV. 31X31 Binary Rank test: In this test, a 31X31 binary matrix is constructed by taking the 31 leftmost bits from random integers of 31-bits to find out the rank of the matrix. Every row of this matrix is a random integer of 31-bit. Now, the rank of the matrix is calculated. It is well known that a rank less than 28 is rare, so any rank less than 28 is kept with rank 28. We calculate the rank on 40000 matrices, and then we perform chi-square test for the rank,<=28, 29, 30, and 31.

V. 6X8 Binary Rank test: In this test, we choose a generator for the test, and from this generator, randomly select six random integers each of 32-bit. Now from these six random integers, one random byte would be chosen from each one. So in this way, we would form a matrix of 6X8. Now, the rank of the matrix is calculated. A rank less than 4 is rare, any rank less than 4 is kept with rank 4. We calculate the rank on one million matrices, and then we perform chi-square test for the rank,<=4, 5, 6, and 7.

VI. Bit Stream test: In this test, the random input sequence is considered as a string of bits. We can call it as a b1b2b3b4.... Here we consider this string as an overlapping combination of 20 letter words, where each letter can have a value either 0 or 1. With this in mind, the first word would be b1b2b3b4b4...b20. Similarly, the second word would be b2b3b4b4...b21. The focus is to find out how many times a 20 letter word is missing from the bit stream of 221 overlapping words of 20 letters. For an ideal random sequence, this missing number should be a standard normal variate, which gives a uniform p-value. We repeat this test 20 times.

VII. OQSO, DNA, and OPSO tests: The OPSO stands for overlapping pair sparse occupancy. The OPSO test chooses only two-letter words out of 1024 letters. In this test, the focus is to produce 221 overlapping two-letter words and check how many times the two-letter word is missing in the entire bit stream. These values are very much nearer to a normal distribution with mean 141909 and sigma 290. Hence it becomes a standard normal variate, which would give uniform p-value. This test chooses s 32-bit at a time, selects any 10-bit set from this, and again restarts to select next 10-bits and so on. The OQSO stands for overlapping quadruples sparse occupancy. This test is very much similar to the OPSO test. The difference is that it takes four-letter words from a 32-letter alphabet. Similar to the previous test, this test also finds the count of missing words, and here, the mean of this would come to 141909, and this time, the sigma would be 295. Now in the DNA test, the alphabet consists of four letters defined by the two bits from random integers. Here also, the focus is to find the count of the missing overlapping 10-letter words.

VIII. Counts 1st Stream test: This test takes the random bit sequence as an input in the form of bytes. Each byte may or may not contain 1s. So the probabilities of each byte having no 1s to all 1s are 1,8,56,..., 1. Now the 5-letter word would be provided, which are overlapping words. So the total number of five letters word is 256000 (overlapping) and this test counts each word's frequencies.

IX. Count 1st Byte test: This test takes the random bit sequence as an input in the form of 32-bit integer and the left 8 bits are selected as a byte. Each byte may or may not contain 1s. As above, the probabilities of each byte having no 1s to all

1s are 1,8,56,…, 1. Now, the specific bytes from consecutive integers generates a sequence of five-letter words. The count of 1s decides the letter. Now similar to the previous test, here also the total number of five letters word is 256000 (overlapping) and this test counts the frequencies of each word from the total five-letter words.

X. Parking lot test: This test is based on an experiment of parking a car in a square, each side of which measures 100. Now, if we try to park the second car and then a third car and so on, there are two possibilities, either one can successfully park the car, or there is a crash. So if the car is successfully parked, the list of cars successfully parked is incremented, and if there is a crash, a new attempt is made to park the car at a different place. Now, if we try to plot the curve between a and s, where a is the total number of attempts to park a car and s is the number of cars successfully parked, interestingly, the curve would be very much closer to that of generated by the random number generator. Since there is no concrete theory behind this phenomenon and also no graphics are available for this test, a simple experiment is performed for this test with s and a = 12000. The experimental results show that if k is 3523 with sigma 21.9 then it would be very nearer to the normal distribution. This provides the input to further KSTEST.

XI. Minimum Distance test: In this test, a set of f=8000 points are chosen on a square, each side of which measures 10000. Now the focus is to calculate the l, which is the minimum distance between the pairs of the randomly chosen points. The minimum distance would be exponentially distributed with a mean of 0.995, if the points chosen are uniformly independent. The value 1 – e (–l2 ÷ 0.995) would be uniform, and the KSTEST on the output becomes the test of uniformity on the chosen random points. This test is repeated 100 times.

XII. 3D Sphere test: In this test, a cube of size 1000 is taken into consideration, and 400 points are chosen on that cube. Taking each point as center and making a sphere, the volume of the smallest sphere would be much closer to the exponential distribution with mean 120pi/3. So,the radius cube becomes exponential, having the mean 30. 4000 such tests were computed 20 times. Every mean radius evaluates p-values, which further is used in KSTEST.

XIII. Squeeze test: In this test, the focus is to find the uniforms on [0,1) by floating random integers. It starts with m = 231 and tries to find out the number of iteration n required to decrease the m to 1. This deduction is made using the function m = ceiling(mxL), where L is given by the input file. The n will be calculated for 1 lac times and the count the number of times the n is less than or equal to 6 or greater than or equal to 48. This then will be used in the chi-square test.

XIV. Overlapping Sums test: In this test, overlapping sums are calculated, which are normal to a specific covariant matrix. These sums are converted into independent standard normal by using linear transformation, which is further converted into a uniform variable for a KSTEST.

XV. Runs test: This test performs the up runs and down runs in the given integers inputs. For example, in the sequence 0.234,0.442,0.835, 0.654,0.345,0.546,0.093, there are up runs of 3 and down runs of length 2. The covariance matrix form with runs up and runs down is best suited to the chi-square test. In this, the runs of length 10000 are counted, and the test is repeated ten times.

XVI. Craps test: This is the gaming of two lac crap games where one tries to find out the number of throws and the number of wins required to complete a game. The number of wins must be a normal, having mean of 200000m, and variance should be 200000m(1-m), given m = 244/495. The number of throws cells count is used to perform the chi-square test.

2.2.2 Memristors and Memristor based TRNGs

A memristor is a nano-scale device with unique features. Various applications of memristors like non-volatile memory, neuromorphic systems, and content-addressable memory (CAM), etc., existed after its first fabrication (Strukov, Snider, Stewart, & Williams, 2008). Though there have been numerous memristor models proposed, the model with a heavily doped TiO2-x layer with oxygen vacancy kept over an almost undoped TiO2 layer is situated between two metallic electrodes, as shown in Figure 2.2(a). This can be considered as a serial combination of two registers, as depicted in Figure 2.2(b). The equivalent resistance can be written as:

$$R_{eq}(t) = \frac{w(t)}{D} \cdot R_{on} + \left(1 - \frac{w(t)}{D}\right) \cdot R_{off} \tag{2.1}$$

where Ron is the low resistance, Roff is the high resistance, w(t) is the instantaneous width of the doped region, and D is the total width of the device (Mathew, Chakraborty, Sahoo, Yang, & Pradhan, 2015).

The memristor's primary feature is that its instantaneous resistance is retained even after the removal of excitation voltage, which makes it unique. The process variations of the device affect the total width D by the magnitude of several orders, which can further affect the electrical characteristics, including the delay of the devices. This variation in the delay can be harnessed to produce randomness, as done in this work. The total width of the device D is highly affected by the process variations. This may be varied by the magnitude of several orders, which affects the electrical characteristics, including the delay of the devices. This delay variation, in turn, can be utilised to generate random bits, as done in this work.

2.2.3 Related Works

Recently, memristor has been widely used in building hardware security primitives such as PUF, TRNG, etc. Energy efficiency, high integration density, and extractable manufacturing process variations make nano-scale devices like memristor more effective than

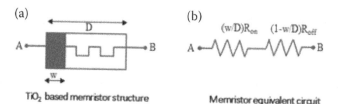

FIGURE 2.2
Memristor: (a) structure, (b) Equivalent circuit.

conventional CMOS devices. Recently, different TRNG design has been proposed by various researchers who use the manufacturing process variation induced in memristor to produce better randomness properties. A CMOS-memristor hybrid TRNG has been proposed in (Hashim, Teo, Hamid, & Hamid, 2016). This work demonstrates a TRNG design, which is a 141 stage ring oscillator based structure, where each stage consists of one memristor and one NMOS (each memristor replaces the PMOS transistor of each inverter stage). The rest of the circuit is inspired by (Sunar et al., 2007). The prime features of this architecture are its regularity and simplicity. In this design, the output signals from a few ring oscillators are XOR-ed, and the final output signal is given to a D flip-flop as data input. A clock reference is used for a clock signal for this D flip flop. The TRNG output would be the output of the D flip-flop. The phase difference between the ring oscillators is the main source of the randomness in this design, which actually happens due to random jitter enhanced due to process variations of the memristor. In this work, it is demonstrated that due to the significant randomness of the memristor, the ring oscillators show better randomness than that of CMOS-inverter based ring oscillator. MTRNG, another memristor-based TRNG, is discussed in (Wang, Wen, Li, & Hu, 2015) and is a compact, fast, and energy-efficient architecture. The random bit stream in this scheme is generated due to the switching between the memristor's binary states. To enhance the entropy of the generated bit string, two MTRNGs are XOR-ed in this design. The TRNG, presented in (Jiang et al., 2017), is based on a diffusive memristor, which exploits the diffusion dynamics of metal atoms. The switching of the device from a low state to a high state is performed after a random time as soon as the applied voltage is removed. This stochasticity is harnessed as a source of randomness to produce random bit streams.

2.3 Proposed Multi-bit Random Number Generator

The schemes discussed in the previous section are various kinds of true random number generators. Some of them use memristors in their architecture, whereas few of them are CMOS-based true random number generators. Although these schemes possess good randomness, they are still not very useful in an IoT oriented environment because of higher resource requirements. In this section, we explain the proposed architecture for an efficient but lightweight true random number generator, which also has good statistical randomness. We also present the modified architecture of the true random number generator by removing the memristors from the design. The results and comparison will be discussed in subsequent sections.

The primary component of proposed TRNG architecture is two quasi-identical ring oscillators, as shown in Figure 2.3. Each ring oscillator has an odd number of inverters. In each ring oscillator, the output of the inverter is given as an input to the memristor. Then, the output of the memristor serves as the input to the next inverter and so on. In the first ring oscillator, the signal after the last inverter is tapped and used as a data input to a positive triggered D flip-flop. Similarly, in the second ring oscillator, the signal after the last inverter is tapped and used as a clock input to the same positive triggered D flip-flop. The D flip-flop acts as an arbiter and produces output depending on the temporal relationship between the clock signal and the data signal. Since the memristor exhibits pronounced process variations and CMOS also has uneven process variations, this shows a good amount of randomness. Hence, random phase shifts would occur between the RO

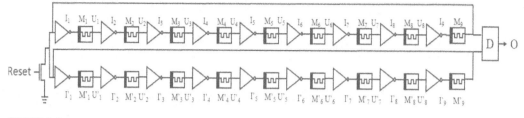

FIGURE 2.3
Single-bit true random number generator.

oscillation waveforms, which in turn can be digitised to generate the random bit streams. The random bit streams are again sampled using a sampling clock signal to produce a more stable bit sequence. We can extend our architecture to a multi-bit TRNG, as shown in Figure 2.4. Here, we capture the signals after each memristor in both ring oscillators. The signal from the top ring oscillator is connected to the data input of the D flip flop. The signal from the bottom ring oscillator is connected as clock input of the D flip-flop. Each arbiter produces the output based on the temporal relationship between the clock signal and the data signal. This design increases the efficiency of the single-bit architecture by producing multi-bit output. Unlike single-bit TRNG architecture, where we would be able to produce a one-bit response, in multi-bit architecture, we would produce an n-bit response. A parallel to serial converter may be used to collect the output. We may reset the main component of the circuit by a global reset signal. We can understand the improvement in the throughput by a simple example: if n=64 in a TRNG, normally, we can generate 1-bit output in single-bit TRNG. But, multi-bit TRNG design can give us the 64-bit output. Suppose, in RSA, we require 1024 bits key, now with only 16 repetitions of this

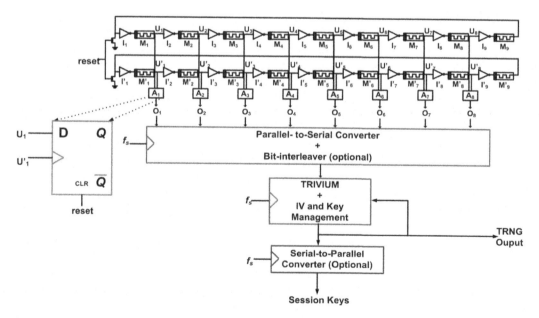

FIGURE 2.4
Proposed true random number generator (multi-bit) circuit.

operation, we can acquire our desired bit length of 1024 bits. This would greatly increase the rate of bit generation.

Finally, for post-processing, the TRIVIUM synchronous stream cipher (Canni`ere & Preneel, 2005) is used. TRIVIUM is a lightweight cipher, and since it is also simpler in implementation, we choose this for post-processing. The starting state is set up using an 80-bit key and a vector IV. The randomness is imposed by TRIVIUM through a mixing of secret key and initialization vector for 1152 cycles, and then it starts producing the output at the rate of one bit per cycle. We have a flexible hardware design of the cipher which can generate one or eight or 64 bits outputs in one clock cycle. If cryptographic keys of desired lengths (e.g. 128-bit) are required, the output of this TRIVIUM cipher has to be processed further by a serial-to-parallel-converter.

2.3.1 TRNG Architecture without Memristor

We have discussed a lot about the memristor and its importance in the background section of this chapter. We explained how a memristor could improve randomness if it is introduced in a circuit by its dynamic behavior. To prove that point here, we slightly modify the original circuit by removing the memristor from the circuit. This circuit is still able to produce the randomness in the output, but the statistical measurements of the randomness would explain the difference between the two circuits. The architectural design of the proposed TRNG without memristor is as shown in Figure 2.5. The primary component of the design has two identical ring oscillators. Each ring oscillator is comprised of nine inverters. The output of one inverter is given as input to the next inverters, and the output of the last inverter is connected to the first inverter, similar to the previous architecture. The nodes after the first inverter from the top and bottom ring oscillators are connected to a positive triggered D-flip flop. The node of the top ring oscillator is connected as data input to the D flip flop, and the node of the bottom ring oscillator is connected as the clock input to the D flip flop. For instance, we can clearly see that the node E1 from the top ring oscillator is connected as a data input to the D flip flop and the corresponding node in the bottom ring oscillator, i.e., node E'1 is connected as a clock input to the D flip flop. Similarly, the nodes E2, E3, ..., E8 are connected to the different D flip flops as data inputs, and the corresponding nodes to these nodes are E'2, E'3,..., E'8 connected as clock inputs to the corresponding D flip flops. Since there are manufacturing process variations in the CMOS inverters, the propagation delay of the signals in both the

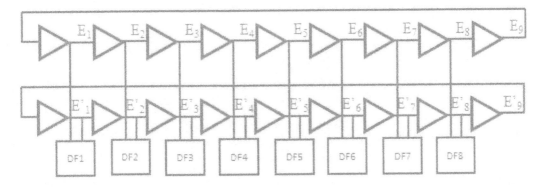

FIGURE 2.5
True random number generator (multi-bit) without memristor.

ring oscillators are different even if we start the signals simultaneously. Due to this difference in the delay, the random phase shift would occur in the both the signals. So, when these two signals would be connected to the D flip flops, it would work as an arbiter. The output of the arbiter would be generated by considering the temporal relationships between both the signals. We use another sampling clock signal to produce a random bit stream using a parallel to serial converter as we collect the random bits from many arbiters in parallel. In the design, we use a global reset signal which can reset the main component of the circuit as needed. Since we remove the memristor in this architecture of true random number generator, there is less variability in the propagation delay in both the ring oscillators, which in turn negatively affects the randomness of the circuit. So, it finally reduces the statistical randomness in the generated bit streams.

2.3.2 Bit Correlation Effect

Though we have proposed a multi-bit random number generator, the binary bits which are collected from the relatively nearer nodes, or nodes in the close neighborhood, has a high chance of correlation. In other words, we can say that these bits may be correlated. So, it affects the randomness of the entire random bit sequence. To test the correlation effects on the random bit stream we collected the five hundred thousand bits generated from each arbiter i.e. from O1, O2, O3, ..., O8. We then performed the bitwise XNOR operation between these random streams collected from each arbiter separately. So, the bit correlation is computed between O1 and O2, O1 and O3, O1 and O4, O1 and O5, O1 and O6, O1 and O7, O1 and O8. Figure 2.6 shows the percentage of the bit correlation between the different combinations of the nodes. The high bit correlation is the indication of a high degree of similarity between the nodes. We interpret from our simulation results that the nodes which are at the distance of $(n-1)/2$ are approximately 50/% correlated, where the n is the number of inverters in a ring oscillator. Thus, these nodes are highly recommended to combine in a single bit stream. However, the nodes which are at n-1 distant away from each other have the highest level of bit correlation due to the "ring structure". This is undesirable for random bit streams. The nodes in close proximity have lowest correlation, which is also not a favorable condition for random bit streams. So, neither high nor low correlation is a good factor for a true random number generator if a sufficient level of statistical randomness in the generated bit streams is desired. This implies a constraint in the multi-bit random number generator in the selection of nodes to

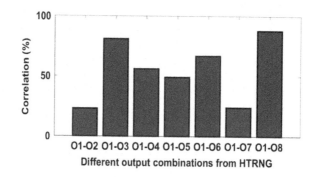

FIGURE 2.6

Correlation between bits generated from different ring oscillator nodes of the proposed true random number generator.

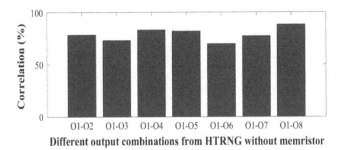

Different output combinations from HTRNG without memristor

FIGURE 2.7
Correlation between bits generated from different ring oscillator nodes of the proposed true random number generator without memristor.

combine together in order to generate a random bit sequence. Following this logic, the best two nodes to combine together to get a satisfactory highly random bit sequence are node 1 and node 5 because they have nearly 50% bit correlation.

Furthermore, we have done the same analysis for the true random number generator without memristor to assess the bit correlation between the different nodes. The result is shown in Figure 2.7. A high bit correlation indicates a higher level of similarity, and a low bit correlation signifies a lower level. We collected five hundred thousand bits, as we did in the previous case, to test the bit correlation. We collected the bits from O1, O2, O3 up to O8 and calculate the bit correlation between O1 and O2, O1 and O3, O1 and O4, O1 and O5, O1 and O6, O1 and O7, O1 and O8 by applying a simple XNOR operation between the bits generated from each node. The results obtained clearly indicate that the bit correlation is high for all these different node combinations; they dictate that no node will get together to generate a high throughput. Thus, it is not recommended for this type of architecture to combine nodes to collect parallel output in order to get a high throughput random number generator.

2.4 Experimental Results

This section of this chapter explains the simulation setup and results for the multi-bit true random number generator. We also discuss the NIST test suite results and Diehard test suite results carried out to test the statistical randomness of the generated bit streams.

2.4.1 Simulation Setup Details

We designed and simulated the proposed true random number generator using Synopsys HSPICE circuit simulator using a 45nm Predictive Technology Model (PTM). We implemented the circuit netlist in HSPICE and collected the data. Then we used some extensions like HSPICE TOOLBOX to import those data to MATLAB for further processing and digitization. There are several memristor models widely accepted and used, but we chose to use the memristor model discussed in (Mathew et al., 2015). We consider 10%-15% variations in the parameters like Ron, memristor resistance in on state, and Roff, memristor resistance in off state, and over the threshold voltage Vth of the transistors.

TABLE 2.1

Parameters for memristor

Parameter	Value
R_{on}	60 Kohm
R_{off}	120 ohm
L	10mm
D	10mm
W	10mm
μ	10^{-14} m^2/(V.s)

The value of Roff is taken as 60 kΩ as the base value, and 120 Ω is taken as Ron as the base value. The memristor parameters are shown in Table 2.1. The sampling frequency used here is 1.8 GHz. Figure 2.8(a) and 2.8(b) show the distributions of resistances Ron and Roff. These variable resistances are the major source of randomness in the design. The variations in the threshold voltage Vth is shown in Figure 2.8(c), which also induced variability in the delay of the inverters. The variation in the delay of the inverters is shown in Figure 2.8(d). So, these variations in the different parameters trigger randomness in the path delay variations of the circuit, which is the basis of our true random number generator design.

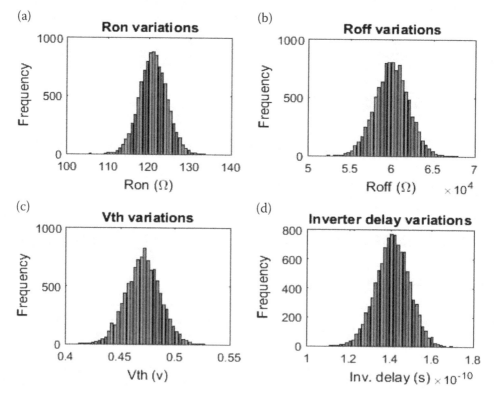

FIGURE 2.8

R_{on}, R_{off}, V_{th} and inverter delay distribution.

2.4.2 Statistical Randomness Testing Results

The randomness of the random bit streams generated by the true random number generator should be measured by some globally accepted testing techniques. So, we used two standard and commonly used and widely accepted testing techniques called the NIST and Dieharder test suites. We implemented these tests over 100 lacs bit, which are generated using the proposed true random number generator. The NIST test results and Dieharder test results are compiled in Table 2.2 and Table 2.3, respectively. Since the P-values obtained in the NIST tests are above 0.001 for all the tests, so the proposed TRNG passed the NIST tests. Similarly, P-values found to be greater than the threshold value in the Dieharder test, so the bit streams passed this test too. Hence, we can say that our design offers satisfactory levels of randomness. The NIST tests for TRNG without memristor are also shown in Table 2.2, which clearly show that it fails some of the tests, like Longest RUNS, Universal test, RUNS test, Serial tests, Approximate entropy test, Random Excursions, and Random Excursions variant test. The results of the Dieharder test suite for TRNG without memristor are also compiled in Table 2.3, which indicates weaker results in RUNS tests and STS RUNS tests etc. We also compare the proposed TRNG design with other TRNG discussed in (Acerbi et al., 2018). The result of the NIST test comparison is given in Table 2.2. The result clearly showed that the proposed TRNG performed better in eleven tests, while two tests could not be compared. In only two tests out of 15, proposed TRNG lagged behind the RNG discussed in (Acerbi et al., 2018). We also compared the proposed TRNG with a TRNG proposed in (Yang, Blaauw, & Sylvester, 2016) against the Dieharder test, and the result is compiled in Table 2.3. It illustrates that the proposed TRNG had better performance in twenty-three tests but lagged in three.

TABLE 2.2

NIST Test Suite Randomness Testing Results comparison between proposed TRNG, TRNG without memristor, and RNG discussed in (Acerbi et al., 2018)

Test	Memristor TRNG (proposed)		TRNG without memristor		RNG in (Acerbi et al., 2018)	
	Proportion of passed test	P-value	Proportion of passed test	P-value	Proportion of passed test	P-value
Frequency	992/1000	0.6124	995/1000	0.1972	988/1000	0.372
Block Frequency	994/1000	0.6983	992/1000	0.5682	983/1000	0.4980
Cumulative Sums	995/1000	0.7219	986/1000	0.4628	988/1000	0.9236
Runs	996/1000	0.6832	64/1000	0.0824	991/1000	0.8684
Longest Run	993/1000	0.4389	46/1000	0.0184	996/1000	0.8940
Rank	997/1000	0.3536	989/1000	0.2264	992/1000	0.2164
FFT	997/1000	0.7292	988/1000	0.5928	984/1000	0.8574
Non Overlapping Template	995/1000	0.7382	987/1000	0.4825	989/1000	0.6214
Overlapping Template	993/1000	0.0658	982/1000	0.0684	978/1000	0.6124
Universal	994/1000	0.4286	70/1000	0.0035	982/1000	0.4252
Approximate Entropy	989/1000	0.4680	135/1000	0.0942	985/1000	0.2516
Random Excursions	994/1000	0.0584	25/1000	0.0784	990/1000	0.4782
Random Excursions Variant	991/1000	0.2082	18/1000	0.0862	988/1000	0.6938
Serial	994/1000	0.6326	150/1000	0.0286	990/1000	0.2064
Linear Complexity	989/1000	0.0924	987/1000	0.3894	994/1000	0.2904

TABLE 2.3

Dieharder Test Suite Randomness Testing comparison between proposed TRNG, TRNG without memristor, and RNG discussed in (Yang et al., 2016)

Test	Proposed TRNG	TRNG without memristor	TRNG in (Yang et al., 2016)
	P-value	P-value	P-value
Birthdays Test	0.8928	0.4206	0.9136
OPERM5 Test	0.7536	0.3580	0.2075
32x32 Binary Rank Test	0.6028	0.5836	0.4905
6x8 Binary Rank Test	0.6382	0.4137	0.4817
Bit stream Test	0.6974	0.9209	0.8935
OPSO	0.0684	0.3974	0.0486
OQSO Test	0.4194	0.3897	0.2795
DNA Test	0.9694	0.9725	0.8473
Test for 1st stream count	0.5896	0.5391	0.2614
Test to count 1st byte	0.9176	0.7402	0.8132
Parking Lot Test	0.6109	0.8205	0.2784
Test for 2d Circle Minimum Distance	0.7294	0.2973	0.2174
Test for Minimum Distance (3D sphere)	0.5894	0.3850	0.3186
Squeeze Test	0.8503	0.7291	0.3482
Sums Test	0.5907	0.5904	0.4385
Runs Test	0.5291	0.4938	0.3461
Craps Test	0.4026	0.2672	0.2965
Test for GCD (Marsaglia and Tsang)	0.7901	0.5162	0.6283
Test for Monobit (STS)	0.8217	0.7093	0.4284
Test for Runs (STS)	0.3906	0.2194	0.1892
Serial Test (STS)	0.9026	0.5092	0.7395
Bit Distribution Test	0.5209	0.3985	0.3495
Test for Minimum Distance Test (Generalised)	0.4297	0.1905	0.3508
Permutations Test	0.7093	0.6194	0.4297
Lagged Sum Test	0.9302	0.8184	0.8826
Kolmogorov-Smirnov Test	0.6385	0.6143	0.5739

2.4.3 Entropy Calculation

The n-bit entropy shows the amount of entropy exhibits by the n-bit output of a random number generator. The entropy analysis is carried out using the bit strings generated from our proposed true random number generator. Entropy is defined by Equation (2.2). We used the Equation (2.2) and the entropy is calculated as 0.9683.

$$H_n = \left(- \sum_{x \in (0,1)^n} P(x) \cdot log_2 P(x) \right) \tag{2.2}$$

TABLE 2.4

Comparison between existing TRNGs based on different parameters

TRNGs schemes	Output bits	Entropy source	NIST Pass
Ring oscillator based (Hashim et al., 2016)	Single	Ring oscillator	No
MTRNG (Wang et al., 2015)	Single	Memristor and transistor	No
Proposed TRNG	Multiple	Memristor and ring oscillator	Yes
TRNG without memrisor	Single	Ring oscillator	No

2.5 Comparison with Existing Memristor Based TRNGs

In this section, we compare the proposed true random number generator architecture with the two memristor based TRNGs discussed in (Hashim et al., 2016) and (Wang, Wen, Li, & Hu, 2015) and also with the TRNG without memristor to find out the benefits of our proposed design. In this comparison, along with statistical randomness, we have used some other parameters to find out the clear differences between these true random number generators. Here, we have taken the parameters like output bits, entropy source, post processing and statistical testing. The results of the comparison are presented in Table 2.4. The main difference between the proposed TRNG and the other TRNGs discussed in Hashim et al. (2016) and Wang et al. (2015) is that proposed TRNG has a high throughput as it may generate multiple output bits in a single cycle, whereas other TRNGs can only generate one bit in a cycle. This is the demonstrated advantage of our design. The proposed architecture is favorable in terms of hardware requirements also. The TRNG presented in Hashim et al. (2016) needs 141 inverters and 141 memristors in the designs. The TRNG discussed in Wang et al. (2015) requires 12 transistors and two memristors in its architecture. However, the proposed design only requires 2 memristors and 2 inverters effectively to produce one bit. Statistical testing is performed for the proposed design, whereas no such testing is carried out in the TRNGS presented in Hashim et al. (2016) and Wang et al. (2015). The bit correlation analysis is done for proposed architecture and also for the TRNG without memristor. The bit correlation analysis for the true random number generator without memristor clearly shows that the bit correlation is high between all the nodes like O1 and O2, O1 and O3, O1 and O4, O1and O5 etc. So, this is an undesirable situation to combine the nodes for multiple bit output. The comparison is also done for the NIST test between proposed TRNG and TRNG without memristor, which clearly indicates that proposed TRNG performed well and passed all the tests while the TRNG without memristor failed in many tests like Longest RUNS, Universal test, RUNS test, Serial tests, Approximate entropy test, Random Excursions and Random Excursions variant.

2.6 Conclusion

A CMOS-memristor based high throughput true random number generator with its complete architecture and post processing is presented in this work. We assess the statistical randomness of the proposed TRNG using two widely accepted statistical randomness testing suite like NIST and Dieharder. We also compare our design with some of the existing TRNGs and find our design to be better in terms of randomness, hardware requirements and throughput. Future work should focus on the analysis of the circuit and maximum limit of the extractable entropy.

References

Acerbi, F., Bisadi, Z., Fontana, G., Zorzi, N., Piemonte, C., Pavesi, L. (2018). A robust quantum random number generator based on an integrated emitter-photodetector structure. *IEEE Journal of Selected Topics in Quantum Electronics*, 24(6), 1–7.

Canni`ere, C. D., & Preneel, B. (2005). TRIVIUM specifications. eSTREAM submitted papers. Accessed: May 2017 [Online]. Available: http://www.ecrypt.eu.org/stream/p3ciphers/trivium/triviump3.pdf

Chari, S. N., Diluo_o, V. V., Karger, P. A., Palmer, E. R., Rabin, T., Rao, J. R., Rohotgi, P., Scherzer, H., Steiner, M., & Toll, D. C. (2010). Designing a side channel resistant random number generator. In Smart Card Research and Advanced Application: 9th IFIP WG 8.8/11.2 International Conference, CARDIS 2010, Passau, Germany, April 14–16. Proceedings. Berlin, Heidelberg: Springer Berlin Heidelberg, 2010 (pp. 49–64) [Online]. Available: https://doi.org/10.1007/978-3-642-12510-2_5

Fischer, V. (2014). Random number generators for cryptography design and evaluation [Online]. Available: https://summerschool-croatia.cs.ru.nl/2014/slides/Random Number Generators for Cryptography.pdf

Fischer, V. (2012). A closer look at security in random number generators design. In W. Schindler, & S. A. Huss (Eds.), *Proceedings of the third international workshop on constructive side-channel analysis and secure design (COSADE 2012)* (pp. 167–182). Berlin, Heidelberg: Springer Berlin Heidelberg.

Gupta, A., Chakraborty, C., & Gupta, B. (2019). Medical information processing using smartphone under IoT framework. *Springer: Energy Conservation for IoT Devices, Studies in Systems, Decision and Control*, 206, 283–308, ISBN 978-981-13-7398-5, https://doi.org/10.1007/978-981-13-7399-2_12.

Haddad, P., Fischer, V., Bernard, F., & Nicolai, J. (2015). A physical approach for stochastic modeling of tero-based trng. In *Workshop on Cryptographic Hardware and Embedded Systems, CHES 2015, st-malo, France* [Online]. Available: https://hal-ujm.archives-ouvertes.fr/ujm-01164105

Hashim, N. A. B. N., Teo, J., Hamid, M. S. A., & Hamid, F. A. B. (2016). Implementing memristor in ring oscillators based Random Number Generator. In 2016 IEEE Student Conference on Research and Development (SCOReD). IEEE (pp. 1–5).

Jiang, H., Belkin, D., Savelev, S. E., Lin, S., Wang, Z., Li, Y., Joshi, S., Midya, R., Li, C., Rao, M.et al. (2017). A novel true random number generator based on a stochastic diffusive memristor. *Nature Communications*, 8(1), 882.

Killmann, W., & Schindler, W. (2011). A proposal for: Functionality classes for random number generators (v 2.0). Bundesamt fur Sicherheit in der Informationstechnik (BSI), Publication, Bonn, Germany. Retrieved May 2017. [Online]. Available: https://www.bsi.bund.de/EN/Home/home node.htm

Lao, Y., Tang, Q., Kim, C. H., & Parhi, K. K. (Apr. 2016). Beat frequency detector-based high-speed true random number generators: Statistical modeling and analysis. *Journal on Emerging Technologies in Computing Systems*, 13(1), 9:1–9:25.

Mandal, R., Mondal, M. K., Banerjee, S., Chinmay, C., & Biswas, U. (2020). *A survey and critical analysis on energy generation from datacenter*. Data de-duplication approaches-concepts, strategies and challenges, New York, NY: Academic Press, (Ch. 11), 203–230, https://doi.org/10.1016/B978-0-12-823395-5.00005-7, ISBN: 978-0-12-823395-5.

Marsaglia, G. (1995). Diehard battery of tests of randomness. Retrieved December 2017 [Online]. Available: https://web.archive.org/web/20160125103112/http://stat.fsu.edu/pub/diehard/

Martin, H., Natale, G. D., & Entrena, L. (2017). Towards a dependable true random number generator with self-repair capabilities. *IEEE Transactions on Circuits and Systems I: Regular Papers*, *PP*(99), 1–10.

Massey, J. (Jan 1969). Shift-register synthesis and BCH decoding. *IEEE Transactions on Information Theory*, *15*(1), 122–127.

Mathew, J., Chakraborty, R. S., Sahoo, D. P., Yang, Y., & Pradhan, D. K. (2015). A novel memristor based physically unclonable function. *Integration, the VLSI Journal*, *51*, 37–45.

Rukhin, A. et al. (2010). A statistical test suite for random and pseudorandom number generators for cryptographic applications. Retrieved May 2017 [Online]. Available: http://csrc.nist.gov/groups/ST/toolkit/rng/documents/SP800-22rev1a.pdf

Strukov, D. B., Snider, G. S., Stewart, D. R., & Williams, R. S. (2008) The missing memristor found. *Nature*, *453*(May), 80–83.

Sunar, B., Martin, W. J., & Stinson, D. R. (2007). A provably secure true random number generator with built-in tolerance to active attacks. *IEEE Transactions on Computers*, *56*(1), 109–119.

Wang, Y., Wen, W., Li, H., & Hu, M. (2015). A novel true random number generator design leveraging emerging memristor technology. In Proceedings of the 25th Edition on Great Lakes Symposium on VLSI, ser. GLSVLSI '15. New York, NY, USA: ACM (pp. 271–276) [Online]. Available: http://doi.acm.org/10.1145/2742060.2742088.

Wold, K., & Tan, C. H. (2008). Analysis and enhancement of random number generator in FPGA based on oscillator rings. In Proceedings of the International Conference on Recon_gurable Computing and FPGAs (ReConFig 2008) (pp. 385–390).

Yang, K., Blaauw, D., & Sylvester, D. (2016). An all-digital edge racing true random number generator robust against pvt variations. *IEEE Journal of Solid-State Circuits*, *51*(4), 1022–1031.

3

Secured Testing of AES Cryptographic ICs for IoT Devices

Kusum Lata

Department of Electronics and Communication Engineering, The LNM Institute of Information Technology, Jaipur, India

3.1 Introduction

The proliferation of the Internet of Things (IoT) devices and the advances in technology are opening the door for business and industry alike to benefit from IoT applications. The basic idea of this concept is that numerous objects such as sensors, cell phones, Radio Frequency Identification (RFID), etc., can communicate with each other to accomplish a common goal via a unique address system (Giusto, Iera, Morabito, & Atzori, 2010). Everything in the IoTs is connected virtually which means every individual and object has a position, a distinct address, and a legible partner through the internet (Atzori, Iera, & Morabito, 2014). The lucrative advantages that IoT can bring also come with lots of security concerns along with their design challenges (Chinmay & Joel, 2020). IoT devices are incorporated more and more often into facets of our everyday lives. The enormous connectivity of these devices and extensive air data communicated through these devices becomes susceptible to various types of attacks. Therefore, these IoT systems demand that high-security features be integral to their implementation. They exchange critical and sensitive data. Therefore, to protect all IoT devices, the proper use of Cryptography by Cryptographic ICs is necessary. Even a cryptographic IC must be checked to ensure its proper functionality. Although various testing methods exist, the most common technique is scan-based testing. When it comes to Cryptographic ICs, these scan-based testing methods can retrieve secret information stored inside them. This happens because scan chain structure provides a backdoor to hackers who can exploit easily sensitive information. Such attacks are known as scan-based side channel attacks. Therefore, the reliability and testing capabilities of these cryptographic ICs used to secure IoT devices need to be assured, to make sure that their functional correctness is maintained.

Cryptographic Algorithms used for security purposes in IoT devices are discussed, along with various side-channel attacks on AES Cryptographic ICs that were reported in the literature. The existing countermeasures for securing the scan chain which is typically inserted in the AES Cryptographic ICs is examined. The design and simulation results of the scan-inserted AES Crypto module, as well as the proposed methods used to enhance the security of the scan chain of the AES Crypto module are presented. The results and analysis of the implemented Crypto module are offered. The conclusion highlights the application of security to the testing architecture of crypto chips.

3.2 Cryptography for Security in IoT Devices

Gartner ("Gartner.Com," n.d.) states that over 80% of organizations have launched IoT, and in the past three years, almost 20% of organizations have identified an IoT-based attack. This happens because the lucrative advantages that IoT can also come with new protection and privacy issues relative to the reliability and credibility of the knowledge sensed, gathered, stored and shared. Such problems mean IoT deployments are particularly susceptible to various kinds of attacks, resulting in unsafe IoT environments. Cryptography has been a security aspect of networks for secure communication and data sharing for a long time.

Many cryptography algorithms are used for various applications in the domain of IoT (Garg, Chukwu, Nasser, Chakraborty, & Garg, 2020). A total of the top 15 most relevant publications from 2016 to 2020 to date were identified from the Scopus database ("Scoups," n.d.). The selected language of the published papers is English.

Citation analysis identified Cryptography, Security, and IoT. The study then employed keyword analysis to identify themes in IoT security by Cryptography research. IoT, Cryptography, Security, Authentication, Public Key Cryptography, and Data Privacy were the themes identified. Figure 3.1 shows the distribution of the 1,576 published research articles on security for IoT devices using cryptography in various subject areas. It shows that theComputer Science and Engineering fields are the major contributors in this domain; however, other fields are also publishing such articles. It is possible to improve IoT network protection using ECC for wireless devices (Gauniyal & Jain, 2019). The security problems that device design engineers face to secure embedded systems are discussed in (Dhillon & Kalra, 2017). It is claimed that ECC is the most suitable crypto algorithm in IoT systems for resource-constrained real-time embedded systems. The use of Physical Unclonable Functions (PUFs) as primitive protection to provide IoT system security is presented in (Aman, Chua, & Sikdar, 2016). This also demonstrates the initial implementation of the PUF-based mutual authentication protocol for IoT applications.

The combined Symmetric and Asymmetric cryptographic model can be used to define secure communication between IoT devices (Henriques & Vernekar, 2017). It is claimed that the combination of cryptographic algorithms reduces the encryption times compared with using only the Asymmetric cryptographic algorithm. Using random keys for Symmetric encryption each time avoids the session-key distribution issue and improves the approach to symmetric encryption.

A hybrid cryptosystem is proposed based on shared-key and public-key cryptography in (Jian, Cheng, & Shen 2019). IoT device MAC address information is used for public key exchanging with the help of asymmetric cryptography, whereas the shared key is exchanged between server and IoT devices. Implementation of the proposed mechanism shows that the hybrid cryptosystem was able to secure the data. HSC-IoT is PUF based security to ensure the hardware-software integrity of IoT devices (Sadkhan & Hamza, 2017). HSC-IoT is based on ECC which provides a lightweight authentication scheme for

Documents by subject area

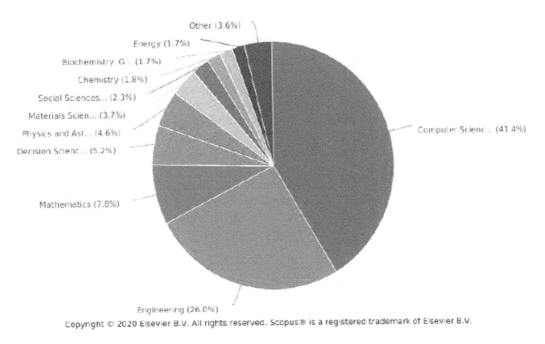

FIGURE 3.1
Distribution of 1,576 relevant publications for "security for IoT using cryptography" in various subject areas from 2016 to 2020.

the resource-constrained IoT devices. Another ECC based ECQV implicit certificates for IoT devices and its usage for authenticated key exchange in DTLS (Ha, Nguyen, & Zao 2016). IoT cryptography-based security solutions of the health care system are suggested (Arunkumar, Vetriselvi, & Thanalakshmi, 2020). IoT systems for environmental monitoring and control are now commonly used every day (Banerjee, Chakraborty, & Chatterjee, 2018). For an IoT based environmental monitoring system, optimum lightweight cryptography was presented and tested for its security purposes in (Jeong, Bajracharya, & Hwang, 2019). Proposed lightweight cryptography was also compared for its performance with the various existing similar approaches. Hardware/Software co-design for ECC would be suitable for lightweight devices (Salman & El-Tawab, 2019). This architecture was implemented on an edge node that could be used in a healthcare system for indoor localization (Chanda et al., 2021). An abstract secure communication model is implemented for IoT using lightweight symmetric block cipher AES (Pattanayak & Amic, 2020). A method to secure communication for IoT devices with the minimum resource usage was proposed in (Anggorojati & Prasad, 2018), loosely/tightly connected network security research and the complex design of IoT devices were also discussed. To include the digital signature and encryption functions to comply with the authenticity and confidentiality of resource-restricted IoT devices, a new simplified certificateless signcryption was suggested (Karati, Fan, & Hsu, 2019). It is proposed that gCLSC can be adopted in case of requirements for confidentiality and authenticity. The IoT application must be lightweight. Well-known cryptographic techniques are used for providing semantic data and reasoning actionable knowledge on context-aware IoT environments

(Jangra & Gupta, 2017). Various cryptographic techniques used in real-life scenarios for IoT applications are compared analytically and proposed in this work. RSA used in many IoT devices is one of the first public-key cryptosystems that are still commonly used for safe data transmission (Rivest, Shamir, & Adleman, 1978). Its protection is based on the complexity of large integers being factorised. Although it is considered that such an algorithm would be unbreakable by general computing devices, it is demonstrated that RSA can be breached by using a quantum computer in polynomial time (Shor, 1999). L Marin et al in (Marin, Pawlowski, & Jara, 2015) have implemented the optimised ECC for protected interaction between heterogeneous IoT devices. Elliptic Curve Diffie-Hellman (ECDH) cryptographic algorithm is designed and implemented for IoT Devices (Akhundov, van der Sluis, Hamdioui, & Taouil, 2020). This key exchange algorithm enables even smaller keys to be used and therefore is the most suitable candidate for wireless IoT devices. RFID authentication schemes based on Elgamal Elliptic Curve Cryptography (EECC) have recently gained a lot of attention and have been used in healthcare applications. EECC-based authentication schemes are analyzed in healthcare environments (Balasubramaniam, Sathya, Ashicka, & Kumar, 2016). Cryptographic algorithms are also used in the Medical Internet of Things (Sun et al. 2018). There are various survey articles (Khattak, Shah, Khan, Ali, & Imran, 2019; Yang, Geng, Du, Liu, & Han, 2011; Yang, Wu, Yin, Li, & Zhao, 2017; Ali, Sabir, & Ullah, 2019; Sun et al., 2018; Perera, Zaslavsky, Christen, Geogakopoulos, 2014; Kane, Chen, Thomas, Liu, & Mckague, 2020; Hameed, Khan, & Hameed, 2019) and many more are published in which the application of cryptographic algorithms in IoT environment and devices are discussed. Therefore, it can be seen that cryptography is becoming the main integral component of the security perspective of IoT devices. Cryptographic algorithms used in IoT devices' security are summarised in Table 3.1 (Adat & Gupta, 2018):

3.3 Advanced Encryption Standard (AES) Algorithm for Security in IoT Devices

IoT system protection is of utmost importance in any IoT network; therefore, security is crucial to ensure the confidentiality, integrity and authenticity of the data. Cryptographic systems are essential solutions for IoT device protection and privacy (S. Khan et al., 2019; Farooq, Hasan, Baig, & Shehzad, 2019; Dhanda, Singh, & Jindal, 2020). Among them, AES is a commonly used cryptographic method that utilises a symmetric cipher to achieve the maximum security possible (Panagiotou, Sklavos, Darra, & Zaharakis, 2020). AES has

TABLE 3.1

Cryptographic algorithms used in IoT devices' security

Cryptography based Security of IoT devices
AES Confidentiality
ECC Resource constrained applications
SHA-1, SHA-256 Integrity
Diffie-Helman Key sharing

robust security features, and deployment of software and hardware is reasonably easy. It is a round-based algorithm and iterative that supports the 128-bit block and 128, 192, 256 bit key sizes. AES hardware implementation needs lots of resources and is generally not recommended for area constrained IoT devices. However, due to the advancement in the technology and optimised design processes, cryptographic algorithms such as AES are becoming an integral part of the IoT devices. In this regard, efficient implementations of AES are proposed (Jung, Fiedler, & Lerch, 2005; Farooq & Aslam, 2017); thus, it is a suitable solution for the inclusion of security to IoT-based systems. An appropriate adaptive system is presented to safeguard IoT devices, where AES implementations are considered (Farooq et al., 2019); it is shown that the IoT-based systems are adaptively protected by the proposed architecture.

As discussed in the previous section, this document search is limited to only the AES algorithm used for security purposes for IoT devices. The language of the selected papers is English only; the keywords used are similar to the previous section with the addition of AES. The distribution of the 182 published research articles on security for IoT devices using AES in various subject areas follows the same pattern as in section 3.2. Out of the 182 most relevant papers, 15 are selected here for discussion. Power information extraction of the AES state is preferred over performing fault injection for practical attackers (Li, Chen, & Wang, 2017). The relationship between the information used and the key recovery results is confirmed by theoretical analysis as well as simulation results. An algebraic side-channel attack is also considered for variant countermeasures to construct the general security evaluation method. Successful key recovery of an algebraic side-channel attack on AES key expansion is explained with some observations. A high-performance security algorithm for data analytics in IoT based devices is proposed (Manikandan & Subha, 2018). It is claimed that through parallel processing of the AES algorithm, data analytics in IoT-based system performance can be improved. The proposed method also has been tested on multi-core processing architectures of Intel. AES key generation based on memristor is proposed for low power hardware modules for IoT (Rady, Hossam, Saied, & Mostafa, 2019). It is suggested that the proposed module could meet the requirement of IoT embedded devices secure communication. An abstract secure communication model using lightweight AES for IoT with two-level security is discussed and implemented (Pattanayak & Amic, 2020). S-box lightweight implementation of AES using dual-edged activated S-box is presented (S. Khan et al., 2019). The proposed architecture ASIC implementation reported low power consumption and reduced hardware by 30%. Secure communication between IoT devices authenticated by the Diffie-Hellman encryption algorithm and encrypted along with verified nodes by AES and MD5 is proposed (Quist-Aphetsi & Xenya, 2019). Encryption and decryption durations and energy consumption results in three variants of AES viz tinyAES, B-con's AES, and Contiki's own built-in AES presented in (Tsao, Liu, & Dezfouli, 2019). A combined architecture of modified lightweight AES using a chaotic system is proposed (Naif, Abdul-Majeed, & Farhan, 2019). The modified lightweight AES reduced the processing complexity of AES. Cryptographic calculation and the power consumption required for the AES algorithm and different payload lengths are analyzed (Hung & Hsu, 2018). Software-based AES-CB, hardware-based AES-ECB, and hardware-based AES-CCM are implemented. It is concluded that the hardware-based AES consumes less power and cycle requirements are also less. Power-efficient implementation of AES core was done for IoT constrained devices (Agwa, Yahya, & Ismail, 2017). AES core implementation was modified by adding a white box and doubling the AES encryption (Javed, Khan, Qahar, & Abdullah,

2017). S-box was also replaced with the white box. It is claimed in this work is that the proposed solution can prevent a DoS attack on IoT and devices. An experimental study is done (Kane et al., 2020) to compare the performance of cryptographic algorithms on three different devices. One of the devices performed best in terms of overall energy cost while giving reasonably good performance. AES-AES cascading is conducted, consisting of 5 rounds instead of 10 (Ritambhara, Gupta, & Jaiswal, 2017). It is also mentioned that the cascaded structure takes half of the time with respect to the simple AES algorithm. Ultra-low power architecture of AES cipher is presented (S. Khan et al., 2019). AES energy consumption and duration was implemented on software and hardware using two resource-constrained IoT edge devices, diverse key and buffer size settings (Munoz, Tran, Craig, Dezfouli, & Liu, 2019). It was noted that the implementation of hardware is more susceptible to buffer size settings and, in all situations, the security premium offered by an increase in key size contributes to increased consumption of resources.

Many of the articles claim that AES cryptographic ICs are being used for enhancing the security of IoT devices. It is also noticed that claims about the security enhancement of AES hardware implementations are being published recently. It is reported that obfuscation techniques enhance the security of AES-128 algorithm implementation at the hardware level (Chhabra & Lata, 2018b; 2018a). These papers also discuss possible attacks and their countermeasures with the help of obfuscation techniques. Obfuscated implementation of AES-128 has proven to be the better solution for image-processing applications in the reported work. Image transfer is the most common application of IoT devices today. This kind of security enhancement gives a high level of confidence to users of heterogeneous IoT devices. Finite field arithmetic-based obfuscation of AES is proposed where obfuscation keys utilize variations of finite field construction (Zhang, Shvartsman, Zhou, & Tawfik, 2019). It is also claimed that the proposed method can be applied to other algorithms involving finite field arithmetic. A graph-based approach for the hardware implementation of AES is created by using various technology-mapping and synthesis options (Swierczynski et al., 2015). Dynamic obfuscation is also discussed with its benefits and shortcomings to make sure that reverse engineering becomes more challenging for the attacker. Hardware obfuscation based on the deflection strategy for the AES algorithm is designed in the SMIC65 nm CMOS process (Zhang, Pan, Wang, Ding, & Liu, 2018. It is shown that the code coverage rate is increased significantly with the trade-off of some area and power consumption. Although various publications claim that inbuilt security is being enhanced by applying such techniques, an engineer still has to test the implementation to give confidence to the organization and the user. This is where secured testing plays its major role.

3.4 Scan-based Side-channel Attack on AES Cryptographic ICs

AES can be implemented either on software or hardware, but hardware implementation is preferred as it provides higher throughput than the software counterpart. Throughput is a performance measurement term that simply means the number of data items processed per unit time. Therefore hardware implementation of the AES is desirable; the circuits that apply AES are known as crypto chips. Although there are no known brute force or cryptanalytic attacks that can hamper the security of the AES, the

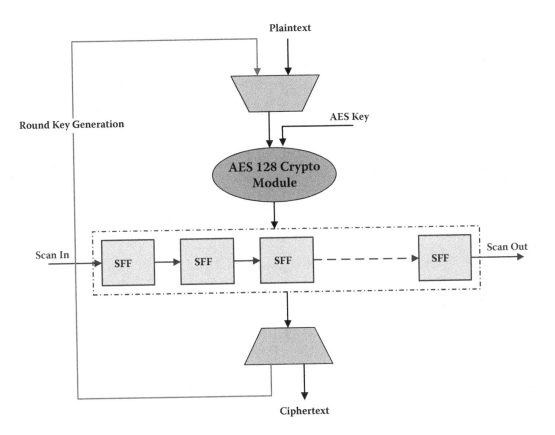

FIGURE 3.2
AES 128 Crypto module with a single scan chain architecture.

hardware is at risk due to side-channel attacks (Yang, Wu, & Karri, 2006). Using scan chains, every node of the circuit is observable and controllable; therefore, internal secret data stored inside the crypto chip can be observed easily and subsequently the secret key can be retrieved. Since testing and security are both of utmost importance, several countermeasures were proposed to maintain the standards of both the security and testing capabilities of the crypto algorithm. Figure 3.2 shows the AES 128 crypto module with single scan chain architecture which consists of AES-128 Crypto module round register Scan Flip-Flops (SFF).

AES one round encryption is shown in Figure 3.3. The attacker aims to break through the value of K0. Register R becomes part of the scan chain. By using this scan chain the content of this register R can be scanned out and then the attacker attempts to find similarities between the scanned out sequences and the user key. This approach makes it possible for attackers to break the designed system. The first step of the attack mentioned in Yang et al. (2006) is to find the register location that holds the result of round one encryption. If there is a shift in one bit in the input then several bits of the output's value are changed; such a phenomenon is called an avalanche effect (Yang et al., 2006). By this means, the location of register R can be identified.

Once the location of the register R has been determined, a known plain text is applied to the input and run for one round operation to get the scanned output results. In the second step, another known plain text is added which differs by only 1 bit from the previously

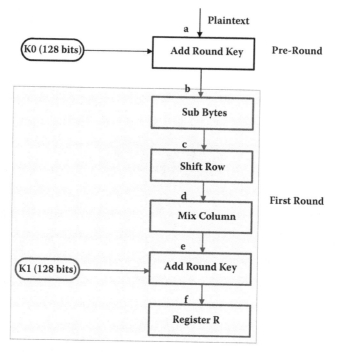

FIGURE 3.3
Round operations of AES encryption.

applied plain text, then run again for the one round operation, in the same way, to get the scanned out results from register R again. By analyzing these two scanned outputs, attackers would be able to recover the values of K (the key of the AES algorithm). This demonstrates that an attack on AES can be done successfully.

The reader is encouraged to go through the work in Kaushik and Lata (2020) with the example of the known plain text to show that using this type of scan-based side-channel attack, the value of $K_{1,1}$ can be retrieved, and then the whole key of AES can be recovered. To recover the AES whole key, Kaushik and Lata (2020) followed all the steps mentioned in Yang et al. (2006). The attack on any cryptographic algorithm by the side-channel was first mentioned in Yang, Wu, and Karri (2004) where they targeted the Data Encryption Standard (DES) crypto algorithm. Later the same authors extended that work to AES to break its security (Yang et al., 2006). The work done in [7],[8] proposes that when crypto algorithms are implemented for testing with scan chain then, due to its structure, access to sensitive information stored in Flip-Flops can be misused. As a consequence, the protection of these crypto algorithms is compromised. For a reasonable period, the attacker will run the crypto chip in normal mode and during this period, internal Flip-Flops store the intermediate results. Then the attacker scans those results and tries to find out the relation between the internal bit and the scanned bits of the Flip-Flops. This procedure is repeated until the algorithm is cracked successfully. This complete process is known as a side-channel attack. This type of encryption algorithms' security depends upon its key which becomes the primary target for the attackers.

Authors in Liu and Huang (2007) have considered advanced DFT structures such as embedded compression architectures for their inherent security purposes. These advanced DFT structures have become the norm of complex designs today. Their integration improves testing time while reducing testing costs. Large scan chains are divided into sub-chains due to restrictions imposed by the available number of input and output pins. To meet these input/output pin requirements, test patterns and output responses are also compressed. Authors in Liu and Huang (2007) show that the design itself becomes self-resistant for scan-based side-channel attacks by using these advanced DFT structures. Initially, these techniques were found to be resistant against side-channel attacks, but failure of these advanced DFT structures against enhanced differential style scan-based attack was reported in DaRolt, Di Natale, Flottes, and Rouzeyre (2011), Da Rolt, Natale, Flottes, and Rouzeyre (2012). BIST (Di Natale, Doulcier, Flottes, & Rouzeyre, 2009; Hetherington et al., 1999) DFT architecture was also used for securing the design, but BIST suffers from low fault coverage and high area overhead requirements.

Another counter measure is recorded when the attacker attempts to monitor the working of the scan chain with the assistance of the probes to gain observability. Hely, Bancel, Flottes, and Rouzeyre (2005) discuss a countermeasure: setting up an alarm that will alert one to malicious intention, while simultaneously all the contents of the scan cell are cleared. These micro probing attacks are also known as invasive attacks because of the physical use of a probe on the hardware by the attacker.

Typically, the attacker tries to deduce the cipher key from the intermediate encrypted results. Therefore, many research papers are published that are based on the obfuscated scan performance known as scan-out pin masking (Paul, Chakraborty, & Bhunia, 2007), or random connections of sub-scan chains known as the scan-chain scrambling technique (Lee, Tehranipoor, Patel, & Plusquellic, 2007). A password-based authentication system to apply the standard test procedure is proposed in Paul et al. (2007). This system requires that the user needs to apply the M vectors which would be of N bits length; afterward, it is compared with the stored authentication key which is put in the chip. In the case of equating both, the scan output pin becomes observable.

Authors of Lee et al. (2007) have proposed the lock and key technique, and have discussed under what circumstances the scan chain would be unlocked to perform normal testing. The whole scan-chain is bifurcated into sub-scan chains for implementation with the order changed. All these sub-chains can be correctly linked back in order after applying the right key. This technique is known as scan-chain reordering. To implement this, we need a Test Security Controller that consists of the Finite State machine (FSM) and the Linear Feedback Shift Registers (LFSRs).

A new test vector application procedure was proposed in Da Rolt et al. (2013) where the authors emphasised the fact that controllability of the SFF does not pose a security leak as compared to the observability of the SFF, due to which the side-channel attack occurs. Hence the authors modified the standard test procedure with some slight variations. Another approach, called key blocking, was proposed in which scanned output of SFF is not compared with the actual responses. Instead, on-chip comparison is done, where the encryption key was not allowed to participate during the test procedure (Yang et al., 2006). The authors presented the Mirror Key Registers (MKR)-based secure scan design, which uses secure and insecure modes to perform operations. In this work, normal encryption mode is considered a secure mode and testing mode is considered an insecure mode. During normal encryption mode, the original cipher key is entered into MKRs, whereas during testing mode a pseudo key is entered from MKRs,then test stimuli are applied. It is ensured that test stimuli cannot be applied during normal operation mode. The major constraint to this procedure is that the encryption key design portion cannot be tested and it has a large overhead area.

Normal Scan Flip-Flops (SFF) are replaced by State-Dependent Scan Flip-Flops by the authors (Atobe, Shi, Yanagisawa, & Togawa, 2012). The SDFF is a combination of historic states and SFF. Historic data is combined with normal test data to get the scanned out data. The authors of Cui, Luo, and Chang (2017) have reported a drawback to this method: it reduces testability and increases the complexity of the design.

The obfuscation of scan data in two different modes, viz, static and dynamic, for scan-based side-channel attack resistance is presented in Cui et al. (2017). To strengthen the key and lock method, static obfuscation of the scanned data is suggested to provide resistance against a signature attack on the crypto chip. This paper also discusses a new type of signature attack which is called a test-mode-only signature attack. To overcome this new attack, the authors also discussed another countermeasure, which is based on dynamic obfuscation of the scanned data.

The scan chain structure is modified by inserting the NOT gates at random positions as proposed by (Sengar, Mukhopadhyay, & Chowdhury, 2007). This enhances the security of the scan chain structure and depends upon the randomness of the NOT gates' locations. Agrawal, Karmakar, Saha, and Mukhopadhyay (2008) reported that it increases the susceptibility of the reset attack, in which attackers can reset the scan chain Flip-Flops and find out the inverters' locations easily. However, it is also reported that in this method the design area overhead is minimum. Another approach is proposed in the same paper (Agrawal et al., 2008), in which XOR gates are inserted in scan chain architecture for the stream cipher Trivium to combat scan-based side-channel attack.

3.5 Design and Simulation of Scan-inserted AES Crypto Module

AES is a block symmetric key encryption algorithm, published on 26 November 2001 in Federal Information Processing Standards Publications (FIPS PUBS), issued by the National Institute of Standards and Technology (NIST) (Standard and others 1977). In 1997 NIST issued fresh bids for a new encryption algorithm as the existing one, 3 Data Encryption Standard (3DES), was having security issues and was not efficient for implementing with emerging technologies. Many researchers from different countries submitted their proposals; 15 made it to the final list and from them, 5 were declared finalists, namely, MARS, RC6, RIJNDAEL, SERPENT, and TWOFISH and finally RIJNDAEL. The last algorithm was declared the winner, and was subsequently known as AES algorithm. AES can use keys of size 128, 192, and 256 bits to encrypt and decrypt input data of 128 bits. AES, being a block cipher, processes the input as a block of data items at one time. Another way of processing the input plaintext is the stream cipher process, in which input data items are processed continuously and produce output one data item at a time. The key used for encrypting the input plaintext is called a cipher key and the scrambled message obtained at the output of the encryption algorithm is called ciphertext. The scan chain structure of the AES Crypto module's security is enhanced thereby. The work is divided into three phases. In phase I the 128 bit pipelined AES architecture has been designed and verified using the Xilinx Vivado 2014.4. Phase II describes the process of inserting the scan chains into the design and thereafter applying the test patterns to test the design. In phase III a side-channel attack on AES is launched, then subsequent measures to protect the design against side-channel attacks are

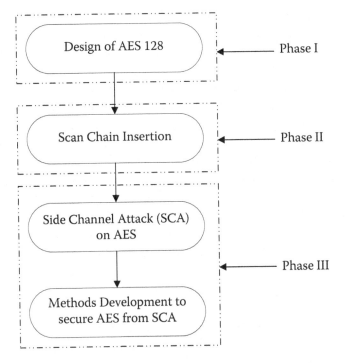

FIGURE 3.4
Flowchart of the presented work.

presented. Phase III is discussed in detail in section 3.6. Figure 3.4 shows the complete flow of the presented work.

3.5.1 Design of AES

The first step involved in any VLSI design flow is to define the functionality or the behavior of the intended design. The AES algorithm functioning is already discussed in the previous chapter. Hardware Description Languages (HDL) such as VERILOG and VHDL define the functionality of the design. The 128 bit AES algorithm is implemented using VHDL. After defining the behavior of the AES design, it is then converted into gate-level representation: the synthesis process. After synthesis, simulation is done to check the correctness of the AES, which can be done by either writing test benches or behavioral simulation. In behavioral simulation, values are forced to the input and correspondingly output is checked as the signal waveform. In test benches, an HDL code is written, in which the desired clock and input stimuli are defined and output according to those stimuli is verified. Top-level round architecture is shown in Figure 3.5. AES consists of ten rounds, so one round is replicated ten times: they are then connected serially to get the encrypted output. The top-level RTL schematic of the AES design is shown in Figure 3.6.

To verify the correctness of the AES algorithm designed, behavioral simulation is done. A standard example is chosen from (Standard and others 1977) for verifying the correctness of our design. Input plaintext chosen is $(00112233445566778899aabbccd-deeff)_{16}$, the applied key is $(000102030405060708090a0b0c0d0e0f)_{16}$, and the encrypted output is $(69c4e0d86a7b0430d8cdb78070b4c55a)_{16}$. The same is obtained in the simulation results shown in Figure 3.7.

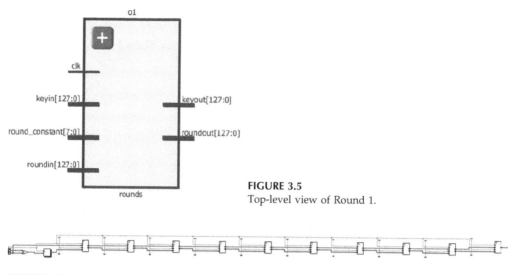

FIGURE 3.5
Top-level view of Round 1.

FIGURE 3.6
RTL Schematic of 128 Bit AES Design.

3.5.2 Design and Simulation of Scan-inserted AES Design

The Scan chain insertion process is done after successful verification of the AES design. AES 128 crypto module is designed using VHDL and is synthesised using the Mentor Graphic tool Leonardo Spectrum, which converts it into the gate-level netlist. Scan chains are needed to test the designs that are inserted into the gate-level netlist generated through the Leonardo tool. These are inserted through the DFTAdvisor tool. The FastScan tool is used for Automatic Test Pattern Generation (ATPG). To test the design, test patterns are applied to it, then fault coverage is checked. ModelSim is used for simulating the scan-inserted AES design to verify its functionality. Simulating the AES design is necessary to verify the working of the scan chains. The top-level schematic of the scan-inserted AES design is shown in Figure 3.8.

The scan in pin is used for changing the mode of the AES design into test mode or normal mode. In normal mode, the scan chain is disconnected and the operation is the same as in the original AES design, i.e., without scan chains. In test mode, test patterns are applied from the scan in pin and the shifted contents of the scan cell are taken out from the scan out port. Figure 3.9 shows the simulation of the scan-inserted AES design with scan en=0.

3.6 Enhanced Protection of AES Crypto Module Scan Chain Structure: A Case Study

The attack discussed in the previous section is termed a differential side-channel attack. The attack analysis reveals that the primary reason for key retrieval is the inclusion of register R in the scan chains: its contents get scanned out during the test mode. One can protect the key by excluding register R in the scan chain, but this will have serious implications on testing capabilities. If it is included, one must make sure its contents are not shifted out during the test mode or do not obfuscate the scanned data so the differential attack fails.

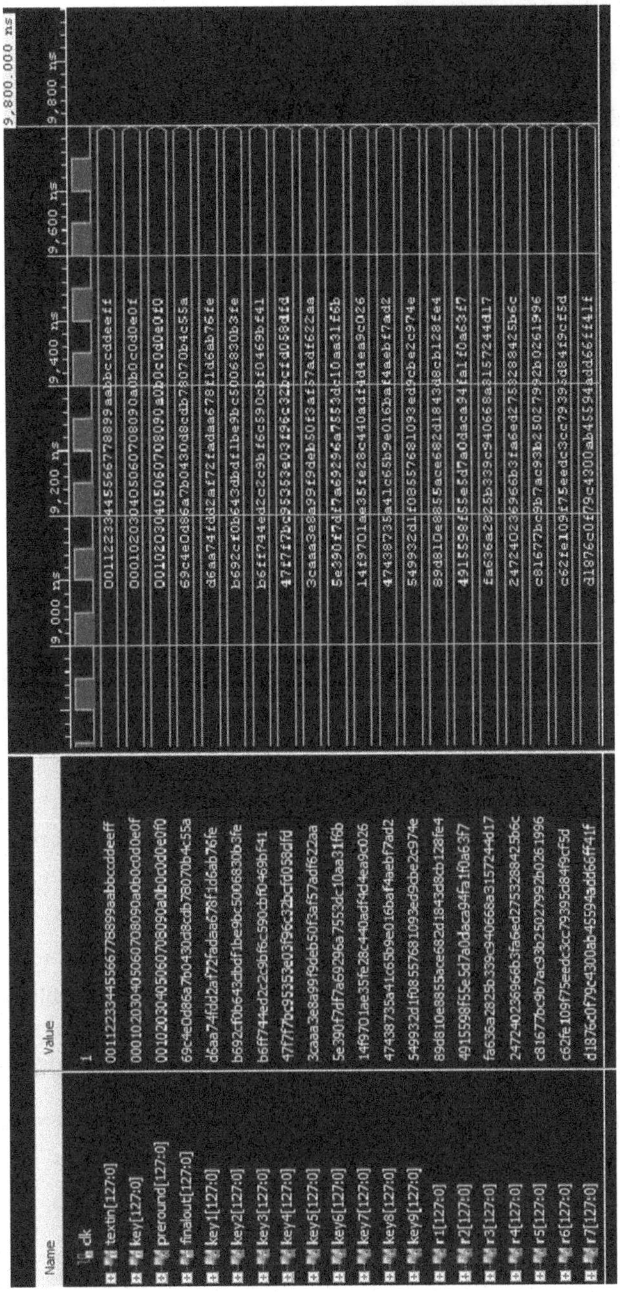

FIGURE 3.7
AES Encryption Simulation Results.

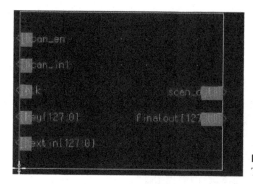

FIGURE 3.8
Top-level view of scan-inserted AES design.

Securing the key by obfuscating the scanned output is recommended in this work. The idea of using XOR gates to obfuscate the scanned out data was first presented in [27] on stream cipher Trivium to successfully prevent the scan-based side-channel attack. Here, first, we applied the same XOR-based obfuscation to the block cipher AES to see if it could resist the side-channel attacks on AES, then we used the hybrid approach to enhance the security testing structure of AES. The reader is encouraged to go through reference (Kaushik & Lata, 2020) that illustrates both approaches towards resisting scan-based side-channel attacks for performing secured testing of AES crypto chips in detail.

3.6.1 XOR Based Obfuscation Technique

Scan-based side-channel attack on AES (Yang et al., 2006) presents cipher key recovery by analyzing the intermediate encrypted results. We present the idea of obfuscating the scanned output so that retrieving back the cipher key becomes next to impossible. To safeguard the AES scan chain structure, XOR gates are placed at random positions and are shown in Figure 3.10.

The modified architecture of the scan flip-flop is shown in Figure 3.11. The Scan-in input of SFF consists of the output of the previous SFF; the other input is the next state of SFF, which is achieved by providing a feedback loop.

3.6.2 Hybrid Obfuscation for the Scan Output

Although the XOR-based obfuscation based scheme can defend itself against a side-channel attack, it is susceptible to SET attack. To overcome this drawback, another secured method is proposed, a hybrid scheme in which NOT gates along with XOR gates are inserted randomly in the scan chain structure of the AES. The proposed architecture is described in Figure 3.12. In this, XOR and NOT gates are combined in the scan chain structure to obfuscate the scan output data.

3.7 Results Analysis

This section discusses various aspects like impact on testability, and security analysis of the proposed schemes.

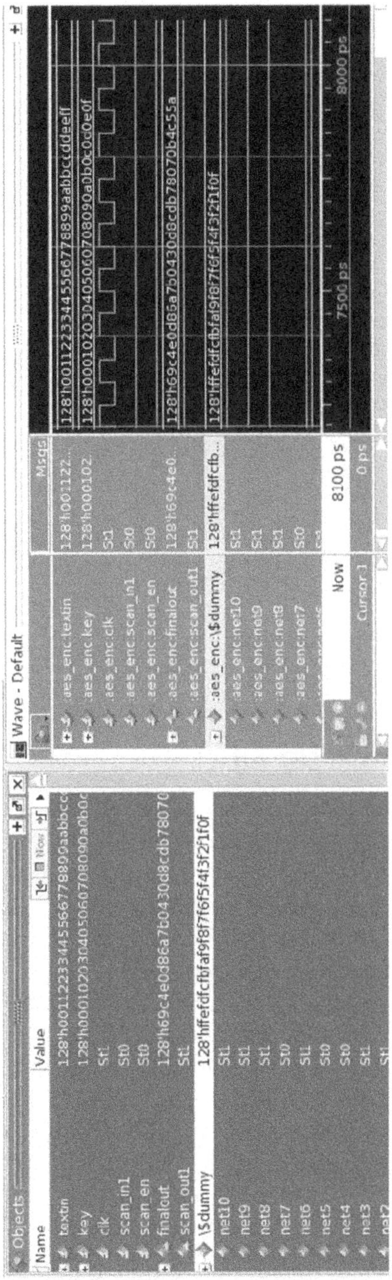

FIGURE 3.9
Simulation of scan-inserted AES design.

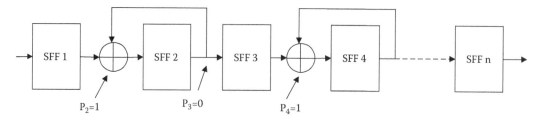

FIGURE 3.10
AES Scan chain structure after insertion of XOR gates.

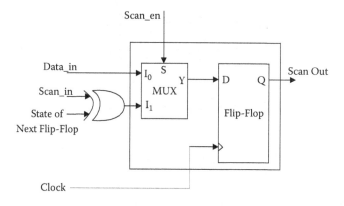

FIGURE 3.11
Modified Scan Flip-Flop.

3.7.1 XOR-based Obfuscation and SET Attack

The overall security of this scheme relies on the random position of the XOR gates inside the scan chain structure. With n Flip-Flops and $m = n/2$ XOR gates, the scan chain structure can be exposed to the probability of $1/2^n$ (Agrawal et al., 2008). AES design has overall 12992 Flip-Flops but there is no need to put 6496 XOR gates as it will create unnecessary overhead in the design. The attack analysis reveals that the key gets retrieved during the one round encryption process, so we need to obfuscate the scanned data only for one round. Register R stores the round one encryption result. The scan Flip-Flops in register R total 128 in number. So, 64 XOR gates are randomly placed, then according to a theorem, the probability of discovering the scan chain structure is $1/2^{128}$, which is computationally unfeasible.

If the position of the XOR gates inside the scan chain is known to the attacker, the security of this scheme gets compromised. An SET attack, as reported in Banik and Chowdhury (2013), can find out the position of XOR gates inside the scan chain structure. Banik and Chowdhury (2013) have proposed that if modern scan Flip-Flops come with a SET/RESET pin, then one can easily find out the position of XOR gates. After setting all the Flip-Flops to 1 then scanning out the contents, the position of the XOR gates can be easily traced. To demonstrate a SET attack, let us consider the scan chain arrangement as shown in Figure 3.13. Setting all the scan Flip-Flops to 1 and scanning out the contents, we get the sequence 110 which indicates the presence of the XOR gate at the first scan flip flop. Similarly, it can be used for finding the position of all the XOR gates present inside the structure.

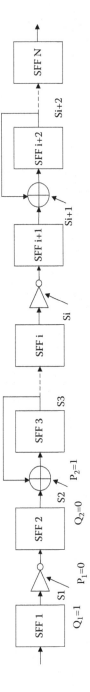

FIGURE 3.12
Hybrid scheme based scan chain obfuscation.

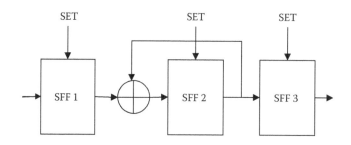

FIGURE 3.13
Demonstration of SET attack.

3.7.2 Hybrid Obfuscation and SET/RESET attack

To overcome the drawback of the SET attack, a hybrid model was proposed. In this scheme, we have used both the NOT and XOR gates. The purpose of using NOT gates is to protect the scan chain structure from a SET attack, which is illustrated in the following example. Consider the scan chain arrangement shown in Figure 3.14 below. Following the attack procedure, set all the Flip-Flops to 1 and scan out the contents. The sequence so obtained is 100. If it was a normal scan chain structure, i.e., without NOT gates or XOR gates, we would get the output of 111. These two zeroes in the scanned out sequence indicate the presence of two XOR gates at position 2 and 1, which is actually only 1. So in this way, it will always give dubious locations for the XOR gates and hence successfully prevents the SET attack.

The RESET attack (Agrawal et al., 2008) is used for finding the position of the NOT gates present in the scan chain structure. In this attack, first we reset all the Flip-Flops to 0, then scan out the data. If the design contains only NOT gates, there will be transitions from 0 to 1 in the scanned out sequence, from which the position of the NOT gates can be known. Now we will show that our hybrid design is resistant to this type of attack. Setting the states of all the scan Flip-Flops to 0 in Figure 3.14 and scanning out the value, we get the sequence of "001". XOR gates become transparent in this type of attack as $0 \oplus 0$ is 0. Hence the design acts like that there are only XOR gates present inside the structure, but actually NOT gates are also present.

The theorem used in the XOR-based obfuscation method can also be applied here to guess the structure of the scan chain. 32 XOR and NOT gates each are randomly placed and then according to a theorem, $1/2^{128}$ is the probability of breaking the scan chain structure of this AES 128 crypto module.

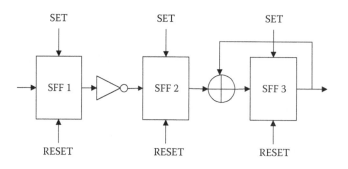

FIGURE 3.14
Example showing resistance against SET/RESET attack.

3.7.3 Signature Attack

Both the proposed approaches, XOR-based obfuscation and hybrid method resistances to signature attack, are recorded in (Nara, Togawa, Yanagisawa, & Ohtsuki, 2009), (Kodera, Yanagisawa, & Togawa, 2012). For this type of attack, as shown in Figure 3.15, the attacker may develop a simulator at the software level to perform simulation of the cryptographic implementation. To perform such an attack, one needs to divide the guesses key of K length into k times sections (Cui et al., 2017): $fk_1, fk_2, fk_3,... fk_k$. Values of the $fk_i (1 \le i \le k)$ can be found out by selecting the p plaintexts. In order to show its effect on the associated intermediate encryption result, it can invoke the function of fk_i in the encryption. Inputs to the simulator are the n plaintexts and figure key fk_i, while the n plaintexts are the input to the original crypto module.

As n plaintexts are applied, each scan cell output will result in the formation of n bit sequence, called the signature* corresponding to the fk_i. Similarly, n bit sequences from each scan cell will be obtained from the real crypto module and the n bit signatures are formed. If the signature obtained from the original crypto module matches with the simulator's signature*, then fk_i is right in the figured key portion. This experimentation procedure is replicated until all the keys in the section have been calculated.

The signature attack fails in the proposed obfuscation methods because the scanned out encrypted results would be different after the application of the known plaintext which forms the signatures. In the proposed methods, the signature formed would be obfuscated in the presence of NOT gates and XOR gates; therefore, those will never match with the received signatures from the simulators.

3.7.4 Impact on Testability

The DFTAdvisor tool is used for performing the DFT on AES-128 implementation first, then XOR and NOT gates are inserted in the design netlist. The Fast scan tool generates the test pattern automatically, viz, ATPG. A total of 277 test stimuli simulate all the faults,

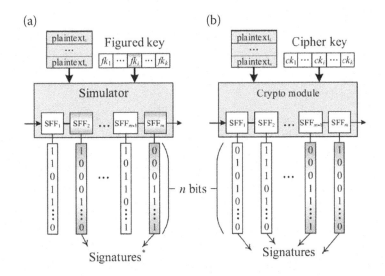

FIGURE 3.15
Signature Attack (Cui et al., 2017).

and the fault coverage achieved is 99.3%. It is to be noted that test stimuli used for true value simulation are applied to the XOR-based obfuscation and hybrid architecture. This can be done without any problem. This also ensures that the testability and the DFT process of the modified AES architecture after scan insertion are still intact.

The key benefit over other current counter-measures is that any additional testing techniques are not required to test the added circuit, which is used to protect the design against side-channel attacks.

3.8 Conclusions

In the last few years, there has been considerable interest among researchers in the study of IoT and its applications. In particular, security and privacy aspects have attracted considerable attention. This will continue for years to come due to various security and privacy challenges raised by connected IoT devices. The potential for unexpected effects of IoT devices' vulnerabilities has a huge impact on overall internet security. Various cryptographic algorithms are being used for IoT connected devices. Cryptography not only protects the devices from hackers, but also offers enormous data confidentiality, integrity, and even authentication to the data. A cryptographic algorithm also needs to be implemented and tested to get the correct functional ICs in these IoT devices. When it comes to the testing of these crypto chips, it is reported in the literature that designs for testable architecture themselves can be hacked and can become the back door for the hackers. These types of attacks are referred to as scan-based side-channel attacks, and we have explored how to secure these crypto chip scan structures.

In this chapter, various scan-based side-channel attacks are described and discussed. As a countermeasure against them, one case study is also presented and discussed, which may provide a solution to the current problem. With the aid of the case study of AES algorithms that are becoming an integral part of IoT devices, this chapter will enhance the understanding of the domain of secure testing of crypto chips.

Acknowledgment

Simulation results are part of the M.Tech. Thesis work carried out by Mr. Nandit Kaushik under my supervision. He is graduated from The LNM Institute of Information Technology Jaipur in July 2019.

References

Adat, V., & Gupta, B. B. (2018). Security in internet of things: Issues, challenges, taxonomy, and architecture. *Telecommunication Systems, 67*(3), 423–441. https://doi.org/10.1007/s11235-017-0345-9.

Agrawal, M., Karmakar, S., Saha, D., & Mukhopadhyay, D. (2008). scan-based side channel attacks on stream ciphers and their counter-measures. *International Conference on Cryptology in India,* 5365, 226–238.

Agwa, S., Yahya, E., & Ismail, Y. (2017). power-efficient AES core for IoT constrained devices implemented in 130nm CMOS. *Proceedings - IEEE International Symposium on Circuits and Systems.* Institute of Electrical and Electronics Engineers Inc. https://doi.org/10.1109/ISCAS.2017.805 0361.

Akhundov, H., van der Sluis, E., Hamdioui, S., & Taouil, M. (2020). Public-key based authentication architecture for IoT devices using PUF. *ArXiv Preprint ArXiv:2002.01277.*

Ali, I., Sabir, S., & Ullah, Z. (2019). Internet of things security, device authentication and access control: A review. *ArXiv Preprint ArXiv:1901.07309.*

Aman, M. N., Chua, K. C., & Sikdar, B. (2016). Physical unclonable functions for IoT security. *2nd ACM International Workshop on IoT Privacy, Trust, and Security, IoTPTS 2016*: Vol. 117583. National University of Singapore, Singapore (pp. 10–13). Association for Computing Machinery, Inc. https://doi.org/10.1145/2899007.2899013.

Anggorojati, B., & Prasad, R. (2018, January). Securing communication in inter domains internet of things using identity-based cryptography. *2017 International Workshop on Big Data and Information Security, WBIS 2017.* Faculty of Computer Science, Universitas Indonesia, Indonesia (pp. 137–142). Institute of Electrical and Electronics Engineers Inc. https://doi.org/10.1109/IWBIS.2017.8275115.

Arunkumar, S., Vetriselvi, M., & Thanalakshmi, S. (2020). Cryptography based security solutions to IoT enabled health care monitoring system. *Journal of Advanced Research in Dynamical and Control Systems,* 12(7), 265–272. https://doi.org/10.5373/JARDCS/V12I7/2 0202008.

Atobe, Y., Shi, Y., Yanagisawa, M., & Togawa, N. (2012). State-Dependent scan flip-flop with key-based configuration against scan-based side channel attack on RSA circuit. *2012 IEEE Asia Pacific Conference on Circuits and Systems,* 607–610.

Atzori, L., Iera, A., & Morabito, G.. (2014). From "Smart Objects" to "Social Objects": The next evolutionary step of the internet of things. *IEEE Communications Magazine,* 52(1), 97–105.

Balasubramaniam, R., Sathya, R., Ashicka, S., & Kumar, S. S. (2016). An analysis of RFID authentication schemes for internet of things (IOT) in healthcare environment using Elgamal elliptic curve cryptosystem. *International Journal of Recent Trends in Engineering and Research(IJRTER),* 2(3), 241–247.

Banerjee, S., Chakraborty, C., & Chatterjee, S., (2018). A survey on IoT based traffic control and prediction mechanism, *Springer: Internet of Things and Big data Analytics for Smart Generation, Intelligent Systems Reference Library,* Ch. 4, 154, 53–75, ISBN: 978-3-030-04203-5.

Banik, S., & Chowdhury, A. (2013). Improved scan-chain based attacks and related counter-measures. *International Conference on Cryptology in India,* 78–97.

Chanda, P.B., Das, S., Banerjee, S., & Chinmay, C. (2021). Study on edge computing using machine learning approaches in IoT framework, CRC: Green Computing and Predictive Analytics for Healthcare, 1st edition, Ch. 9, 159–182.

Chhabra, S., & Lata, K. (2018a). Design and analysis of logic encryption based 128-Bit AES algorithm: A case study. *INDICON 2018 - 15th IEEE India Council International Conference.* https://doi.org/10.1109/INDICON45594.2018.8987098.

Chhabra, S., & Lata, K. (2018b). Hardware software co-simulation of obfuscated 128-Bit AES algorithm for image-processing applications. *Proceedings - 2018 IEEE 4th International Symposium on Smart Electronic Systems, ISES 2018.* https://doi.org/10.1109/iSES.2018.00049.

Chinmay, C., & Joel, J. P. C. R. (2020). A comprehensive review on device-to-device communication paradigm: Trends, challenges and applications. *Springer: International Journal of Wireless Personal Communications,* 114, 185–2017. doi: 10.1007/s11277-020-07358-3

Cui, A, Luo, Y., & Chang, C. (2017). Static and dynamic obfuscations of scan data against scan-based side-channel attacks. *IEEE Transactions on Information Forensics and Security,* 12, 363–376.

Cui, A., Luo, Y., Li, H., & Qu, G. (2017). Why current secure scan designs fail and how to fix them?, *Integration, 56*, 105–114.

DaRolt, J., Di Natale, G., Flottes, M.-L., & Rouzeyre, B. (2011). Scan attacks and countermeasures in presence of scan response compactors. *2011 Sixteenth IEEE European Test Symposium*, 19–24, doi: 10.1109/ETS.2011.30.

DaRolt, J., Di Natale, G., Flottes, M. -L., & Rouzeyre, B. (2012). Are advanced DfT structures sufficient for preventing scan-attacks? *2012 IEEE 30th VLSI Test Symposium (VTS)* (pp. 246–251), doi: 10.1109/VTS.2012.6231061.

DaRolt, J., Di Natale, G., Flottes, M. -L., & Rouzeyre, B. (2013). Thwarting scan-based attacks on secure-ICs with on-chip comparison. *IEEE Transactions on Very Large Scale Integration (VLSI) Systems, 22*(4), 947–951.

Dhanda, S. S., Singh, B., & Jindal, P. (2020). Lightweight cryptography: A solution to secure IoT. *Wireless Personal Communications, 112*(3), 1947–1980. https://doi.org/10.1007/s11277-020-07134-3.

Dhillon, P. K., & Kalra, S. (2017). Elliptic curve cryptography for real time embedded systems in IoT networks. *5th International Conference on Wireless Networks and Embedded Systems, WECON 2016*. Department of Computer Science and Engineering, Guru Nanak Dev University, Regional Campus, Jalandhar, Punjab, India. Institute of Electrical and Electronics Engineers Inc. https://doi.org/10.1109/WECON.2016.7993462.

Di Natale, G., Doulcier, M., Flottes, M. -L., & Rouzeyre, B. (2009). Self-test techniques for crypto-devices. *IEEE Transactions on Very Large Scale Integration (VLSI) Systems, 18*(2), 329–333.

Farooq, U., & Aslam, M. F. (2017). Comparative analysis of different AES implementation techniques for efficient resource usage and better performance of an FPGA. *Journal of King Saud University-Computer and Information Sciences, 29*(3), 295–302.

Farooq, U., Hasan, N. U., Baig, I., & Shehzad, N. (2019). Efficient adaptive framework for securing the internet of things devices. *EURASIP Journal on Wireless Communications and Networking, 2019*(1), 210. https://doi.org/10.1186/s13638-019-1531-0.

Garg, L., Chukwu, E., Nasser N., Chakraborty C., & Garg G. (2020). Anonymity preserving IoT-based COVID-19 and other infectious disease contact tracing model, *IEEE Access, 8*, 159402–159414, 10.1109/ACCESS.2020.3020513, ISSN: 2169-3536.

"Gartner.Com." n.d. https://www.gartner.com/en/doc/iot-security-primer-challenges-and-emerging-practices.

Gauniyal, R., & Jain, S. (2019). IoT security in wireless devices. *3rd International Conference on Electronics and Communication and Aerospace Technology, ICECA 2019*, Institute of Electrical and Electronics Engineers Inc, Amity University, Noida, India (pp. 98–102). https://doi.org/10.1109/ICECA.2019.8822124.

Giusto, D., Iera, A., Morabito, G., & Atzori, L. (2010). *The Internet of Things: 20th Tyrrhenian Workshop on Digital Communications*. Springer Science & Business Media.

Ha, D. A., Nguyen, K. T., & Zao, J. K. (2016, December 8-9). Efficient authentication of resource-constrained IoT devices based on ECQV implicit certificates and datagram transport layer security protocol. 7th Symposium on Information and Communication Technology, SoICT 2016. Association for Computing Machinery, National Chiao Tung University, 1001 University Road, Hsinchu, Taiwan (pp. 173–179). https://doi.org/10.1145/3011077.3011108.

Hameed, S., Khan, F. I., & Hameed, B. (2019). Understanding security requirements and challenges in internet of things (IoT): A review. In D. F. H. Sadok (Ed.), *Journal of Computer Networks and Communications, 2019*, 9629381. https://doi.org/10.1155/2019/9629381.

Hely, D., Bancel, F., Flottes, M.-L., & Rouzeyre, B. (2005). Test control for secure scan designs. *European Test Symposium (ETS'05)*, 190–195, doi: 10.1109/ETS.2005.36.

Henriques, M. S., & Vernekar, N. K. (2017). Using symmetric and asymmetric cryptography to secure communication between devices in IoT. *2017 IEEE International Conference on IoT and Its Applications, ICIOT 2017*. Departmentof Computer Engineering, Goa College of Engineering,

Farmagudi, Goa, India. Institute of Electrical and Electronics Engineers Inc. https://doi.org/10.1109/ICIOTA.2017.8073643.

Hetherington, G., Fryars, T., Tamarapalli, N., Kassab, M., Hassan, A., & Rajski, J.. (1999). Logic BIST for large industrial designs: Real issues and case studies. *Proceedings of the International Test Conference 1999. (IEEE Cat. No. 99CH37034)*, 358–367.

Hung, C.-W., & Hsu, W.-T.(2018). Power consumption and calculation requirement analysis of AES for WSN IoT. *Sensors (Switzerland), 18*(6), 1675. https://doi.org/10.3390/s18061675.

Jangra, P., & Gupta, M. (2017, January). Expositioning of cryptography techniques in IoT domain. In M. Sood & S. Jain (Eds.), *4th IEEE International Conference on Signal Processing, Computing and Control, ISPCC 2017.* Department of Electronics and Communication Engineering, University Institute of Engineering and Technology, Kurukshetra University, Kurukshetra, India (pp. 414–419). Institute of Electrical and Electronics Engineers Inc. https://doi.org/10.1109/ISPCC.2017.8269714.

Javed, Y., Khan, A. S., Qahar, A., & Abdullah, J. (2017). Preventing DoS attacks in IoT using AES. *Journal of Telecommunication, Electronic and Computer Engineering, 9*(3–11), 55–60. https://www.scopus.com/inward/record.uri?eid=2-s2.0-85041748882&partnerID=40&md5=e46c6f7bd45f97b6c9cd71788ed28ca7.

Jeong, J., Bajracharya, L., & Hwang, M. (2019). Optimal lightweight cryptography algorithm for environmental monitoring service based on IoT. In K. J. Kim, K. J. Kim, & N. Baek (Eds.), *International Conference on Information Science and Applications, ICISA 2018 .* Department of Eco-Friendly Offshore Plant FEED Engineering, Graduate School, Changwon National University, Changwon, South Korea (pp. 641–773). Springer Verlag. https://doi.org/10.1007/978-981-13-1056-0_37.

Jian, M. -S., Cheng, Y. -E., & Shen, C. -H. (2019, February). Internet of things (IOT) cybersecurity based on the hybrid cryptosystem. *21st International Conference on Advanced Communication Technology, ICACT 2019 .* Cloud Computing and Intelligent System Lab., National Formosa University, Department of Computer Science and Information Engineering, Taiwan (pp. 176–181). Institute of Electrical and Electronics Engineers Inc. https://doi.org/10.23919/ICACT.2019.8701957.

Jung, M., Fiedler, H., & Lerch, R. (2005). 8-Bit microcontroller system with area efficient AES co-processor for transponder applications. *Ecrypt Workshop on RFID and Lightweight Crypto*, 32–43.

Kane, L. E., Chen, J. J., Thomas, R., Liu, V., & Mckague, M. (2020). Security and performance in IoT: A balancing act. *IEEE Access, 8*, 121969–121986.

Karati, A., Fan, C. -I., & Hsu, R. -H.(2019). Provably secure and generalized signcryption with public verifiability for secure data transmission between resource-constrained IoT devices. *IEEE Internet of Things Journal, 6*(6), 10431–10440. https://doi.org/10.1109/JIOT.2019.2939204.

Kaushik, N., & Lata, K. (2020). An approach towards resisting side-channel attacks for secured testing of advanced encryption algorithm (AES) cryptochip. *ISEA-ISAP 2020 - Proceedings of the 3rd ISEA International Conference on Security and Privacy 2020.* https://doi.org/10.1109/ISEA-ISAP49340.2020.235014.

Khan, S., Gupta, N., Vishvakarma, A., Singh Chouhan, S., Pandey, J. G., & Kumar Vishvakarma, S. (2019). Dual-edge triggered lightweight implementation of AES for IoT security. *International Symposium on VLSI Design and Test*, 298–307. https://doi.org/10.1007/978-981-32-9767-8_29.

Khan, S., Gupta, N., Raut, G., Rajput, G., Pandey, J. G., & Vishvakarma, S. K. (2019). An ultra low power AES architecture for IoT. In V. Singh, R. Sharma, S. K. Vishvakarma, A. Sengupta, & S. Dasgupta (Eds.). *Communications in Computer and Information Science, 1066*, 334–344. https://doi.org/10.1007/978-981-32-9767-8_29.

Khattak, H. A., Shah, M. A., Khan, S., Ali, I., & Imran, M. (2019). Perception layer security in internet of things. *Future Generation Computer Systems, 100*, 144–164. https://doi.org/https://doi.org/10.1016/j.future.2019.04.038.

Kodera, H., Yanagisawa, M., & Togawa, N. (2012). Scan-based attack against DES cryptosystems using scan signatures. *2012 IEEE Asia Pacific Conference on Circuits and Systems*, 599–602.

Lee, J., Tehranipoor, M., Patel, C., & Plusquellic, J.. (2007). Securing designs against scan-based side-channel attacks. *IEEE Transactions on Dependable and Secure Computing, 4*(4), 325–336.

Li, Y., Chen, M., & Wang, J. (2017). Another security evaluation of spa countermeasures for AES key expansion in Iot devices. *Journal of Information Science and Engineering, 33*(4), 1085–1100. https://doi.org/10.6688/JISE.2017.33.4.15.

Liu, C., & Huang, Y. (2007). Effects of embedded decompression and compaction architectures on side-channel attack resistance. *25th IEEE VLSI Test Symposium (VTS'07)*, 461–468, doi: 10.1109/VTS.2007.29.

Manikandan, N., & Subha, S. (2018). Parallel AES algorithm for performance improvement in data analytics security for IoT. *International Journal of Networking and Virtual Organisations, 18*(2), 112–129. https://doi.org/10.1504/IJNVO.2018.10012669.

Marin, L., Pawlowski, M. P., & Jara, A. (2015). Optimized ECC implementation for secure communication between heterogeneous IoT devices. *Sensors (Basel, Switzerland), 15*(9), 21478–21499. https://doi.org/10.3390/s150921478

Munoz, P. S., Tran, N., Craig, B., Dezfouli, B., & Liu, Y. (2019). Analyzing the resource utilization of AES encryption on IoT devices. *2018 Asia-Pacific Signal and Information Processing Association Annual Summit and Conference, APSIPA ASC 2018 - Proceedings* (pp. 1200–1207). Institute of Electrical and Electronics Engineers Inc. https://doi.org/10.23919/APSIPA.2018.8659779.

Naif, J. R., Abdul-Majeed, G. H., & Farhan, A. K. (2019). Secure IOT system based on chaos-modified lightweight AES. *2019 International Conference on Advanced Science and Engineering, ICOASE 2019* (pp.12–17). Institute of Electrical and Electronics Engineers Inc. https://doi.org/10.1109/ICOASE.2019.8723807.

Nara, R., Togawa, N., Yanagisawa, M., & Ohtsuki, T. (2009). A scan-based attack based on discriminators for AES cryptosystems. *IEICE Transactions on Fundamentals of Electronics, Communications and Computer Sciences, 92*(12), 3229–3237.

Panagiotou, P., Sklavos, N., Darra, E., & Zaharakis, I. D. (2020). Cryptographic system for data applications, in the context of internet of things. *Microprocessors and Microsystems, 72*, 102921. https://doi.org/https://doi.org/10.1016/j.micpro.2019.102921.

Pattanayak, B. K., & Amic, S.(2020). Modified lightweight AES based two-level security model for communication on IoT. *Test Engineering and Management, 82*(1–2), 2323–2330. https://www.scopus.com/inward/record.uri?eid=2-s2.0-85080083441&partnerID=40&md5=7ef79ef687f153a424efc671a141afab.

Paul, S., Chakraborty, R. S., & Bhunia, S. (2007). Vim-Scan: A low overhead scan design approach for protection of secret key in scan-based secure chips. *25th IEEE VLSI Test Symposium (VTS'07)*, 455–460, doi: 10.1109/VTS.2007.89

Perera, C., Zaslavsky, A., Christen, P., & Georgakopoulos, D. (2014). Context aware computing for the internet of things: A survey. *IEEE Communications Surveys & Tutorials, 16*, 414–454.

Quist-Aphetsi, K., & Xenya, M. C. (2019). Node to node secure data communication for IoT devices using Diffie-Hellman, AES, and MD5 cryptographic schemes. *Proceedings - 2019 International Conference on Cyber Security and Internet of Things, ICSIoT 2019* (pp. 88–92). Institute of Electrical and Electronics Engineers Inc. https://doi.org/10.1109/ICSIoT47925.2019.00022.

Rady, H., Hossam, H., Saied, M. S., & Mostafa, H. (2019, August). Memristor-based AES key generation for low power IoT hardware security modules. *Midwest Symposium on Circuits and Systems* (pp. 231–234). Institute of Electrical and Electronics Engineers Inc. https://doi.org/10.1109/MWSCAS.2019.8885031.

Ritambhara, Gupta, A., & Jaiswal, M. (2017, January). An enhanced AES algorithm using cascading method on 400 bits key size used in enhancing the safety of next generation internet of things (IOT). In V. Sharma, A. Swaroop, M. Singh, P. N. Astya, & K. Gupta (Eds.), *Proceeding - IEEE International Conference on Computing, Communication and Automation, ICCCA 2017* (pp. 422–427). Institute of Electrical and Electronics Engineers Inc. https://doi.org/10.1109/CCAA.2017.8229877.

Rivest, R. L., Shamir, A., & Adleman, L. (1978). A method for obtaining digital signatures and public-key cryptosystems. *Communications of the ACM, 21*(2), 120–126.

Sadkhan, S. B., & Hamza, Z. (2017). Cryptosystems used in IoT-current status and challenges. *2017 International Conference on Current Research in Computer Science and Information Technology, ICCIT 2017*. Information Network Dept., IT College, Babylon University, Babylon, Iraq (pp. 58–62). Institute of Electrical and Electronics Engineers Inc. https://doi.org/10.1109/CRCSIT.2017.7965534.

Salman, A., & El-Tawab, S. (2019). Efficient hardware/software co-design of elliptic-curve cryptography for the internet of things. *2019 International Conference on Smart Applications, Communications and Networking, SmartNets 2019*. College of Integrated Science and Engineering, James Madison University, United States. Institute of Electrical and Electronics Engineers Inc. https://doi.org/10.1109/SmartNets48225.2019.9069777.

"Scoups." n.d. https://www.scopus.com/.

Sengar, G., Mukhopadhyay, D., & Chowdhury, D. R. (2007). Secured flipped scan-chain model for crypto-architecture. *IEEE Transactions on Computer-Aided Design of Integrated Circuits and Systems, 26*(11), 2080–2084.

Shor, P. W. (1999). Polynomial-time algorithms for prime factorization and discrete logarithms on a quantum computer. *SIAM Review, 41*(2), 303–332.

Standard, Data Encryption, and others. (1977). Federal information processing standards publication 46. National Bureau of Standards, US Department of Commerce 23.

Sun, W., Cai, Z., Li, Y., Liu, F., Fang, S., & Wang, G. (2018). Security and privacy in the medical internet of things: A review. H. Chen (Ed.), *Security and Communication Networks, 2018*, 5978636. https://doi.org/10.1155/2018/5978636.

Swierczynski, P., Fyrbiak, M., Paar, C., Huriaux, C., & Tessier, R. (2015). Protecting against cryptographic trojans in FPGAS. *Proceedings - 2015 IEEE 23rd Annual International Symposium on Field-Programmable Custom Computing Machines, FCCM 2015* (pp. 151–154). https://doi.org/10.1109/FCCM.2015.55.

Tsao, B., Liu, Y., & Dezfouli, B. (2019). Analysis of the duration and energy consumption of AES algorithms on a contiki-based IoT device. *ACM International Conference Proceeding Series* (pp. 483–491). Association for Computing Machinery. https://doi.org/10.1145/3360774.3368202.

Yang, B., Wu, K., & Karri, R. (2004). scan-based side channel attack on dedicated hardware implementations of data encryption standard. *2004 International Conferce on Test* (pp. 339–344).

Yang, B., Wu, K., & Karri, R. (2006). Secure scan: A design-for-test architecture for crypto chips. *IEEE Transactions on Computer-Aided Design of Integrated Circuits and Systems, 25*(10), 2287–2293.

Yang, G., Geng, G., Du, J., Liu, Z., & Han, H. (2011). Security threats and measures for the internet of things. *Journal of Tsinghua University Science and Technology, 51*(10), 1335–1340.

Yang, Y., Wu, L., Yin, G., Li, L., & Zhao, H. (2017). A survey on security and privacy issues in internet-of-things. *IEEE Internet of Things Journal, 4* (5), 1250–1258.

Zhang, X., Shvartsman, P., Zhou, J., & Tawfik, E. (2019). Hardware obfuscation of AES through finite field construction variation. *Proceedings - IEEE International Symposium on Circuits and Systems*: Vol. 2019-May. https://doi.org/10.1109/ISCAS.2019.8702285.

Zhang, Y., Pan, Z., Wang, P., Ding, D., & Li, G. (2018). Design of hardware obfuscation AES based on state deflection strategy. *Dianzi Yu Xinxi Xuebao/Journal of Electronics and Information Technology, 40*(3), 750–757. https://doi.org/10.11999/JEIT170556.

4

Biometric-based Secure Authentication for IoT-enabled Devices and Applications

J. Mahesh, M. Bodhisatwa, and D. Somnath
Indian Institute of Technology Indore, India

4.1 Internet-of-Things (IoT) Impacting our Livelihood

The Internet has played a significant role in recent innovations and advancements in information technology. Internet-of-Things is referred to as a network of smart hardware devices (things) communicating with each other using the Internet to provide specific functionality for individuals, industries, and organizations. With the advent of numerous IoT applications, human lives are more comfortable, as such applications employ intelligent programmable everyday objects to interact with people around them and amongst themselves. A few years back, what seemed to be an electronic device has now become a smart device by employing IoT as a backbone. The present-day world is witnessing the impact of IoT products and systems on everyday life as humans are increasingly becoming more reliant on the Internet for routine activities. Some sectors that employ IoT applications include healthcare, infrastructure, agriculture, logistics, manufacturing, automation industries and many others (Glaroudis, Iossifides, & Chatzimisiors, 2020).

IoT applications employ smart hardware, such as smart sensors, and intelligent software. The hardware is capable of sensing its environment to collect data and communicate the data to a local gateway (a device that enables communication over the Internet to the cloud-based server). The gateway acts as a common medium of contact for local devices, and the remote software and hardware get connected through the Internet. The gateway can also perform local analytics on the received data and take informed decisions. The server most often applies machine learning and data mining algorithms to the incoming data from the gateways to extract essential information (Sinche et al., 2020). Analysis reports are available for stakeholders through web portals or smartphone apps, which can also act as a controlling interface for the IoT infrastructure. IoT systems provide authentication mechanisms for administrators and consumers to prevent unauthorised access to any component within the system.

A smart speaker wakes us up with our favorite music in the morning; a smart wristband keeps the record of our health, and suggests a healthy diet; a smart assistant reminds us about our daily schedule and appointments: the list continues to grow. With the advent of smart cities, facilities, such as a smart home, smart grid, smart cars, smart street lights and traffic signals, smart meters, and smart TVs, are influencing human lives in some or the other way. The vendor provides a smartphone app for smart consumer appliances to connect and control the device conveniently through the Internet. Such devices employ sensors, whereas the smartphone app performs local analytics. If reports of analytics need to be shared with multiple parties, the system uses a cloud-based server for analytics. A smart wristband gives health updates using different sensors to smartphone applications. In such cases, local analytics in the application are sufficient. A health-monitoring system for an older person living alone may require data processing at the cloud server since a physician, a caretaker, or a relative may be sharing the reports.

Existing IoT systems, such as smart home, smart grid, smart cities, smart car, and smart farming employ several different types of sensors. These systems have multiple goals to achieve, which comprise security, smooth functioning, load balancing, emergency handling, and dynamic decision making. IoT systems most often require user authentication to authorise a user. An IoT-based smart home includes entry and exit locking systems for family members and relatives, fire and burglar alarms, or switching lights and fans on or off based on the presence or absence of an individual. A smart grid may employ sensors for monitoring and distributing power based on domestic and industrial usage, predicting future demand, or locating a fault in the grid. The smart city project, already a reality in developed countries, makes use of IoT for controlling traffic signals and street lights, monitoring criminal activities, and managing vehicle parking, among other applications (Ross, Banerjee, & Chowdhury, 2020). Many tech giants are ready to make self-driving cars equipped with sensors and GPS available for commercial sale in the upcoming years. Farmers now have the technology at their doorsteps for maintaining and watering crops based on environmental conditions.

The automation and transportation industries extensively use Radio-frequency identification (RFID) tags. The tracking of assets, paying toll charges, and managing inventory are some other application areas of these tags. These tags contain digital information and require a reader device to collect it through radio waves. Near-Field Communication (NFC) tags have also found their place in IoT infrastructure. These tags offer low-range communication over a low-speed channel. The vendors employ NFC tags for labeling and electronic payment. The application of IoT to the specific field is named accordingly. For instance, industrial IoT (IIoT) (Munirathinam, 2020), Internet of Medical Things (IoMT) (Joyia et al., 2017), Internet of Battlefield Things (IoBT) (Farooq & Zhu, 2018),and Environmental Internet of Things (EIoT) (Zheng, Wang, & Zhao, 2019) are some examples of explicit IoT applications. The IoT-based home and consumer appliances include HomePod, HomeKit, Siri, Apple Watch from Apple Inc., Amazon's Echo and Alexa, Google's Nest and Google Home.

The advancement and innovations in information and communication technologies (ICT) have significantly influenced the development of IoT and similar technologies. Since most of the IoT devices are battery operated and require continuous communication, the standard protocols and encryption algorithm are optimised for IoT infrastructure. There is a growing need for embedding multiple functionalities in a single device. As IoT systems are inherently resource-constrained, lightweight ciphers, low power consuming electronic circuits, and short-range and secure communication protocols and gateways are employed in IoT devices and systems. There are a plethora of machine learning, artificial intelligence, and deep learning-based algorithms, which are extensively studied by the research community to render them more suitable for the

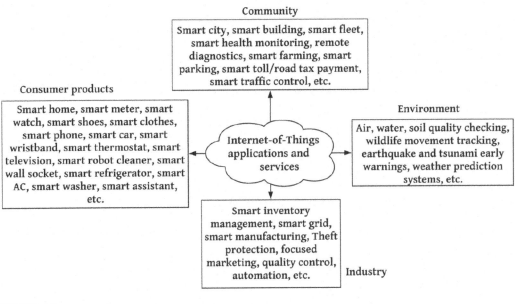

FIGURE 4.1
IoT-enabled applications and services.

IoT environment. Cloud computing and fog computing technologies are evolving continuously to fulfill IoT requirements. Data mining, big data analytics techniques deployed on server systems, also gained significant attention from the IoT industry. Cyber-Physical Systems (CPS) is also closely related to IoT (Sobin, 2020). Figure 4.1 depicts a glimpse of IoT-enabled applications and services in the present-day world.

4.2 IoT Ecosystem

Smart-connected consumer devices employing IoT as the backbone are becoming a part of day-to-day human life. These products are fascinating to everyone but bear a dark side of becoming potential threats to consumers and vendors. With enormous expanse of reachability over the Internet, users and smart device manufacturers are losing finances as well as confidential data to the adversary mainly due to employing insecure authentication methods. Even though there are alliances to brainstorm the specific problem and standardization of protocols for IoT infrastructure, there is a slow growth in incorporating the most secure and cost-effective solution to security issues in IoT. We must understand the vulnerabilities and loopholes in IoT infrastructure and correctly design mitigation techniques to build a robust system.

We should be aware of the functioning inside a typical IoT system to figure out the vulnerabilities and threats associated with it. Figure 4.2 shows major components in the IoT ecosystem. We can roughly classify the elements and communication channels in the IoT system into six categories; firstly, the environment and things directly sensing and responding to the environment. Furthermore, low-range LAN communication protocols employed for intranet data transfer, the local control centre cum the IoT gateways, the Internet, and lastly, remote servers and user interfaces.

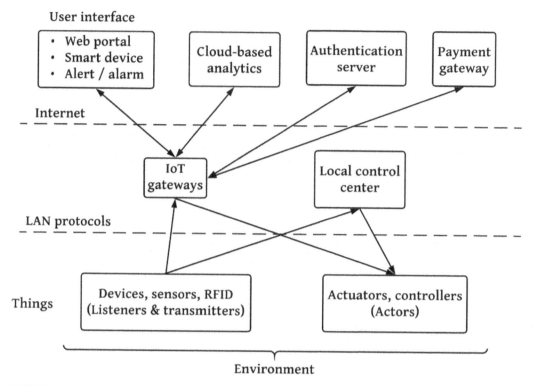

FIGURE 4.2
Major components in the IoT ecosystem.

A system or device that can sense and respond to the environment promptly and accurately is indeed desirable. Most IoT systems listen to their environment in one or the
other way: a voice assistant senses voice commands; a smart car employs several sensors
to sense the climate, the traffic and obstacles in the way; a smoke detector categorises
harmful gases in the surroundings, etc. The significant achievement of IoT systems and
smart consumer devices lies in automation. The manufacturers and service providers for
IoT devices and applications develop convenient, comfortable, and alert solutions for
their customers.

The things in an IoT ecosystem include hardware-based embedded systems, such as
listeners and responders to the environment. The interface with the environment comprises sensors (nerves), RFID tags, controllers, and actuators (muscles). In general, sensors exist for capturing physical observables, such as temperature, ambient light,
humidity, dust, fire, motion, smoke, color, water, and many more (Ross et al., 2020). These
sensors continuously collect data for the desired parameter and send it to the IoT gateway
and the local control system without delay. Sensors, typically, use LAN protocols for
transmitting data. RFID tags use a similar approach to send smart barcodes embedded
on them through radio frequency technology. An RFID reader can sense, collect, and read
the information from the tag. In both these cases, the communication is one-way. Hence,
these devices are referred to as listeners and transmitters. Furthermore, certain things
in the IoT system intend to respond to the environment based on collected data or the
instructions received over the Internet. Such devices are termed actuators and controllers.
They act by following the preprogrammed actions based on the commands received.

These devices can only receive data through LAN protocols and usually do not respond to the source of information. Such devices are therefore termed actors. Communication within the local IoT subsystem among things, gateways, and local control centers uses low-range wireless protocols. Some of these protocols include Bluetooth, ZigBee, Wi-Fi, BLE, NFC, etc. In general, these technologies provide low-range communication.

IoT gateways provide the communication link between the local IoT subsystem and the remote cloud-based servers, user interfaces, and data analytics (Banerjee, Chakraborty, & Paul, 2019). These devices can also employ local data analysis for instant decision making on the incoming data stream from the sensors. The data analysis and control capabilities of IoT gateways can sometimes be shared using an explicit local control centre. Local network management, system diagnostics and device configuration management are some additional functions of IoT gateways and local control centers. The Internet has been a boon to IoT-like technologies and services. In a typical IoT environment, remote user authentication and data analytics servers connect with the IoT infrastructure with the help of the TCP/IP based Internet. The system usually provides a user interface in the form of a web portal, smartphone application, and alert/alarm application for customers and system administrators to process raw data and visualise data analysis. A dedicated authentication server performs security checks, such as user authentication. Some IoT applications, such as smart parking, food ordering, paying toll charges, and periodic subscription-based services also employ a payment gateway.

Despite so many features and utilities of IoT systems, such systems also face vulnerabilities and threats associated with various components and communication channels in the entire IoT ecosystem. Experts in multiple applications have reported several incidences of financial losses due to credit and debit card fraud in recent times (Orme, 2019). In IoT systems too, as we use the Internet for data transmission, users have faced threats to data confidentiality and data integrity aspects of communication between devices themselves and IoT gateway. Furthermore, security breaches in IoT other than financial fraud exist, which include insecure mobile interface, privacy issues, and insecure cloud and web interfaces. Such threats must be addressed and monitored closely to understand the risk intensity and accordingly, mitigation mechanisms against them must be provided.

4.3 Classification of IoT-powered Applications and Services

In the present-day world, a considerable number of IoT-enabled consumer products and services are in use. With the emerging evolution of IoT technology, we can find several such devices around us. We classify these smart devices into seven categories based on their working mechanism, as follows.

1. *Sensing only:* There is a class of IoT devices wherein sensors located at different places collect data and send them to the server. The server stores the data, then a system administrator retrieves the data whenever needed and presents the data in its original form. Smart cities that monitor temperature, oxygen, humidity, dust, CO, etc., can use sensor-enabled infrastructure and display the same figures periodically to let citizens know the best and worst neighborhoods in the city.

Health-monitoring applications for a bed-ridden patient at home or hospital can also use sensors to record various health parameters, such as pulse rate, blood pressure, temperature, and other critical health parameters, and a display device shows variations in such parameters per second.

2. *Sensing and real-time local analysis:* IoT systems wherein the sensors collect data from the environment and perform real-time analysis of the data at the local gateway fall under the second category. Weather-prediction systems employ various sensors to collect data about temperature, humidity, and other environmental factors, and perform local data analysis to predict rainfall or forecast temperature variations in the locality. A vehicle maintenance system similarly uses sensors to monitor the condition of critical spare parts and shares the analysis report of the respective vehicle with the driver and maintenance staff.

3. *Sensing and cloud-based data analytics:* Occasionally, data analysis can occur at a centralised cloud server for sensors deployed in a geographically large area, such as a state or a country. Train track-monitoring systems use sensors at a short distance over all the tracks in a country. An unexpected event on the tracks may lead to a major accident or interrupt rail traffic for a longer duration until the damage gets repaired. Hence, to divert trains to an alternate route and stop trains approaching the faults, cloud analytics embedded in the system collect the data, and provide reports to different authorities to the station and onboard train staff based on real-time locations of trains.

4. *Sensing and analysis of data with automated control:* IoT systems can use rule-based automatic control mechanisms in case of an alarming situation. In smart lighting and home systems, the actuators play a significant role in responding to the environment based on rules and instructions according to the underlying algorithm. Electrical appliances switch on and off based on the presence of an individual sensed by motion detectors. Consequently, the air-conditioning system can adjust the temperature based on the number of individuals present in a room perceived by the CO_2 level inside the room.

5. *Sensing and analysis of data with manual control:* It may not always be possible for the system itself to deal with every unexpected situation and thus it may need manual interventions. Sensors and analytics can show a damaged node within a network that needs replacing or repairing urgently. In another scenario, a smartphone app can also provide an interface to display a probable malicious node and allow the consumer or administrator to either restart the node or stop transmitting the sensed data. Suppose the system detects unusual data collected at the gateway or cloud server, and the system cannot decide to identify the actual hacked sensor. In such a case, the network expects a manual intervention.

6. *Sensing and analysis of data with automatic and manual control:* We usually provide intelligence to smart devices so that they can respond quickly to certain situations. But they are still not intelligent enough to differentiate and respond to a more severe event that is within their capability. Smart cameras installed at traffic signals can automatically zoom in on registration numbers of vehicles violating traffic rules and send images to the server. In case there is an accident, the traffic police sitting in the control room can control the camera to zoom in on the offender, rather than allowing the cameras to operate normally.

7. *Smart devices (artificial intelligence (AI) + machine learning (ML) + rule-based and manual control):* Smart consumer appliances usually embed the latest hardware and software technology, like intelligent sensors, artificial intelligence, machine learning algorithms, etc., to provide a state-of-the-art experience to the consumer. These devices can learn by experience and perform better as the consumer interacts with them. They can place orders and pay on behalf of the consumer, automatically sense the consumer's mood and play songs accordingly, read the news, receive calls, set reminders, among other functions.

4.4 IoT Security Breach

It is possible to mount a zero-day attack for compromising devices, such as smart TV, printers, smart security cameras, etc. (Srinivas, 2020). Bitdefender discovered that Ring Doorbell cameras from Amazon were allowing hackers to access a user's Wi-Fi as well as other devices connected through Wi-Fi. It was also possible to mount a distributed denial-of-service (DDoS) attack via Blink XT2 security camera systems from Amazon (Srinivas, 2020). Additionally, the cameras provided access to footage from the camera and audio output. Check Point researchers demonstrated that fax machines are vulnerable to hacking using the fax number and the telephone line (Srinivas, 2020). They also illustrated the exploitation of a security bug in HP all-in-one printers during a conference (Srinivas, 2020). The FBI stated that the camera and microphone integrated into most of the smart TV manufacturers could be hacked to control volume and even channels (Srinivas, 2020). Researchers also illustrated information leakage based attacks on smart lights controlled using infrared (Maiti & Jadliwala, 2020).

A News channel in the US reported an incident of hacking a smart home and a thermostat of a couple (Srinivas, 2020). Researchers from academic institutions developed malware to steal confidential data from a smartphone via a hacked microphone from the device (Shumailov, Simon, Yan, & Anderson, 2019). They also demonstrated the acoustic side-channel attack on a touch-screen device which unveils everything a user types. Smart coffee machines possess high-security risks as their vendors give minimal efforts to its security aspects while designing. The apps for these machines are vulnerable to reveal information about the consumer's bank and cards (Srinivas, 2020). It is also observed that LAN printers pose a risk for cyber attacks into organizations (Srinivas, 2020). Researchers hacked the smart speaker, Amazon Echo, during a live demonstration at a security conference (Srinivas, 2020). As per Trend Micro researchers, IoT-based cyber attacks are possible via Internet-connected gas stations (Srinivas, 2020).

During a demonstration by researchers from the University of Central Florida, a Nest Learning thermostat was hacked within fifteen seconds when the hackers were allowed to access the device physically (Gemalto, 2016). The hacker then employed the thermostat to expose the Wi-Fi credentials, spy on the consumer, and attack other devices connected with the same wireless network (Gemalto, 2016). It was also possible to track geographically the movement of fitness trackers by employing a customised Raspberry Pi (Gemalto, 2016). Further, experts successfully managed to send spam and phishing emails using an Internet-connected refrigerator (Gemalto, 2016). There was also an incident of hacking the in-flight entertainment system of an airplane (Gemalto, 2016). A professor reported that a hi-tech train signaling system was prone to hacking, prompting severe consequences in the UK (Gemalto, 2016). It was also claimed that more than 0.45 million connected vehicles are vulnerable to intrusion attacks if the attacker gets access to their IP addresses (Gemalto, 2016).

Insurance firms employing IoT-enabled services and devices possess a high risk of cyberattacks (Leong & Chen, 2020). More than 34% of present-day Internet traffic accounts for the cyber threat. Average financial loss due to IoT related hacks passed 8 million USD in 2019 (Leong & Chen, 2020). Most of the security breaches reported in the past are the consequences of some critical misconceptions about IoT security. Some of them include: constrained-IoT things pose no risk to the system as they transfer unimportant raw data; mere authentication and authorization are sufficient for securing the entire system; and it is enough to employ threat detection mechanism as a device reset can stop such attempt immediately. (Leong & Chen, 2020)

There are several reported incidents of such security breach in IoT-enabled systems. The IoT industry finds it difficult to catch up with fast development and innovation in lightweight protocols, hardware devices, and authentication mechanisms proposed explicitly for IoT-based products. Furthermore, they are reluctant to embed the evolution with a fear that the consumers may not find the new product budget-friendly. Smart cities, smart homes, smart cars, smart grids, etc., are gaining attention from various sections of society with no knowledge about the vulnerabilities that are associated with these emerging systems.

A well-aware consumer sometimes complains publically (on social media platforms) about security and privacy invasions through an electronic device and sometimes ignores or replaces the instrument, considering it faulty. The casual approach towards minor financial or data loss due to a smart product which provides a little comfort should change to a more profound perspective. Data interception over an insecure LAN or the Internet is mostly beyond the scope of the consumer, who is seldom aware of any such event over the communication network. Consumer behavior plays a significant role in improving product quality. If every consumer is overly concerned about the quality of products and services received, and carefully points out the pros and cons associated with them, then vendors and manufacturers will take care of every minute detail about the presumable issues and deliver robust, secure, and well-designed products and services.

The most severe, vulnerable, and easily targetable loophole into any system or device forms the basis for the direction of research into a field. As IoT devices and services are achieving new milestones every passing year with more innovative and creative solutions, adversaries and hackers have plenty of options to choose their probable target. Mostly, startups become the victim as they lack sufficient expertise and background knowledge to visualise all possibilities of their product or service glitches. Security experts, cryptanalysts, and researchers from various fields closely associated with IoT-powered devices and services to be launched by a company must play the lead role, along with the core development team, to visualise and mitigate any attempt to break into the system.

4.5 Current Scenario of Security in IoT Infrastructure

The device-to-user authentication while accessing connected consumer devices is the area in IoT systems that needs serious attention from academia and the biometric industry. The evidence of IoT security breaches reveals that the incidence of financial fraud is increasing every passing year due to security vulnerabilities associated with payment gateways and user authentication. Data loss through communication channels and leakage in cloud-based servers is also common. Several issues need to be addressed before the world of IoT smart devices and services is made free from all forms of threats and vulnerabilities.

A standard TCP/IP protocol stack exists for the Internet. The security of the protocols had been well studied and researched before everyone could accept it for data transmission over the globe. In the case of IoT systems, there exists no common standardization with which every service provider and consumer product vendor agrees. Multiple versions of lightweight protocols designed and developed explicitly for IoT infrastructure are mainly applicable to the LAN environment. The cryptanalysts and researchers brainstorm these proposals to pinpoint the flaws in them. For such systems, some proprietary security mechanisms (hardware and software-based) do exist, whose internals remain secret. Still, analysts claim the robustness for all existing threats to generate revenue and increase their business. Manufacturers of such systems need to ensure security features, such as device identification, device configuration, data protection, logical access to interfaces, software and firmware updates, and cyber security event logging. The manufacturers then use such proprietary products to build their smart devices without much effort on verifying their security concerns.

As predicted by many leading global agencies, there will be a vast demand for IoT-based appliances and services in the coming years. So, market leaders regularly launch new innovative solutions applicable to day-to-day problems employing IoT and similar technologies. This competition has made a significant impact on increasing security vulnerabilities. The product designers and developers heavily rely on third-party proprietary software and hardware to meet launch deadlines for new IoT products. Moreover, due to existing security associated with the Internet protocols, there is minimal attention given to securing data during transmission from a device to the server. The consumer is rarely aware of the technical specifications of such a device and out of curiosity, comfort or security, buys these products and enjoys them until they notice some security flaw.

The automated-payment system implemented in smart consumer products faces a threat from malicious attackers. A smart refrigerator that automatically places orders for vegetables, eggs, ice creams, etc., and pays the amount through credit or debit card credentials of its owner is one of the probable targets of the adversary. The attacker can hack the device through the Internet. In such a scenario, card details can be misused to order unnecessarily for the owner, or the attacker may buy something with the owner's money. The consumer may be unaware of the payment gateway server leaking bank details or card information through a security flaw, and therefore would be surprised to see the debited amount, without consent or knowledge. An authorised user becomes the victim of such incidents due to the negligence of the service provider.

Public-key authentication can be used within the IoT infrastructure. In this mechanism, the device securely stores a private key and announces a public key for communicating with the device. The main hurdle again turns out to be the strength of the device's security, which assures that no intruder is allowed to access or modify the stored key in any unauthorised way. One more problem with device-to-device communication lies in identifying genuine components of the system once a request is received (Chakraborty & Rodrigues Joel, 2020). A malicious unknown source can also generate such a request and forward it to an IoT sensor through the Internet. Hence, the problem of detecting a genuine system component and authenticating only such devices to access and process the information remains the major challenge.

The local network of IoT sensors, actuators, and gateways is the best target to access data for the adversary since there is a large possibility that such systems use proprietary wireless LAN communication protocols. The hackers are intelligent and well- equipped to break into such protocols, whose security is verified by a limited set of experts within an organization. Once such a flaw opens the door, the attacker becomes an integral part of the system and accesses, modifies, and deletes data from the devices without any hint to the system administrators, until they come across some significant evidence. The smart home, smart grids, smart lighting, and smart traffic signaling systems are a few examples that may be the victim of unauthorised access to network communication.

There are multiple incidents of hacking IoT sensors and actuators. The adversary targets the source of data into an IoT infrastructure. The smart home system, smart healthcare system, and smart farming are some of the most vulnerable systems falling under this category. Once the adversary gets access to the data generated by sensors within these systems, attacks can be triggered on other IoT-enabled devices and services connected to the same LAN. Attacks can also be mounted using a Trojan horse to destroy the existing network. The pre-installed Trojan horse can also harm the consumer and the infrastructure. As there is no single vendor who designs, develops, manufactures, and performs security analysis of every software and hardware component in an IoT system, we can expect some flaws in these smart devices.

4.6 IoT Threat Model and Mitigation Approaches

A wireless Sensor Network (WSN) forms the basis, and the Internet provides the backbone for any IoT-based smart devices and services. Hence, the vulnerabilities these technologies exhibit may also apply to the IoT infrastructure. Denial-of-Service (DoS), Hello flooding, Sybil, and Sinkhole are some attacks that likely target a WSN (Aufner, 2020). The wireless network connecting the sensors, actuators, IoT gateways in an IoT infrastructure may possess similar risk. The proprietary protocols employed in such systems may not be secure and robust enough to resist all the probable existing attacks on the network. A bug in a smart device such as a web camera, smart speaker, etc., may become a threat to the entire system using the same infrastructure.

The incidences of IoT security breaches indicate that IoT-enabled consumer appliances, smart devices, and services are still vulnerable to attacks, such as hacking and data interception. As we have several components and communication channels connecting them, the chances to get exposed to any attack attempt remains high. So, we should identify all the possibilities for a probable attack and categorise them such that a set of mitigation techniques can protect and secure the IoT infrastructure. Based on the major components involved in an IoT ecosystem discussed in Figure 4.2, we propose an IoT threat model with ten vulnerabilities. Figure 4.3 shows the proposed threat model, and the naming convention V_1 to V_{10} depicts the ten vulnerabilities.

The class of threats associated with listeners and transmitters falls under vulnerability V_1. Adversaries hack these devices and divert traffic to the fraudulent server. Malicious content may also be injected into the network to harm the infrastructure. A "hardware Trojan" is a type of deliberate insertion into hardware design (Sengupta & Kundu, 2017). Usually, an act of a rogue designer or vendor can lead to such a security breach in the system. A denial-of-service attack can also be mounted through the IoT nodes. These

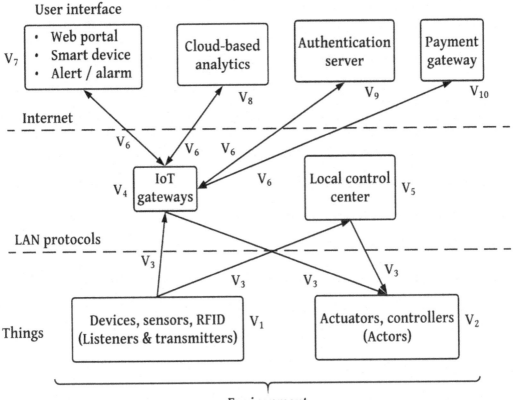

FIGURE 4.3
IoT threat model.

memory-constrained devices are also vulnerable to side-channel attacks. The attacker analyses computation time, electric emission etc., to collect confidential information such as encryption keys or authentication secrets. As these constrained devices are usually battery-powered, it is not feasible to provide additional software or hardware just for security motives.

The possible attacks on nodes that directly interact with the IoT environment, such as actuators and controllers, fall under vulnerabilities V_2. The hacker can target devices such as actuators, controllers, etc. responsible for acting on behalf of the user or the automated control mechanism in the IoT system to perform an unintended activity. Tools such as oscilloscope, logic analyzer, and ChipWhisperer are often employed to figure out the vulnerabilities in the target node. Such an incidence is often noticeable, and the attacker has very little to achieve in the long term except destroying or damaging the components on the IoT infrastructure. Hence, an intelligent adversary invests the least time and effort on such attempts. However, security experts must give necessary consideration to probable threats at such nodes.

Typically, nodes in the IoT infrastructure communicate via technologies, such as Wi-Fi, fiber, Ethernet, ZigBee, 3G, 4G, Bluetooth LE, etc. We categorise the threats to these communication media as V_3. The adversary can retrieve secret information or perform traffic analysis by eavesdropping over the communication medium. Once the attacker obtains some confidential data, it can be used to mount a replay attack at a later time.

Packet flooding in such LAN is a form of denial-of-service attack. If the hacker succeeds in controlling the IoT nodes, several other attacks can be mounted, such as man-in-the-middle (MITM) and Sybil attack, as the entire network will be at the attacker's fingertips (Chen et al., 2018). Consequently, vulnerabilities in the communication protocol and the nodes can be exploited to destroy or damage the whole system.

In a distributed denial-of-service attack scenario, a set of geographically dispersed computers target different nodes within a network infrastructure to deny ultimately any services to the authorised users. During a Sybil attack in a WSN, the adversary employs the compromised node to mislead a victim node by presenting multiple identities. The victim node, in turn, executes the same instruction or operation redundantly. Sniffer instruments are heavily employed to collect network-related information, communication patterns, physical locations of various wireless access points, and the protocols used in the network (Chen et al., 2018). A Sinkhole attack is another way to target the network infrastructure in the IoT system. Here, a malicious node gathers data from its nearby nodes and bypasses all other communication links without any hint to the system (Chen et al., 2018).

Usually, IoT gateways possess high processing power compared to other nodes in the LAN, which is sufficient to execute critical and intensive applications (Garg, Chukwu, Nasser, Chakraborty, & Garg, 2020). This empowerment, on a darker side, provides more opportunities for the attacker to succeed in mounting an attack. IoT gateways located at the intersection of LAN and the Internet thus become an entry point into the entire system if found vulnerable to any threat at software or hardware level. IoT gateways are susceptible to data leakage and topology disclosure. In data leakage, the adversary manages to collect the data from the local storage at a node or by diverting the traffic from the victim node. As every sensor, actuator, and controller device within the network communicates with the gateway, a malicious gateway can consequently disclose the location and identity of these nodes. We label the attacks at IoT gateways as V_4. The attacks corresponding to a local control center within an IoT LAN fall under the category of vulnerability V_5. The threats to the IoT gateways also apply to the local center. Additionally, they both can be a possible target of a Trojan horse attempt.

Communication to various remote servers, cloud-based services, and user interfaces from the gateway occurs via the Internet. One must be concerned about the possible threats to IoT infrastructure that can be mounted via the Internet. Even though we have a well-established Internet service throughout the world, it is still vulnerable to specific risks and so cannot be labeled a fully-secure communication medium. An adversary can execute a DoS attack, man-in-the-middle attack, eavesdropping, selective forwarding, Sybil attack, channel congestion, or collision attack on Internet services (Makhdoom, Abolhasan, Lipman, Liu, & Ni, 2019). We categorise all such threats through Internet traffic under V_6.

The service provider and smart product vendor usually provide a user interface for the consumer and various teams working at the cloud server. Web portals, smartphone apps, and plug-in for alarm or alert mechanism are a few examples of such interfaces. The vulnerabilities associated with them fall under category V_7. The hacker can use another app or operating system bug to capture the smartphone screen or read the app data from system memory. An insecure web application portal or APIs can disclose secret user information or delete some files.

Almost every IoT application requires data analytics, and vendors prefer cloud-based services for the same. But the application, in turn, draws in all the risks associated with cloud servers to its environment. Several attacks have been mounted on cloud servers, which comprise flooding attack, cloud malware injection, SQL injection attack, and signature wrapping attack, etc. (Chen et al., 2018). The adversary tries to control the cloud services by injecting malware, a malicious service instance, or virtual machine on the cloud. The attacker can also modify the XML signature commonly employed by cloud servers for ensuring service integrity. A malicious SQL query code for updating, deleting or reading the database contents also poses a potent threat to cloud infrastructure. The adversary uses the Internet for mounting a DoS attack on the cloud through flooding. Such risks to cloud-analytics servers are categorised as V_8.

An authentication server in an IoT environment ensures that only authorised individuals are allowed to access the system resources and services. Usually, these servers are maintained separately away from the LAN for security issues. Such remote servers are vulnerable to leaking user credentials, false data injection, eavesdropping, record delete or update, identity theft, password, or key or session token disclosure. These types of threats to the remote authentication server fall under vulnerabilities V_9.

Smart devices allow consumers to place orders and pay the merchant via stored card credentials. The IoT service vendor of such consumer appliances also maintains a payment gateway remotely for providing a secure payment process. The adversary targets these gateways for performing financial fraud, intercepting traffic to such servers to collect user credentials and utilise them, or placing unnecessary orders on behalf of the smart device. DoS, DDoS, or false data injection attempts on the gateway can also be executed. We classify such threats to the payment gateways under the class of vulnerabilities V_{10} in our threat model.

The security mechanisms at different levels in the IoT infrastructure ensure that the system components and communication links connecting them are least vulnerable to known threats. Building an IoT application or service free from all existing and unforeseen risks cannot be guaranteed. However, the best consumer product or system should be resistant to a large number of the most common threats and incur the least loss in terms of finances, data, and user privacy. As we can broadly categorise the IoT components into three categories—namely hardware, software, and communication medium—we present countermeasures for the IoT system in these categories to enhance security and thus build a robust application or service from scratch.

Hardware devices should be tamper-proof and employ code signing to avoid any possibility of a Trojan horse attack. Hardware-security features such as ARM TrustZone ensure secure data flow within the devices (Sikder, Petracca, Aksu, Jaeger, & Uluagac, 2018). Designing ICs with active shields protects them from probable side-channel attacks. To provide identity to each node, an X.509 Digital Certificate should be employed. Subsequently, communication between the nodes will use HTTPS or NTLS protocols; each node can verify the identity of every other trusted node. Additional security measures include embedding a Trusted Platform Module (TPM) device, using of a PUF (Physical Unclonable Function), randomising instruction execution cycles, using lightweight hardware implementation of a cipher, employing network segmentation, or implementing a device registry.

The strategy to secure communication channel includes encryption using a hash function (Meneghello, Calore, Zucchetto, Polese, & Zanella, 2019). The authentication mechanism should use message authentication codes (MAC), digital signatures, and hash functions. A pseudo-random number generator that satisfies a majority of randomness tests can further enhance the security of communication by generating asymmetric keys and lower the chance of a replay attack. We can reduce the risk of most software-based threats by adapting lightweight intrusion detection methods, using software-defined networking (SDN), ensuring software integrity during updates, and auditing (log management for each update). Cloud-based IoT systems should implement homomorphic encryption and Cloud Access Security Broker (CASB) for providing security and privacy protection in cloud-based services. Block-chain technology can also protect against replay attack, ransomware, or malware.

The threats to various communication links, software modules, and hardware components have different intensities in terms of financial and data loss to the consumer and the service provider. Smart devices have touched almost every aspect of human life, and subsequently, we are inviting their probable dangers into our lives, exposing everything we own. Since not every consumer is technologically savvy, they believe most news item and review available online and form some misconceptions. We addressed the possible risks associated with smart devices and also specified the ways to mitigate them. In today's digital world, a consumer excited about smart devices must become smart enough to understand the know-how about every such device in their possession.

4.7 Authentication Using Biometric Systems

The conventional ways for user authentication include password, token, patterns, security question, personal identification number (PIN), and identity card (ID), among others. We have seen rapid growth in usage of the Internet in recent times. So a large number of services are available to everyone, and they require any of the conventional authentication approaches to secure accounts created for accessing these services. The main drawbacks with these approaches are that a user has to remember them and a brute-force or dictionary attack can crack them with minimal efforts. Additionally, for each attempt when a user opts for "forgot password" or "reset PIN" option, the service provider incurs a financial loss. Hence, we need an authentication mechanism for which the user incurs small effort overhead, the service provider faces minimal expenses, and authentication is less prone to attacks. A biometric system has emerged as an alternative to these methods that provides more convenient and secure authentication. A system employing biometric-based authentication assures that only legitimate users control the smart device and access the system through the user interface. It also guarantees that every transaction is signed using consumer biometric data. Such a mechanism can mitigate vulnerabilities V_7, V_9, and V_{10} from Figure 4.3.

The biometric system employs an individual's biological traits for user recognition. These systems possess high accuracy and are available for commercial and personal use. Several government and private organizations rely on a biometric system rather than conventional token or password-based systems. smartphones, home security applications, university attendance systems, cash dispenser machines, and e-voting are a few examples employing biometric user authentication in the current scenario. These systems are easy to operate even for a layperson, and most importantly, the user is free from remembering passwords or tokens for verifying his identity. In general, a system administrator enrolls a new user to the biometric-system database with two or more samples of his biometric traits. The system uses these enrolled traits from the database to recognise the user whenever biometric data is submitted to the biometric system.

Most often, an administrator enrolls the user in the biometric system. There are two ways to provide access control using a biometric system. The device can be an identification system or an authorization system. During identification, the system compares recently-acquired biometric input to all the samples stored in the database. The system identifies the input with the identity of the database entry that generates the highest similarity. In this case, the unknown user does not reveal any other information about identity to the system other than biometric data. The user claims identity while submitting traits to a biometric authentication system. The system then compares the user's features with the database entries and with the claimed user's samples to conclude if the claimed identity is true or false.

Biometric systems employ physiological and behavioral traits of an individual. The physiological characteristics include fingerprint (Maltoni, Cappelli, & Meuwly, 2017), palmprint (Teoh & Leng, 2015), iris (Rathgeb, Uhl, & Wild, 2013), ear (Abaza, Ross, Hebert, & Harrison, 2013), voice (speech) (Duarte, Prikladnicki, & Calefato, 2014), face (Subban & Mankame, 2013), palm, veins and hand geometry (Christo, 2017), DNA (Hicks & Coquoz, 2015), and electroencephalographic (EEG) signals (de Albuquerque, Damasevicius, Tavares, & RogérioPinheiro, 2018). Some behavioral features capable of identifying an individual include signature, gait (Connor & Ross, 2018), keystroke dynamics (Zareen, Matta, Arora, Singh, & Jabin, 2018), and posture (Burda & Chudá, 2018). A biometric system extracts unique features from submitted biometric data to create an *encrypted biometric template* (Jain, Uludag, & Ross, 2003) for storing in the database. A biometric system may employ more than one trait to improve recognition accuracy. Such designs fuse features from biometric data to create a highly unique template. We term such systems *multimodal biometric* (Toli & Preneel, 2014) recognition systems. Due to advancement in technology, biometric systems evolved during the past two decades. Traditional biometric systems store user templates in a remote database (i.e., match-in-database). Meanwhile, the advanced biometric application keeps an individual's template in a smart card dedicated to the user. Card-based biometric systems include *template-on-card (ToC)* (Joshi, Mazumdar, & Dey, 2020), *match-on-card (MoC)* (Benhammadi & Bey, 2013), and *system-on-card (SoC)* (Joshi et al., 2020).

Figure 4.4 shows various components of a typical match-in-database fingerprint biometric system. The sensor in the input device accepts the user's finger and generates a grayscale or RGB image. The feature-extraction module tries to find out specific patterns in these images. During the enrollment phase, the template-creation module uses these patterns to generate an encrypted template for storing in the remote database. However, while identifying or authenticating an individual through fingerprints, the template-creation module communicates the template to the matcher module. The matcher module then fetches the templates from the database to compare with the newly-generated template. The matcher module uses similarity scores between the compared templates to confirm either the identity of the unknown user or verify the user's claim.

The biometric field has evolved into diverse applications over the past fifty years. The system that initially gained popularity as a mere verification or identification system was found suitable for various problems. Present-day uses, including border security, access control, forensic investigations, child trafficking control, monitoring infant vaccination, channelizing government schemes for underprivileged citizens, home security, university attendance system and banking services such as cash dispenser machines, employ biometric systems. These systems found scope in security and privacy applications along with identity management. Present-day smartphones, laptops and other handheld devices, and smart homes are a few recent examples that incorporate biometric recognition. Researchers are now looking forward to figuring out the possibility of integrating biometrics into IoT-based devices and systems.

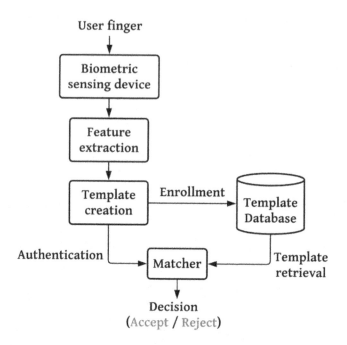

FIGURE 4.4
Block diagram for match-in-database fingerprint biometric systems.

Recent research outcomes prove that the biometric field is marching in diverse directions. Researchers are focusing on multimodal biometric-based authentication, wearable biometrics (Ross et al., 2019), poor-quality data, post-mortem biometrics (Ross et al., 2019), human-computer interaction, scalability (Ross et al., 2019), template protection, personal privacy, and data integrity. Also, there is scope in detecting and mitigating presentation attack and soft biometrics (Ross et al., 2019), improving the accuracy and response time for modalities that use 3D images. Manufacturers of biometric systems employing face, gait, and iris recognition using high-resolution 3D images are targeting the development of pocket-friendly devices that have high precision even with noisy data. Figure 4.5 depicts a glimpse of biometric-enabled applications.

In the case of IoT infrastructure, we can provide access to raw IoT data and system components via biometric user- authentication. We can replace the traditional authentication server from Figure 4.3 with a biometric cancellable-template database. A multimodal biometric authentication would further enhance the security of the system. The system becomes more trustworthy and acceptable by a large community. We can even add smartphone-based biometric authentication and associate the hardware identity to further lower the chances of replay attacks. In addition, there are minimal chances of vulnerability V_6 occurring at the authentication server.

4.8 Authentication in IoT System

We can provide secure user authentication within an IoT infrastructure with user interfaces which recognize individuals and ensure the authenticity of a payment process.

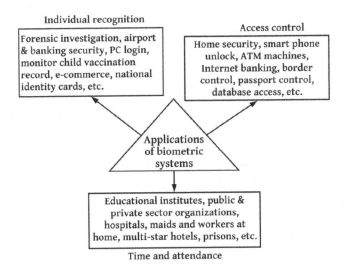

FIGURE 4.5
Applications areas of biometric systems.

User authentication and authorization are receiving significant attention from researchers across academia and industry. When it comes to commercial IoT products, vendors still prefer to stick with the traditional user-authentication methods. However, academic research efforts try to figure out more innovative and IoT-enabled product and system-specific solutions. Multifactor remote-user authentication is shown to be secure from existing threats to the system (Geeta & Kalra, 2020; Lee, Kim, Yu, Park, & Park, 2019). IoT medical applications can be secured using mutual authentication achieved by an elliptic curves cryptosystem (ECC) (Elngar, 2019). An Efficient Anonymous User Authentication (E-AUA) protocol that reduces computation and communication costs for mobile IoT devices using multiple servers proves to address network congestion in the system (Zeng, Xu, Zheng, Xiang, & Zhou, 2019).

FlexPass (Belk, Fidas, & Pitsillides, 2019) employs a one-time text and graphical secret chosen by the user for authentication in the IoT system. In this paradigm, the user has the flexibility to select the kind of secret desired for authentication, which can be changed if found inconvenient. The WSN and IoT-powered greenhouse-monitoring system that utilises a two-factor user- authentication mechanism including a confirmation code was demonstrated useful for practical application (Akhtar, Hussain, Arshad, & Ahmad, 2019). It is also possible to secure an IoT-enabled Farm Management Information Systems (FMIS) using RFID technology (Bothe, Bauer, & Aschenbruck, 2019). An innovative authentication protocol based on hashing and XOR-ing operations that is free from time stamping is suitable for the Industrial IoT (IIoT) technology (Eldefrawy, Ferrari, & Gidlund, 2019). A lightweight key agreement and user-authentication protocol facilitate session-key negotiation to ensure mutual authentication between communicating parties for WNS within IoT infrastructure (Turkanovic, Brumen, & Holbl, 2014). The approach implements simple operations such as hashing and XOR-ing.

The way users interact with IoT devices within the system may also help in user authentication with above 85% accuracy (Ongun et al., 2018). The approach performed with 97% accuracy for classifying five users interacting with the IoT devices and nodes. The use of smartcards for multi-factor remote user authentication proves to be secure against several existing threats to IoT and cloud infrastructure (Sharma & Kalra, 2018; Song, 2010;

Sood, Sarje, & Singh, 2010; Chen, Kuo, & Wuu, 2014). Recent research has also focused on user authentication into IoMT. It is possible to use three-factor authentication, including password and smartcard, to allow medical practitioners to assess a patient's health updates securely (Dhillon & Kalra, 2018). There is evidence of using Block-chain technology for secure and safe user authentication in the IoTsystem (Almadhoun, Kadadha, Alhemeiri, Alshehhi, & Salah, 2018). The cryptographic notion of elliptic-curve cryptography (ECC) can also be employed to provide session-key security as well as user anonymity (Li, Wu, Chen, Lee, & Chen, 2017).

Level-Dependent Authentication (LDA) is a resource-efficient key-sharing mechanism employing the organizational hierarchy for allowing the user to censor communication (Patel & Doshi, 2020). A signature scheme requiring no certificate and performing unidirectional communication for anonymous authentication during mobile payments in IoT-enabled smart devices proves to be lightweight, efficient, and practical (Chen et al., 2019). A hardware fingerprint from the sensor can be obtained within an IoT infrastructure and further used to compare against reference measurements collected before deployment for a secret-free authentication (Lorenz et al., 2020). It is also feasible to generate strong keys dynamically to provide secure communication and authentication between IoT gateways and edge nodes within an IoT system (Sathyadevan, Achuthan, Doss, & Pan, 2019). User-privacy preservation under the IoT environment requires an anonymous user-authentication approach (Xie, He, Xu, & Gao, 2019).

Smartphones' SMS facility can act as a means of device authentication for constrained nodes in IoT systems (Perkovic, Cagalj, & Kovacevic, 2019). A novel approach employing ZigBee technology proposes node-to-node anonymous and mutual authentication suitable for smart home systems (Alshahrani, Traore, & Woungang, 2019). It is feasible to generate a unique ID for each IoT device from its SRAM to authenticate the device securely (Mahmod & Guin, 2020). An artificial-intelligence-powered, lightweight and fast authentication approach can reduce communication latency between IoT devices (Fang, Qi, & Wang, 2020). CirclePIN is a mechanism for authorizing a user to IoT nodes via a smartwatch (Guerar et al., 2020). It is possible to mitigate data extraction and similar attacks on communication channels within an IoMT using random numbers and session keys for hashing and to encrypt the data (Bae, 2019). The IoT-powered WSN (Wireless Sensor Network) (Chakraborty, Gupta, & Ghosh, 2013) is vulnerable to DoS attack, along with stolen verifier and user traceability attempts. An approach to authenticating a remote user via symmetric-key protocol secures the communication channel against such attempts (Anwar et al., 2019).

Device-to-device authentication will be more promising if it is feasible to exploit specific unique properties from the hardware device. The uniqueness, unclonability, and tamper-evident characteristics of Physical Unclonable Functions (PUFs) can play a vital role in securing communication between two IoT nodes (Barbareschi, Benedictis, Montagna, Mazzeo & Mazzocca, 2019). Advancements in software technologies also open new doors for providing security and privacy features to communicate over different mediums. The algorithms that evolve on their own and can adapt to variable conditions make the most suitable choices for the IoT environment. A combination of deep-neural network and machine-learning algorithms at communicating nodes within an IoT environment enhances security during authentication (Chatterjee, Das, Maity, & Sen, 2019).

Most existing approaches rely on traditional user-authentication approaches for user authentication to assure that only legitimate individuals can use IoT-powered smart devices and systems. However, data-sensing devices and intermediate nodes within the

IoT infrastructure are equally at risk of being hacked by an adversary. Researchers inadvertently overlooked this domain in their work. It is not enough to have a centralised authentication server to maintain the database of legitimate users and assure that the system in its entirety is free from all sorts of external risks. The more secure a system becomes, attackers try to figure out some new loophole to access the services illegally.

We can link an authentication mechanism to a user interface, authentication server, and payment gateway to thwart vulnerabilities V_7, V_9, and V_{10}, respectively in the threat model shown in Figure 4.3. Thus, biometric authentication will be required to approve automated financial transactions within the IoT environment. Also, the mechanism will guarantee access to resources by legitimate users and authorities. We are in the initial stages of approving banking services using consumer biometric data. It will be a massive achievement once we cross this milestone in most countries.

4.9 Biometrics for IoT Security

The advantage of biometric-based user authentication to the IoT system lies in its uniqueness. The user gets relieved from worries of forgetting his credentials for authentication virtually forever. If biometrics are stolen by some means, we have *cancellable biometrics*, which generate another authentication data from the same biometric. Since there are multiple traits that an individual possesses, the system can be made more secure by employing a combination of these traits. Also, the *template protection mechanisms* for storing encrypted biometric data on the server assure that information about the biometric features or patterns of a user is entirely confidential. Hence, even if the attacker gets access to the template, it will be unfeasible to know the biometric details of the owner.

There exists a high possibility to employ biometric-based user authentication for IoT devices and systems. However, as we have multiple characteristics of an individual for unique identification and authentication, we must decide the most feasible trait for a specific application. Facial recognition systems may not differentiate identical twins; in some cases, even face masks can fool such systems (Orme, 2019; Klare, Paulino, & Jain, 2011). Since social media has entered our everyday lives, we are becoming less hesitant to post our own and family photos on such platforms. Hence, we should avoid facial authentication for financial transactions or critical applications. Voice-based authentication may not distinguish a recorded voice and thus fail to provide the expected level of security. We have modalities such as voice, ear, and gait, among others, which are still in the experimental stage (Orme, 2019).

The fingerprint of each individual is unique, and to date, no record exists for two people with similar fingerprint patterns. Moreover, research in this domain began fifty years ago, and we have highly accurate fingerprint biometric systems available in the market (Jain, Nandakumar, & Ross, 2016). Additionally, compared to other modalities, the fingerprint biometric system requires the least assistance from the user, and it is comfortable to operate even for a layperson. Hence, being completely mature, efficient, and convenient among all the characteristics, these systems are employed for user-identification and user-authentication systems in various government, private, and even high-security applications. Fingerprint-based biometric authentication thus wins the race for being the most suitable and practical option for the IoT environment.

A cancelable biometric authentication can help in preserving security and privacy for IoT-based applications (Punithavathi & Geetha, 2019). Similarly, the multimodal biometric system was proved to render enhanced security and improved accuracy while authorizing an individual for accessing an IoT network (Olazabal et al., 2019). The biometric-powered anonymous user authentication scheme performing lightweight operations in IoT infrastructure proves to be efficient (Shayan, Naser, & Hossein 2019). The mechanism is best suited for smart homes, and it assures that the rest of the smart devices inside the house remain unaffected even if the hacker breaks into the security of the smart home. The user may choose to use ECG (electrocardiography) as a biometric trait for authentication into the IoT system (Barros, Rosário, Resque, & Cerqueira, 2019; Zhang, 2018). The facial-recognition approach can provide user authentication on smartphone devices (Lee, Sa, Cho, & Park, 2017).

A novel combination of system-level obfuscation methods along with electro-cardiogram (ECG) and photoplethysmograph (PPG) biometrics solves the issue of unauthorised access to IoT nodes (Guo, Karimian, Tehranipoor, & Forte, 2016). The approach also helps in preventing any possible tampering or reverse engineering. The research also shows the use of behavioral biometrics to provide secure and authorised access to the IoT system components (Janik, Chudá, & Burda, 2020; Abuhamad, Abusnaina, Nyang, & Mohaisen, 2021). Also, keystroke biometrics has demonstrated promising results when targeted for user authentication (Wang, Wu, Zheng, & Wang, 2019). We can employ these approaches, especially for smartphone authentication. Smart health care systems also need authorised access exclusively by medical practitioners and family members. The biometric-based system can be a feasible solution in such scenarios (Hamidi, 2019). A bimodal biometric authentication provides enhanced security on smartphone devices (Buriro, Crispo, & Conti, 2019).

The use of the smart card for storing the biometric template mitigates several attacks associated with templates and requires no additional remote template database server. A similar approach can also help to secure and authorise access to IoT resources (Banerjee, Chunka, Sen, & Goswami, 2019; Cui, Sui, Zhang, Li, & Cao, 2018). Additionally, if the system employs a SoC, then most vulnerabilities associated with the biometric system will be eliminated. The consumer would find it convenient to carry his smartcard and get rid of several issues related to traditional authentication mechanisms. Biometric-based access-control systems would benefit vendors and buyers in multiple folds. Hence, sooner or later, IoT-enabled smart devices and services would offer such solutions to the consumers.

The Fast Identity Online (FIDO) Alliance provides solutions for replacing conventional approaches to identity management (IM) in a more secure, convenient, and feasible manner (Chadwick et al., 2019). The alliance includes more than 200 industry leaders in software and hardware sectors. Their first framework provides smart devices with a password-free authentication. In another protocol, the coalition presents a small hardware token for implementing two-factor authentication. An asymmetric encryption method forms the basis for authentication under both approaches. The Verifiable Credentials (VC) data model proposed by the World Wide Web Consortium (W3C) is a similar initiative to identify better user-centric solutions for the identity ecosystem (Laborde et al., 2020). These solutions are decentralised digital-identity systems and are presently available for real-time use in smart devices and IoT-based user authentication. As there is a profound demand for a standard protocol across various consumer products and services, different market leaders are collaborating with the alliance and incorporating strong authentication standards as a security mechanism for the multiple products and services they offer.

We can employ a fingerprint-based biometric system over other modalities to solve the problem of user-to-device authentication as well as allowing only authorised individuals to access data and services concerning vulnerability V_9 from the threat model proposed in Figure 4.3. It is also possible that consumers appreciate such a move, as most of them might have used a fingerprint-based biometric system in the past. We are living in a rapidly-progressing world where privacy and security need utmost attention to ensure that the future generation finds it safe to use any smart devices and services. A fingerprint biometric system would meet the requirements of the time and become a solution to authentication-related queries of industry innovations.

4.10 Conclusion

We are living in a world of information and communication technology (ICT). The advancements and innovations powered by ICT have made a significant impact on our daily routine. We are so connected with this new Internet world that people feel uncomfortable if their Internet goes down even for a few minutes. Hence, Internet service providers (ISP) worldwide are exercising every possible step to provide uninterrupted and high-speed connectivity to their consumers, even in a remote area. The notion of Internet of Things is the outcome of a sufficiently mature ICT. The chapter provides a broad review of the current scenarios of Internet-of-Things technology. It covers various types of IoT devices and services and presents differences between them in terms of their functionalities. The chapter also talks about security issues in these devices and suggests various corrective and mitigative steps against probable threats to such systems. The study also addresses the need for a highly-secure authentication mechanism like a biometric system to authorise and authenticate an individual. We emphasise the benefits of employing a fingerprint-based biometric system in an IoT infrastructure. We tuned the contents such that the readers understand the significance of every term specific to IoT and biometric systems. In a nutshell, the chapter introduces the reader to IoT and biometric systems from a security perspective and encourages them to address various threats to them.

References

Abaza, A., Ross, A., Hebert, C., Harrison, M. A. F., & Nixon, M. S. (2013). A survey on ear biometrics. *ACM Computing Surveys*, 45(2), 22:1–22:35.

Abuhamad, M., Abusnaina, A., Nyang, D. H., & Mohaisen, D. (2021). Sensor-based continuous authentication of smartphones' users using behavioral biometrics: A survey. *IEEE Internet Things Journal*, 8(1), 65–84.

Akhtar, M., Hussain, M., Arshad, J. A., & Ahmad, M. (2019). User authentication scheme for greenhouse remote monitoring system using WSNs/IOT. In *Proceedings of the 3rd International Conference on Future Networks and Distributed Systems, ICFNDS 2019*, Paris, France, July 01–02, 2019 (pp. 47:1–47:8). ACM.

de Albuquerque, V.H.C., Damasevicius, R., Tavares, J.M.R.S., & RogérioPinheiro, P. (2018) EEG-based biometrics: Challenges and applications. *Computational Intelligence and Neuroscience*, *2018*, 5483921:1–5483921:2.

Almadhoun, R., Kadadha, M., Alhemeiri, M., Alshehhi, M., & Salah, K. (2018). A user authentication scheme of IoT devices using blockchain-enabled fog nodes. In *15th IEEE/ACS International Conference on Computer Systems and Applications, AICCSA 2018*, Aqaba, Jordan, October 28 - Nov. 1, 2018 (pp. 1–8). IEEE Computer Society.

Alshahrani, M., Traoré, I., & Woungang, I. (2019). Anonymous mutual IoT inter device authentication and key agreement scheme based on the ZigBee technique. *Internet Things*, 7.

Anwar, G., Mansoor, K., Mehmood, S., Chaudhry, S. A., Rahman, A. U., & Najmus Saqib, M. (2019). Security and key management in IoT-based wireless sensor networks: An authentication protocol using symmetric key. *International Journal of Communication Systems*, 32(16), e4139.

Aufner, P. (2020). The IoT security gap: a look down into the valley between threat models and their implementation. *International Journal of Information Security*, 19(1), 3–14.

Bae, W.-S. (2019). Verifying a secure authentication protocol for IoT medical devices. *Cluster Computing*, 22(1), 1985–1990.

Banerjee, S., Chakraborty, C., & Paul, S. (2019). Programming paradigm and Internet-of-Things, CRC: A handbook of Internet-of-Things & big data. *A Handbook of Internet-of-Things and Bigdata*, First edition, 148–164, ISBN 9781138584204.

Banerjee, S., Chunka, C., Sen, S., & Goswami, R. S. (2019). An enhanced and secure biometric-based user authentication scheme in wireless sensor networks using smart cards. *Wireless Personal Communication*, 107(1), 243–270.

Barbareschi, M., Benedictis, A. D., Montagna, E. L., Mazzeo, A., & Mazzocca, N. (2019). A PUF-based mutual authentication scheme for Cloud-Edges IoT systems. *Future Generation Computer Systems*, 101, 246–261.

Barros, A., Rosário, D. do, Resque, P., & Cerqueira, E. (2019). Heart of IoT: ECG as biometric sign for authentication and identification. In *15th International Wireless Communications & Mobile Computing Conference, IWCMC 2019*, Tangier, Morocco, June 24–28, 2019 (pp. 307–312). IEEE.

Belk, M., Fidas, C., & Pitsillides, A. (2019) FlexPass: Symbiosis of seamless user authentication schemes in IoT. In *Extended Abstracts of the 2019 CHI Conference on Human Factors in Computing Systems, CHI 2019*, Glasgow, Scotland, UK, May 04–09, 2019. ACM.

Benhammadi, F. & Bey, K. B. (2013). Embedded fingerprint matching on smart card. *International Journal of Pattern Recognition and Artificial Intelligence*, 27(2), 1350006.

Bothe, A., Bauer, J., & Aschenbruck, Nils (2019). RFID-assisted continuous user authentication for IoT-based smart farming. In *IEEE International Conference on RFID Technology and Applications, RFID-TA 2019*, Pisa, Italy, September 25–27, 2019 (pp. 505–510). IEEE.

Burda, K. & Chudá, D. (2018) Influence of body postures on touch-based biometric user authentication. In *SOFSEM 2018: Theory and Practice of Computer Science - Proceedings of the 44th International Conference on Current Trends in Theory and Practice of Computer Science*, Krems, Austria, January 29 - February 2, 2018, (pp. 459–468). Springer.

Buriro, A., Crispo, B., & Conti, M. (2019). AnswerAuth: A bimodal behavioral biometric-based user authentication scheme for smartphones. *Journal of Information Security and Applications*, 44, 89–103.

Chadwick, D. W., Laborde, R., Oglaza, A., Venant, R., SamerWazan, A., & Nijjar, M. (2019). Improved identity management with verifiable credentials and FIDO. *IEEE Communications Standards Magazine*, 3(4), 14–20.

Chakraborty, C., & RodriguesJoel, J.P.C. (2020). A comprehensive review on device-to-device communication paradigm: Trends, challenges and applications. *Springer: International Journal of Wireless Personal Communications*, 114, 185–207. doi: 10.1007/s11277-020-07358-3

Chakraborty, C., Gupta, B., & Ghosh, S. K. (2013). A review on telemedicine-based WBAN framework for patient monitoring. *Mary Ann Libert Inc.: International Journal of Telemedicine and e-Health*, 19(8), 619–626. ISSN: 1530-5627, 10.1089/tmj.2012.0215

Chatterjee, B., Das, D., Maity, S., & Sen, S. (2019). RF-PUF: Enhancing IoT security through authentication of wireless nodes using in-situ machine learning. *IEEE Internet Things Journal*, 6(1), 388–398.

Chen, B.-L., Kuo, W.-C., & Wuu, L.-C. (2014) Robust smart-card-based remote user password authentication scheme. *International Journal of Communication Systems*, 27(2), 377–389.

Chen, K., Zhang, S., Li, Z., Zhang, Y., Deng, Q., Ray, S., & Jin, Y. (2018). Internet-of-Things security and vulnerabilities: Taxonomy, challenges, and practice. *Journal of Hardware and Systems Security*, 2(2), 97–110.

Chen, Y., Xu, W., Peng, L., & Zhang, H. (2019). lightweight and privacy-preserving authentication protocol for mobile payments in the context of IoT. *IEEE Access*, 7, 15210–15221.

Christo, L. E. de (2017) Multimodal biometric-system for identity verification based on hand geometry and hand palm's veins. In *Communication Papers of the 2017 Federated Conference on Computer Science and Information Systems, FedCSIS 2017*, Prague, Czech Republic, September 3–6, 2017 (pp. 207–212).

Connor, P. & Ross, A. (2018). Biometric recognition by gait: A survey of modalities and features. *Computer Vision and Image Understanding*, 167, 1–27.

Cui, J., Sui, R., Zhang, X., Li, H., & Cao, N. (2018). A biometrics-based remote user authentication scheme using smart cards. In *Cloud Computing and Security - 4th International Conference, ICCCS 2018*, Haikou, China, June 8–10, 2018, Revised Selected Papers, Part IV (pp. 531–542). Springer.

Dhillon, P. K. & Kalra, S. (2018). Multi-factor user authentication scheme for IoT-based healthcare services. *Journal on Reliable Intelligent Environments*, 4(3), 141–160.

Duarte, T., Prikladnicki, R., Calefato, F., & Lanubile, F. (2014). Speech recognition for voice-based machine translation. *IEEE Software*, 31(1), 26–31.

Eldefrawy, M. H., Ferrari, N., & Gidlund, M. (2019). Dynamic user authentication protocol for industrial IoT without timestamping. In *15th IEEE International Workshop on Factory Communication Systems, WFCS 2019*, Sundsvall, Sweden May 27-29, 2019 (pp. 1–7). IEEE.

Elngar, A. A. (2019). An efficient user authentication model for IOT-based healthcare environment. International Journal of Information and Computer Security, 11(4/5), 431–446.

Fang, H., Qi, A., & Wang, X. (2020). Fast authentication and progressive authorization in large-scale IoT: How to leverage AI for security enhancement. *IEEE Network*, 34(3), 24–29.

Farooq, M. J. & Zhu, Q. (2018). On the secure and reconfigurable multi-layer network design for critical information dissemination in the internet of battlefield things (IoBT). *IEEE Transactions on Wireless Communications*, 17(4), 2618–2632. doi:10.1109/TWC.2018.2799860, https://doi.org/10.1109/TWC.2018.2799860.

Garg, L., Chukwu, E., Nasser, N., Chakraborty, C., & Garg, G. (2020). Anonymity preserving IoT-based COVID-19 and other infectious disease contact tracing model. *IEEE Access*, 8, 159402–159414. 10.1109/ACCESS.2020.3020513, ISSN: 2169-3536.

Gemalto (2016). A safer Internet-of-Things: Gemalto's guide to making the Internet-of-Things a safe place to connect, https://www.thalesgroup.com/sites/default/files/gemalto/iot-security-ebook.PDF. Retrieved: January, 2016.

Glaroudis, D. P., Iossifides, A. C., & Chatzimisios, P.(2020). Survey, comparison and research challenges of IoT application protocols for smart farming. *Computer Networks*, 168, 107037.

Guerar, M., Verderame, L., Merlo, A., Palmieri, F., Migliardi, M., & Vallerini, L. (2020). CirclePIN: A novel authentication mechanism for smartwatches to prevent unauthorized access to IoT devices. *ACM Transactions on Cyber-Physical Systems*,4(3), 34:1–34:19.

Guo, Z., Karimian, N., Tehranipoor, M. M., & Forte, D. (2016). Hardware security meets biometrics for the age of IoT. In *IEEE International Symposium on Circuits and Systems, ISCAS 2016*, Montréal, QC, Canada, May 22–25, 2016 (pp. 1318–1321). IEEE.

Hamidi, H. (2019). An approach to develop the smart health using Internet-of-Things and authentication based on biometric technology. *Future Generation Computing Systems*, 91, 434–449.

Hicks, T. & Coquoz, R. (2015). *Forensic DNA Evidence*. Springer US.

Jain, A. K., Uludag, U., & Ross, A. (2003). Biometric template selection: A case study in fingerprints. In *Audio-and Video-Based Biometrie Person Authentication, Proceedings of the 4th International Conference, AVBPA 2003*, Guildford, UK, June 9–11, 2003 (pp. 335–342). Springer.

Jain, A. K., Nandakumar, K., & Ross, A. (2016). 50 years of biometric research: Accomplishments, challenges, and opportunities. *Pattern Recognition Letters*, 79, 80–105.

Janik, L., Chudá, D., & Burda, K. (2020). SGFA: A two-factor smartphone authentication mechanism using touch behavioral biometrics. In *Proceedings of the International Conference on Computer Systems and Technologies 2020, CompSysTech 2020*, Ruse, Bulgaria, June 19–20, 2020 (pp. 35–42). ACM.

Joshi, M., Mazumdar, B., & Dey, S.(2020). A comprehensive security analysis of match-in-database fingerprint biometric-system. *Pattern Recognition Letters*, *138*, 247–266. issn: 0167-8655. doi: https://doi.org/10.1016/j.patrec.2020.07.024. http://www.sciencedirect.com/science/article/pii/S0167865520302701.

Joyia, G. J. et al. (2017). Internet of medical things (IOMT): Applications, benefits and future challenges in healthcare domain. *Journal of Communication*, *12*(4), 240–247. doi: 10.12720/jcm.12.4.240-247, https://doi.org/10.12720/jcm.12.4.240-247.

Klare, B., Paulino, A. A., & Jain, A. K. (2011). Analysis of facial features in identical twins. In *2011 IEEE International Joint Conference on Biometrics, IJCB 2011*, Washington, DC, USA, October 11–13, 2011 (pp. 1–8). IEEE Computer Society.

Laborde, R., Oglaza, A., SamerWazan, A., Barr`ere, F., Benzekri, A., Chadwick, D. W., & Venant, (2020). A user-centric identity management framework based on the W3C verifiable credentials and the FIDO universal authentication framework. In *IEEE 17th Annual Consumer Communications & Networking Conference, CCNC 2020*, Las Vegas, NV, USA, January 10–13, 2020 (pp. 1–8). IEEE.

Lee, S., Sa, J., Cho, H., & Park, D. (2017). Energy-efficient biometrics-based remote user authentication for mobile multimedia IoT application. *KSII Transactions on Internet and Information Systems*, *11*(12), 6152–6168.

Lee, J. Y., Kim, M. H., Yu, S. J., Park, K. S., & Park, Y. (2019). A secure multi-factor remote user authentication scheme for cloud-IoT applications. In *28th International Conference on Computer Communication and Networks, ICCCN 2019*, Valencia, Spain, July 29 - August 1, 2019 (pp. 1–2). IEEE.

Leong, Y., & Chen, Y. (2020). Cyber risk cost and management in IoT devices-linked health insurance. *The Geneva Papers on Risk and Insurance - Issues and Practice*, *45*(4), 737–759. https://doi.org/10.1057/s41288-020-00169-4.

Li, C.-T., Wu, T.-Y., Chen, C.-L., Lee, C.-C., & Chen, C.-M. (2017). An efficient user authentication and user anonymity scheme with provably security for IoT-based medical care system. *Sensors*, *17*(7), 1482.

Lorenz, F., Thamsen, L., Wilke, A., Behnke, I., Waldmüller-Littke, J., Komarov, I., Kao, O., & Paeschke, M. (2020) Fingerprinting analog IoT sensors for secret-free authentication. 29th International Conference on Computer Communications and Networks, Honolulu, HI, USA, August 3–6, 2020 (pp. 1–6). IEEE.

Mahmod, M. J. & Guin, U. (2020). A robust, low-cost and secure authentication scheme for IoT applications. *Cryptography*, *4*(1), 8.

Maiti, A. & Jadliwala, M. (2020). Smart light-based information leakage attacks. *GetMobile: Mobile Computing and Communications*, *24*(1), 28–32.

Makhdoom, I., Abolhasan, M., Lipman, J., Liu, R. P., & Ni, W. (2019). Anatomy of threats to the Internet-of-Things. *IEEE Communications Surveys and Tutorials*, *21*(2), 1636–1675.

Maltoni, D., Cappelli, R., & Meuwly, D. (2017). *Automated fingerprint identification systems: From fingerprints to fingermarks*. Springer.

Meneghello, F., Calore, M., Zucchetto, D., Polese, M., & Zanella, A. (2019). IoT: Internet of Threats? A survey of practical security vulnerabilities in real IoT devices. *IEEE Internet-of-Things Journal*, *6*(5), 8182–8201.

Munirathinam, S. (2020). Chapter Six - Industry 4.0: Industrial Internet-of-Things (IIOT). *Advances in Computers*, *117*, 129–164. doi: 10.1016/bs.adcom.2019.10.010, https://doi.org/10.1016/bs.adcom.2019.10.010.

Olazabal, O., Gofman, M. I., Bai, Y., Choi, Y., Sandico, N., Mitra, S., & Pham, K. (2019). Multimodal biometrics for enhanced IoT security. In *IEEE 9th Annual Computing and Communication Workshop and Conference, CCWC 2019*, Las Vegas, NV, USA, January 7–9, 2019 (pp. 886–893). IEEE.

Ongun, T., Spohngellert, O., Oprea, A., Nita-Rotaru, C., Christodorescu, M., & Salajegheh, N. (2018). The house that knows you: User authentication based on IoT data. *Proceedings of the 2018 ACM SIGSAC Conference on Computer and Communications Security, CCS 2018*, Toronto, ON, Canada, October 15–19, 2018 (pp. 2255–2257). ACM.

Orme, D. (2019). Can biometrics secure the Internet-of-Things? *Biometric Technology Today, 2019*(5), 5–7.

Patel, C. & Doshi, N. (2020). A level dependent authentication for IoT paradigm. *IACRCryptol. ePrint Arch. 2020*, 686.

Perkovic, T., Cagalj, M., & Kovacevic, T. (2019). LISA: Visible light based initialization and SMS based authentication of constrained IoT devices. *Future Generation Computer Systems*, 97, 105–118.

Punithavathi, P. & Geetha, S. (2019). Partial DCT-based cancelable biometric authentication with security and privacy preservation for IoT applications. *Multimedia Tools and Applications*, 78(18), 25487–25514.

Rathgeb, C., Uhl, A., & Wild, P. (2013). *Iris biometrics - from segmentation to template security* (Vol. 59). Springer.

Ross, A., Banerjee, S., & Chowdhury, A. (2020). Security in smart cities: A brief review of digital forensic schemes for biometric data. *Pattern Recognition Letters*, 138, 346–354.

Ross, A., Banerjee, S., Chen, C., Chowdhury, Anurag, Mirjalili, V., Sharma, R., Swearingen, T., & Yadav, S. (2019). Some research problems in biometrics: The future beckons. In *2019 International Conference on Biometrics, ICB 2019*, Crete, Greece, June 4–7, 2019 (pp. 1–8). IEEE.

Sathyadevan, S., Achuthan, K., Doss, R., & Pan, L. (2019). Protean authentication scheme - a time-bound dynamic KeyGen authentication technique for IoT edge nodes in outdoor deployments. *IEEE Access*, 7, 92419–92435.

Sengupta, A. & Kundu, S.(2017). Guest editorial securing IoT hardware: Threat models and reliable, low-power design solutions. *IEEE Transactions on Very Large Scale Integration Systems*, 25(12), 3265–3267.

Sharma, G. & Kalra, S. (2018). A lightweight multi-factor secure smart card based remote user authentication scheme for cloud-IoT applications. *Journal of Information Security and Applications*, 42, 95–106.

Sharma, G. & Kalra, S. (2020). Advanced lightweight multi-factor remote user authentication scheme for cloud-IoTapplications. *Journal of Ambient Intelligence and Humanized Computing*, 11(4), 1771–1794.

Shayan, M., Naser, M., & Hossein, G. (2019) IoT-based anonymous authentication protocol using biometrics in smart homes. In *16th International ISC (Iranian Society of Cryptology) Conference on Information Security and Cryptology, ISCISC 2019*, Mashhad, Iran, August 28–29, 2019 (pp. 114–121). IEEE.

Shumailov, I., Simon, L., Yan, J., & Anderson, R. (2019). Hearing your touch: A new acoustic side channel on smartphones. CoRR abs/1903.11137.

Sikder, A. K., Petracca, G., Aksu, H., Jaeger, T., & Uluagac, A. S. (2018). A survey on sensor-based threats to Internet-of-Things (IoT) devices and applications. CoRR abs/1802.02041.

Sinche, S., Raposo, D. M. G., Armando, N., Rodrigues, A., Boavida, F., Pereira, V., & Silva, J. S. (2020). A survey of IoT management protocols and frameworks. *IEEE Communications Surveys & Tutorials*, 22(2), 1168–1190.

Sobin C. C. (2020). A survey on architecture, protocols and challenges in IoT. *Wireless Personal Communication*, 112(3), 1383–1429.

Song, R. (2010). Advanced smart card based password authentication protocol. *Computer Standards & Interfaces*, 32(5-6), 321–325.

Sood, S. K., Sarje, A. K., & Singh, K. (2010) An improvement of Xu et al.'s authentication scheme using smart cards. In *Proceedings of the 3rd Bangalore Annual Compute Conference, Compute 2010*, Bangalore, India, January 22–23, 2010 (pp. 15:1–15:5). ACM.

Srinivas, R. (2020). 10 IoT security incidents that make you feel less secure, CISO MAG. https://cisomag.eccouncil.org/10-iot-security-incidents-that-make-you-feel-less-secure/ (accessed January10, 2020).

Subban, R., & Mankame, D. P. (2013). Human face recognition biometric techniques: Analysis and review. In *Recent Advances in Intelligent Informatics - Proceedings of the Second International Symposium on Intelligent Informatics, ISI 2013*, August 23–24 2013, Mysore, India (pp. 455–463). Springer.

Teoh, A. B. J. & Leng, L. (2015). *PalmprintMatching*. Springer US.

Toli, C.-A. & Preneel, B. (2014). A survey on multimodal biometrics and the protection of their templates. In *Privacy and Identity Management for the Future Internet in the Age of Globalisation - 9th IFIP WG 9.2, 9.5, 9.6/11.7, 11.4, 11.6/SIG 9.2.2*, International Summer School, Patras, Greece Springer, September 7–12, 2014, Revised Selected Papers (pp. 169–184).

Turkanovic, M., Brumen, B., & Hölbl, M. (2014). A novel user authentication and key agreement scheme for heterogeneous ad hoc wireless sensor networks, based on the Internet-of-Things notion. *Ad Hoc Networks, 20*, 96–112.

Wang, Y., Wu, C., Zheng, K., & Wang, X. (2019) Improving reliability: User authentication on smartphones using keystroke biometrics. *IEEE Access, 7*, 26218–26228.

Xie, R., He, C., Xu, C., & Gao, C. (2019). Lattice-based dynamic group signature for anonymous authentication in IoT. *Annales des Télécommunications, 74*(7-8), 531–542.

Zareen, F. J., Matta, C., Arora, A., Singh, S., & Jabin, S. (2018). An authentication system using keystroke dynamics. *International Journal of Biometrics, 10*(1), 65–76.

Zeng, X., Xu, G., Zheng, X., Xiang, Y., & Zhou, W. (2019). E-AUA: An efficient anonymous user authentication protocol for mobile IoT. *IEEE Internet Things Journal, 6*(2), 1506–1519.

Zhang, Q. (2018). Deep learning of electrocardiography dynamics for biometric human identification in era of IoT. In *9th IEEE Annual Ubiquitous Computing, Electronics & Mobile Communication Conference, UEMCON 2018*, New York City, NY, USA, November 8–10, 2018 (pp. 885–888). IEEE.

Zheng, R., Wang, H., & Zhao, J. (2019). A unified management framework for EIoT systems based on metadata and event detection. *IEEE Access, 7*, 112629–112638. doi:10.1109/ACCESS.2019.2930290, https://doi.org/10.1109/ACCESS.2019.2930290.

5

An Improved Verification Scheme Based on User Biometrics

N. Ambika

Department of computer Applications, Sivananda Sarma Memorial RV r College, Bengaluru, India

5.1 Introduction

Remote sensor systems (Akyildiz, Su, Sankarasubramaniam, & Cayirci, 2002; Arampatzis, Lygeros, & Manesis, 2005) contain an assembly of sensors (Ambika, 2020; Nagaraj, 2021) for explicit purposes. Modest sensor nodes in the system process data they get in the waking state of detecting their environment. Such systems can be applied to countless applications (Boginski, Commander, Pardalos, & Ye, 2011) including basic health observation (Abbate, Avvenuti, Corsini, Light, & Vecchio, 2010), ecological control (Mukhopadhyay, 2012), agriculture (Mukhopadhyay, 2012), and combat zone reconnaissance (Lee, Lee, Song, & Lee 2009). In the majority of these applications, the client can get information straightforwardly through a portal since demands are handled on this gateway. Be that as it may, getting information from a gateway is sometimes troublesome or even incomprehensible. Along these lines, information is acquired directly through sensors. Detected information may be confidential, and unlawful clients can undoubtedly get to delicate information by sending a request to a sensor gateway. Client verification is necessary to oppose unlawful utilization of system information and assets.

A remote-sensor arrangement is conveyed in a bound territory, which is separated into various zones. Approved clients can get to the system's Wireless Sensor Network (WSN) utilizing a cellphone or Notebook (Chakraborty, Gupta, & Ghosh, 2013). The cellphone is then able to speak with the sensors inside the network. Before asking any questions in the framework, a client must enroll with a name and secret word, likely at the sensor gateway. Upon successful enrollment, the client can present an inquiry to the sensor-organised framework once inside a definite or authoritative structured epoch. It is regulatable timeframe that arranges distinctively, relying upon the idea of utilizations. During a specific questioning procedure, the client needs to stay set up, log in to the closest device-access station in a region, make the inquiries and recover the outcome. Once the regulatable timeframe terminates, the client initiates another sequence by enrollment in case more queries should be performed.

In the proposal, the drawbacks of the system (Riaz, Gillani, Rizvi, Shokat, & Kwon 2019) are listed and suggestions to strengthen the system are provided. The framework (Riaz et al., 2019) proposes two stages. In the first stage, clients register at with the sensors. The clients give their biometric (Das, Chatterjee, & Sing, 2015) and system computes a hash on them. This is trailed by imparting unique characteristics. After effective enrollment, the hash is produced. In the verification stage, clients again catch their new biometric data and compute a hash on it. This data is sent alongside the mentioned data, unique identity, and the client's present timestamp. The message is verified for its freshness and its value. A gathering credential is shared between the client and the tiny device. The disadvantage is that the system shares its identity with the trusted node. Attacks, such as the eavesdropping attack, can monitor the packets and are prone to various assaults in the system.

In the proposal, the sink node ensures that the user is provided with access to a public key (Ambika & Raju, 2010) when a request is received from the user. The user after receiving the encryption key encrypts the unique identity, location information, and biometric extract. The same is dispatched to the gateway for further validation. The organization has two phases of processing. In the registration period, the sensors and the users enroll themselves sharing the key credentials with the Sink node. On affirmation, the user is permitted to communicate with the sensors. During the authentication phase, the user requests the required data. The sensors after validating the data for its freshness, a hash value is shared by the sink node with the user and the generated biometric hash extract acknowledges with the requested data. The received data is also shared with the sink node via Adhoc request by the base station. Hence the proposal increases confidentiality to the system by minimizing eavesdropping attacks and energy consumption.

The chapter is organised into 8 sections. Segment 2 provides hardware details. Section 5.3 summarises the contributions made by various authors in the domain. The previous system and its drawbacks are explained in section 5.4. The notations used in the proposal are listed in section 5.5. Section 5.6 contains the list of assumptions used in the suggested methodology. The proposed system is elaborated and explained in section 5.7. The work is simulated in NS2. The analysis w.r.t different parameters are estimated and represented in the form of graph in section 5.8. The summary of the proposal with results is offered in section 5.9.

5.2 Working of the Hardware (Biometric Sensor)

The IR illuminators (Lee, 2008) are kept on the finger dorsum, and IR light infiltrates the finger. The camera takes both mirrored and infiltrated light. In the framework, the finger position inside the captured picture is significant; there are no extra picture arrangement techniques. Subsequently, the gadget has a finger dorsum and fingertip guide. The device has a charge-coupled web camera that embraces a general sequential-transport interface. The finger vein designs use NIR light, and have a unique noticeable light-passing channel of the web camera. The NIR-passing channel incorporates inside the camera, which permits just NIR light with frequencies higher than 750 nm to pass. Five extra NIR-light-emitting diodes (LEDs) join to the upper aspect of the gadget.

5.3 Literature Survey

Methodologies are suggested by various authors to tackle different kinds of attack. These are discussed in this section.

The framework (Riaz et al., 2019) proposes two stages. In the first stage, clients register at first with the sensors. The clients give biometric and system compute a hash on them. This is trailed by imparting unique characteristics. After effective enrollment, the hash is produced. In the verification stage, clients again catch their new biometric data and compute a hash on it. This data is sent alongside the mentioned data, unique identity, and the client's present timestamp. The message is verified for its freshness and its value. An assembly credential is made mutual between the client and the device.

Luk, Perrig, and Whillock (2006) offer an improvement of the μTESLA communication-verification convention to give proficient validation to rare messages. The RPT convention (Regular-Predictable TESLA) accomplishes prompt verification for messages sent customarily. The sender makes a single direction-key chain and assigns each at a time. The collector checks the main message, and is given its present time and the most extreme time-synchronization errors. While accepting the subsequent message, the collector initially confirms the key.

Two client-confirmation conventions (Vaidya, Rodrigues, & Park, 2010) that are varieties of an ongoing solid secret-key-based arrangement are proposed. They utilise single-direction hash capacities and XOR tasks to accomplish lower computational and correspondence overheads. In our first proposal, it is expected that one-bounce correspondence between the client gadget and sensor hub happens. The second proposal considers shared verification between the sensor passage and sensor hub. It is comprised of three segments—enrollment, access, and authorization. The above convention is impervious to assaults; for example, legitimate userID, counterfeit secret word, invalid userID, substantial/counterfeit secret key and repeat access demand independent of altered access note.

Wong, Zheng, Cao, and Wang (2006) propose a secret phrase-based answer for this entrance control issue and adjust it into a remote-sensor organised condition. The proposed secret-word-verification approach creates a light computational burden and requires basic activities; for example, single direction hash capacity and selective OR tasks. The User Authentication issue with regard to a WSN where the genuine client is permitted to inquire and gather information at any sensor hub (Chakraborty & Rodrigues 2020). A client enrolls by providing a label and a secret phrase at the device-portal hub. On successful enrollment, the client presents an inquiry to the framework once inside a predefined or authoritative configurable period. During a specific questioning procedure, the client needs to stay set up, login to the closest device access hub in a region, provide the questions and return an outcome. If the regulated timespan elapses, the client may need to rerun the series by enrolling, in case more inquiries should be performed.

He et al. (2015) propose a vigorous unknown validation convention for medical-services applications utilizing remote clinical- sensor systems. The method contains four phases. In the expert enrollment stage, a health professional becomes a legitimate client of the WNSN by enrolling in the entryway hub. The patient should enroll in the medical clinic area during the patient enrollment stage. In the login and verification stages, a health expert gets the patient's physiological data from the WMSN. In the secret-phrase-change stage, the health professional is given the option to change his secret key.

Das, Chatterjee, and Sing (2012) underpin dynamic-hub expansion after the underlying sending of hubs in the system. The proposal doesn't require one to refresh data for new sensors' expansion in the client's card. It supports changing the client's secret word locally without the assistance of the base station. It gives unequivocal protection from hub catch assaults. The group head doesn't uncover any mystery data of other group heads and neither it leads to bargaining of some other secure correspondence between the client and the non-traded off hubs in the system. It sets up a secret meeting key between the client and a group head for future secret correspondence between them of the information inside WSN, utilizing the built-in meeting key.

Nyang and Lee (2009) propose that verification is achieved via the enrollment and validation segments. During the enrollment segment, the hash is determined and XOR-ed. During the confirmation stage, the client presents his identity and secret phrase to validate and encode the information.

The authenticated-query issue is addressed in Benenson, Gedicke, and Raivio (2005). The client verification convention depends on public-key cryptography which is said to be practical for WSNs even without exceptional equipment support. The key framework for elliptic curve cryptography is embraced. ECCM library is used to execute an advanced mark technique with message recuperation called the Nyberg-Rueppel plot. The confirmation-authority key is changed intermittently and circulated to all sensors in the WSN. Any element which has a key marked with the present key of the affirmation authority can get to the system. The clients can renew their keys at the site of the WSN maintainer.

Identity-based multi-user broadcast authentication (IMBAS) (Cao, Kou, Dang, & Zhao, 2008) is an effective identity-based multi-client communication-confirmation scheme with solid refuge and sturdy versatility. The partitions communicate by dual classifications and utilise diverse code natives. Clients' communications are made secure by a novel blending-free identity-based mark with decreased mark size. The procedure accomplishes security, versatility, and effectiveness. The sink communicated uses Schnorr indication with incomplete communication recuperation to upgrade the productivity. Secret-key-based clients insuranceconfirmation is used to oppose active the trade-off assaults.

The methodology of Vaidya, Silva, and Rodrigues (2009) suggests using two phases. During the access run, the client submits user information. The node cross-verifies the same with the lookup table. On affirmation, the trusted node computes value XOR-ing with the received user information. The same is transmitted to the gateway. The hub checks for user validity in the authorization stage. The gateway also verifies the newness of the acknowledged communication (Garg, Chukwu, Nasser, Chakraborty, & Garg, 2020). The methodology minimises replay attack.

Turkanovic, Brumen, and Hölbl (2014) focus on similar areas. They suggest an innovative customer-validation credential scheme for assorted sensing organization. It authorises a customer to arrange a meeting credential with the gateway, using a trivial credential principle. It assures public authorization among the assembly members. Two separate enrollment stages are required after the organization. The first is an enrollment stage between the gateway node and a relating standard sensor hub. The subsequent enlistmentenrollment stage happens between a client and the passage. The plan utilises a card for the client to enroll and authenticate. In the verification stage, the arrangement of the meeting key is made. After effective affirmation, it can be utilised safely to convey any encoding issue. It adjusts to the gratified sketch; it uses straightforward jumble and XOR computation.

Yussoff, Hashim, and Baba (2012) assert that trust-zone address spaces are partitioned into secure and non-secure locales. Standard four Identity Based Encryption stages are diminished to three phases in the execution. The previous two phases are set up, separated, then reconsolidated. The mix of the two phases was made conceivable due to the proposed execution method. The delivery stage disconnects to furnish the system with complete data. When arranged with vital data, the sensor hubs will experience a boot-up process under a controlled situation to produce a unique value. In the sending stage, the sensor hub will boot up safely. In the verification stage, the goal is to enroll the sensor's unique identification with the base station for additional correspondence. The effective boot-up hub will report its trust an incentive to the confided in power.

Chang and Le (2015) offer a four-stage proposal. The stages are: pre-sending; enrollment; validation; and secret-word evolving. The portal hub is a long haul mystery generated randomly. The sensor is appointed a character. The sensor processes the value utilizing the long haul mystery key and its character. In the enrollment stage, the client registers with the sensor. During the validation stage, the client confirms with the sensor.

The Wong et al. (2006) solution has three stages: enrollment; access; and authorization. During the enrollment stage, the identity hash and secret key are used. In the access stage, the client's hash credential and timestamp are shared. In the authorization stage, currency is checked.

Tseng, Jan, and Yang (2007) present a solution which verifies a secret key to a device passage to enroll. The hub figures the pair of credentials to enroll the client. When a client posts an inquiry about sensor information, the user logs in to the hub. The client initially presents the identification and secret word for accessing hub. After gaining data access, the login hub checks for existence of the identification in rundown database. In the validation stage, the hash credential, userID and timestamp are confirmed.

Bohge and Trappe (2003) offer another sort of verification called a TESLA endorsement, that can be utilised by low-controlled sensors to perform element validation. This structure verifies approaching sensors, keeps up trust connections during topology changes through an effective handoff plot, and gives information-inception confirmation to the sensor. Further, the structure allocates validation errands to sensors as indicated by their computational assets; asset-rich access points perform computerised marks and keep up most of the security parameters. Every sensor joining the system must have an individual declaration, given by the system. When a passage or a sensor needs to join the system, the sensor presents its declaration, which eventually will be checked by the application. If the authentication is legitimate, the sensor sets up a common credential for additional handling purpose. The application empowers them to do this by occasionally giving runtime testaments for each passageway and sensor of the system.

Arikumar and Thirumoorthy (2011) uses two-factor confirmation and limits the client logins w.r.t identity, verification, speculation and replay dangers. Two-factor verification is used to validate the instrument, where more than one-factor is required to verify the imparting client. This strategy requires no open-key activities; it uses only cryptographic hash work. By utilizing single direction cryptographic hash work, the convention achieves proficiency. The client gets a personalized certificate from the gateway during the enrollment procedure. With the help of the client's secret key and certificate, the user gains access to the sensor or gateway. The method has three stages: enrollment; confirmation; and secret-phrase alteration. The enrollment stage is performed just a single time; the verification run is executed when the customer signs into the system.

Das and Bruhadeshwar (2013) offer another biometric-based client validation component in heterogeneous remote-sensor systems. The proposed convention gives solid verification, contrasted with customary related secret-word-based plans and accomplishes much; for example, it works without a synchronised clock, unreservedly changes the secret key, has low calculation costs, and uses common validation. The plan builds up a symmetric mystery assembly of credentials that are shared between the client and the hub. The secret meeting credential is used later for secure correspondence. It gives unlimited protection from hub catch and other assaults.

The methodology of Das, Sutrala, Odelu, and Goswami (2017) is an improvement (Li et al., 2016). The procedure is secured against known assaults. Proposed is a three-factor client confirmation methodology utilizing the card, secret phrase and biometrics of a health professional. Thorough casual and formal security examinations utilizing BAN rationale and irregular prophet strategy are performed.

Juang (2006) proposes a unique customer-authorization scheme appropriate for isolated devices. Confirmation is reasonable for such isolated systems. Any two members can verify one another. It can produce a meeting key agreed to by any two members. The plan likewise has low calculation and correspondence costs for client confirmation and key understanding. It utilizes single direction hash capacities and symmetric cryptosystems.

Kumar, Gurtov, Ylianttila, Lee, and Lee (2013) propose another solid confirmation scheme with client protection for WSNs. The proposed plot not only accomplishes end-party shared verification between the client and the sensor hub, but also builds a powerful meeting key. The proposed scheme represents reasonable protection against threats to security.

Yasmin, Ritter, and Wang (2010) propose an effective secure system for verified communication/multicast by sensor hubs for outside client verification. It uses identity-based cryptography and web-disconnected mark plans. The essential objectives of this system are to empower all sensor hubs in the system, communicate or potentially multicast a verified message rapidly, check communication substance, and confirm the authenticity of the external client. Online/Offline Signature (OOS) separates process marking into dual stages. The unplugged stage is performed before the communication to be marked opens up. It calculates mark age, resulting in a fractional mark. For identified communication, the connected stage is used. It recovers the incomplete mark determined during the unplugged stage and plays out a few small brisk calculations to acquire the last certification. The plugged stage is extremely quick, comprising small calculations, while the unplugged stage performs another clever operation. OOS empowers the acknowledged hub to mark communication quickly during its basic reporting.

Benenson, G̈artner, and Kesdogan (2004) offer an n-verification methodology that is a strong variant of basic validation. To protect from failure, this new type of validation succeeds if the client can effectively verify with any subcategory out of a superset. The strategy has three phases—the end organises where all the validated sensors partner with other confirmed hubs. In the legitimacy arrangement, the hubs are verified for their entry. In the understanding stage, the hubs experience understanding after identity certification.

Jiang, Li, and Xu (2007) offer a plan which depends on a self-certified key cryptosystem (SCK), and have altered it to use Elliptic Curve Cryptography (ECC). It builds up paired credentials that are used in client validation methodology. The proposal aims for extremely light computational and correspondence overhead. It is an appropriated client-confirmation methodology reasonable for sensor systems dependent on self-guaranteed key cryptosystems.

The proposal of Ko (2008) stands comparative suspicions (Tseng et al., 2007) and utilises hash capacities and elite or activities as fundamental cryptographic natives.

These two cryptographic natives have low computational overhead and are viewed as moderate to the normal sensor stages.

Lee (2008) offers two straightforward client-confirmation conventions which are varieties of an ongoing secret-code-oriented arrangement. The estimation burden, correspondence price, and safety of the proposal are undergoing assessment. In the principal convention stage, the confirmation procedure decreases the computation heap of the sensor adhering to security of the system. In the subsequent convention stage, the gateway hub checks the passage from itself to the sensor to keep ill-conceived clients away. On access demand, the estimation heap is given higher priority over gateway hub (Sanjukta & Chinmay, 2020).

Liu and Chung (2017) propose a client confirmation plan and information-transmission instrument that assure security; such assurance empowers the clinical workforce to screen quickly patients' health, and give care collectors brief and far-reaching clinical consideration. Utilizing smartcards and passwords, the methodology allows only healthcare workers to access quiet data; for example, temperature, pulse, and circulatory strain.

Quan, Chunming, Xianghan, and Chunming (2015) propose a protected client-validation convention utilizing character-based cryptology. The shared credential is used. A private credential, called a secret-credential producer, is created by an outsider using its main key and the client's uniqueness. In character-oriented cryptology, the key must make an ace open key and an ace private key. At this point, any client can use the main shared credential or client's credential. The client's secret credential is made by using a generator using any of the client's keys.

Fan, Ping, Fu, and Pan (2010) propose a productive and Denial-of-Service safe client-verification methodology for two-layered WSNs, which makes a light computational burden and requires basic activities such as single direction hash capacity and restrictive OR tasks. Every client will get a warrant information table at the time of enrollment. With the assistance of a brilliant card, the system client legitimately accesses trusted nodes from the cell or a hub.

There are four levels in the WBSN framework (Chen, Lee, Chen, Huang, & Luo, 2009) with connections between each. A PC in the emergency clinic comprises the control and information tracks for the WBSN frame. The control walk incorporates capacity control and forces control signals. The application layer does the sensor gatherings and manages by transmitting orders and getting information. The group tier incorporates a few sensor hubs and commands the hubs in the tier by transmitting orders and getting information. The sensing tier is a solitary hub that gets orders from the detector to gather and send information to the assembly. In WBSN framework, every remote sensor hub arranges itself with a unique ciphered ID. The group is the mathematical hub of a Voronoi graph. These sensors group sends orders and characterises got information in its Voronoi cell, and it additionally implies no other sensor gathering can associate it. The unique distinguishing proof code for every individual and Voronoi graph for sensor organise territory make individuals moving openly in an enormous zone to be conceivable by changing sensor arrange a gathering. The control order or distinguishing information is conveyed through Bluetooth, WiFi. It is a worldwide framework for versatile communication of remote buyer electric items.

The removed highlights of ECG signal (Miao, Jiang, & Zhang, 2009) are then quantised and planned to give twofold portrayal to include focussing coordination. Delivered paired highlights and the randomly produced key from ECG limitation by fluffy vault scheme guarantee the security of the cryptographic key. A recurrence-area examination of ECG signals creates the highlights. The entities are executed by two sensors, by testing ECG signal, examining the rate for a fixed time. The fuzzy vault guarantees the security of an arbitrary parallel string.

The procedure depends on a symmetric cryptosystem, which accepts that a powerful and confirmed credential dissemination plot is accessible. In the transmission terminal, biometric attributes are used to submit the key. The other biosensor catches its identity property, variation form and uses them to uncommit the credential at the accepting terminal end. The creator's information is gathered beforehand from two tests, the first of which continuously captures electrocardiogram and photoplethysmogram information to check the pulse. Internal circuits capture information through tempered anodes and infrared visual detectors. In the examination, sound focus enrollment and two PPGs are obtained using the forefingers, and an ECG captured from the middle finger was recorded for 2–3 min. They partitioned the ECG and PPG sets into portions with the goal that the comparing ECG section contained precisely 68 R-waves. An aggregate of 838 information portions are acquired from 99 subjects.

The proposed intelligent Wireless Biometric Smart Home plan (El-Basioni, El-Kader, & Abdelmonim, 2013) joins remote sensor organization to biometrics in building smart homes. It has seven subordinate frameworks: thief location; criminal misdirection; observing and controlling home segments; checking home structure wellbeing; home plant care; and Internet access. The microcontroller, the Atmel AVR ATmega128, is a low-power CMOS 8-cycle microcontroller. The handset is a CC2420 radio handset. The hub may be controlled by a battery or be mains-fueled by the hub place, and the sensors/actuators types will be as indicated by the sensor hub work. An augmentation interface is utilised in hubs to consider associating with the outer sensors and actuators supporting the different establishment of the various subsystems. The Base Station (BS) has only a handset, memory, and processor; it speaks with the worker through a USB association and is controlled by it. The versatile Remote Control (RC) unit is inside the hub without sensors. It has a keypad for directing the microcontroller to do additional activities. The Remote Control is battery-powered.

An alternate EDPS (Miao, Bao, & Li, 2013) concerns the residues of DCT of the AC succession of the anatomical signs. They speak to the cos function segment to force band to the arrange period. The AC/DCT strategy includes four phases. In the principal stage, both the transmitter and beneficiary examine for a fixed span. The assessment of the standardised AC can smother the impact the sufficiency of indication. It offers notable data in recognized topics, later directed to implant closeness highlights in-between proceedings of a similar theme. DCT is applied to the AC residue to distinguish the vitality circulation of materials and documented signs. Like the SWFT technique, the primary purpose of DCT coefficients contain a significant recurrence space data of physiological sample. It separates the highlights for recognizable proof.

The procedure (Poon, Zhang, & Bao, 2006) is dependent on an asymmetric cryptosystem, which expects that a vigorous and confirmed credential-dissemination plot is accessible. Since biometric characteristic is caught at various areas of the body having slight varieties, the system uses fuzzy duty plan to guarantee that the mistakes in the recouped cipher credential decreases to a suitable amount. In the communicating station, fingerprint quality is used to submit the credential. If the trademark is chosen, the encryption key should undergo recouping using attribute. They utilised information gathered beforehand from two trials, the first object of which was to catch electrocardiogram and photoplethysmogram information continuously for the assessment of pulse. They separated the ECG and PPG sets into sections with the goal that the comparing ECG portion contained precisely 68 R-waves. A sum of 838 information portions are created from 99 subjects. They explored the exhibition of the framework when the quantity of IPIs utilised for each example was decreased from 67 IPIs - 34 IPIs.

Biomedical signals were gathered (Mahendran & Velusamy, 2020) with the assistance of sensors placed outside a body. The signs are changed into some identifying factors aided by fuzzy encoder procedure. Every individual sign is encoded with some particular estimation of the same parameters, with some refuse focuses made by polynomials. These encoded outcomes are handled by fuzzy decoder set at the base stations. The existing result is stored during client enrollment. The plot of the vault has limit w.r.t esteem, which has to be balanced to address errors happening during transmission. These two outcomes are contrasted and utilised with the two gadgets. At first, input information is gathered from the patient using bio-sensors associated with the BSN. From the BSN, the removed data is stored in the base-worker part. If an individual needs to recover the stored information, at that point utilizing the proposed fluffy extractor, makes sure of data access is prepared and the fluffy vault produces to actuate the verification key for getting to the information put away in the base station worker with much security.

The proposed encryption plot (Miao et al., 2009) dependent on the stream figure method of AES for BSN. The code created from physiological signs is used to create keystream-by-stream-figure method of AES to scramble and unscramble messages. The uprightness key used to produce keystream-by-stream-figure mode of AES to create Messages Authentication Code is the underlying vector of the AES stream code. The transmitter end scrambles the messages using stream code of AES and key to produce message validation code (MAC). Macintosh connects to eData. A similar cycle is done on the beneficiary's conclusion to create XMAC. Before the decoding cycle, the MAC coordination finishes by deciding whether MAC is the same as XMAC. Macintosh coordination shows whether data changes during the transmission.

5.4 Previous System

The proposed framework (Riaz et al., 2019) has two stages. In the first stage, the client registers with the sensors. The clients give their biometric and the system calculates the hash for the same. This is tracked by imparting unique characteristics. After effective enrollment, the hash is produced. In the verification stage, clients again catch their new biometric data and compute a hash on it. This data is sent alongside the mentioned data, unique identity, and the client's present timestamp. The message is approved for freshness and the hash value. A meeting credential is mutual between the client and the sensor.

One of the disadvantages of this proposal is (Riaz et al., 2019):

- The system shares its identity with the trusted node. Attackers can eavesdrop on the packets; it is also liable to various assaults.

5.5 Notation Employed in the Proposal

Table 5.1 explains the symbols employed in the proposal.

TABLE 5.1

Symbols employed in the study

Notations	Explanation
N_i	i^{th} node of the network
S	Sink node/base station/gateway
U_i	i^{th} user
id_u	Unique identity of the user
L_u	Location of the client
Bio_u	fingerprint extract of the client
id_i	Unique identity of the ith node of the network
L_i	Location of the ith node of the network
R_{join}	Request sent by the user to the sink node for joining the network
L_s	Location of the sink node
T_s	Timestamp estimated by the sink node
R_u	Request transmitted by the user
T_u	Timestamp generated by the user device
T_i	Timestamp generated by the ith node of the network
D_i	Data transmitted by the ith node of the network
C_i	Message count estimated by the ith node of the network
h(...)	Hashing the message (extracted time/biometric/location information/unique identity)
En(...)	Encryption algorithm used to encrypt the input

5.6 Assumptions of the Proposed System

- The sink node is the trusted device of the system. All the nodes rely on it for confidential information sharing.
- The base station/sink node is responsible for embedding algorithms and key credentials in the deployed nodes.
- The adversary is capable of introducing eavesdropping or other attacks in the network.

5.7 Proposed System

The suggested system minimises eavesdropping attacks (Callegati, Cerroni, & Ramilli, 2009). It utilises two phases in order to counter the attack.

  a. *Registration phase*
     After the deployed nodes configure with each other, the nodes register themselves with the sink node using the hash code generated by their identity. As the sink node is aware of the sensor identities, the base station will be able to map with the identity stored in their database. In Equation (5.1) the node N_i generates

the hash value of its identity id_i and the location L_i, concatenates and transmits to the sink node S.

$$N_i \rightarrow S: h(id_i) \mid \mid h(L_i) \tag{5.1}$$

Similarly, the users also register themselves with the base station. The user sends a request to the gateway. In notation (2) the user U_i is sending a request R_{join} to sink node S to join the group.

$$U_i \rightarrow S: R_{join} \tag{5.2}$$

The sink node sends the public key to the user to encrypt the identity, location and biometrics values. In the case of the user, the new database is created by regenerating the identity, location information, and biometrics values. In Equation (5.3) the user U_i is transmitting encrypted unique identity id_u location L_u and biometric bio_u to the base station S.

$$U_i \rightarrow S: En(id_u, L_u, bio_u) \tag{5.3}$$

The base station regenerates the identity and location values of the node by using the decryption key. It stores the same in its database. It generates the hash value of the identity, location information and timestamp and dispatches the same to the respective node. In the Equation (5.4) the sink node uses the identity of the node N_i, location information of the node L_i, location information of the sink node L_s and timestamp T_s to generate the hash value which is later dispatched to the node N_i. This hash code is used as identification by the node.

$$S \rightarrow N_i: h(L_i, L_s, T_s) \tag{5.4}$$

In Equation (5.5) the sink node uses the location information of the user L_u, location information of the sink node L_s and timestamp T_s to generate the hash value which is later dispatched to the user U_i. This hash code is used as identification by the user device.

$$S \rightarrow U_i: h(L_s, L_u, T_s) \tag{5.5}$$

b. *Authentication phase*

During the authentication phase (Riaz et al., 2019) the user communicates his identity, along with the hashed biometric value, request and timestamp to the respective sensor node. If the adversary is able to get hold of the dispatched packets, the identity of the user is at risk.

The user uses this hash code as its identity in the proposed work. This identity along with the request, the hash value of freshly-collected biometrics and the timestamp are transmitted to the sensor for authorization. In Equation (5.6) the user U_i transmits the hash identity of itself, request Ru, timestamp T_u and freshly-collected hashed biometric value bio_u to the sensor node N_i.

$$U_i \rightarrow N_i: h(L_i, L_s, T_s) \mid \mid R_u \mid \mid T_u \mid \mid h(bio_u) \tag{5.6}$$

After receiving the request, the timestamp is evaluated for its currency and the sensor transmits the requested readings to the user. In Equation (5.7) after affirmation of the received message, the sensor node N_i transmits its hashed identity, requested encrypted data D_i and hashed timestamp T_i to the user U_i.

$$N_i \rightarrow U_i: h\left(L_s, L_u, T_s\right) \mid\mid En\left(D_i\right) \mid\mid h\left(T_i\right) \tag{5.7}$$

The sensor is randomly requested by the sink node to transmit the received messages for verification. In Equation (5.8) the node N_i shares a received message with the sink node along with the number of messages C_i received from the user U_i.

$$N_i \rightarrow S: h\left(L_i, L_s, T_s\right) \mid\mid R_i \mid\mid T_i \mid\mid h\left(bio_u\right) \mid\mid C_i \tag{5.8}$$

5.8 Security Analysis

The work is simulated in NS2. The parameters used in the simulation are listed in Table 5.2.

c. *Eavesdropping attack*

Eavesdropping is one type of attack. The assailant conveys in believable manner. It adjusts the interchanges in the assembly. The attacker is vigorously snooping silently. The aggressor makes independent relations with the populace in query and transmits communication. The procedure creates a illusion to others that two parties are over a secret meeting, when the entire conversation is concealed by the assailant. The assailant will snoop every single pertinent communication going between two genuine parties.

The previous framework (Riaz et al., 2019) proposes two stages. In the first stage, the clients register with the sensors. The clients give their biometric and the system computes

TABLE 5.2

List of considerations made in the simulation

Parameters used	Description
Dimension of the network	200m * 200m
Number of devices installed in the system	10 devices
Number of users	3
Time duration	60ms
Length of hash extract (location)	8 bits
Length of hash extract(biometric)	10 bits
Length of hash extract (unique identity)	12 bits
Request message length	10 bits
Length of hash extract (timestamp)	8 bits
Length of count	4 bits
Length of data	256 bits

the hash value for the same. This is trailed by imparting the identity and the hash value. After effective enrollment, the hash is produced. In the verification stage, clients again catch their new biometric data and compute a hash on it. This data is sent alongside the mentioned data, unique identity, and the client's present timestamp. The message is approved for freshness and the hash value. The meeting credential is mutual among the client and the detector. The identity of the user is transmitted is shared with the trusted node and the information if eavesdropped can provide the adversary the chance to execute various assaults. The proposal uses the hash value to keep the rest of the message secure.

The suggested proposal minimises the eavesdropping attack as the user uses the public key to encrypt his identity, location, and biometric value before transmission. The proposal secures the identity and other information by 5.28% compared to (Riaz et al., 2019). The same is represented in Figure 5.1.

d. *Power utilization*

Power is one of the rarest possessions for devices. If identity is compromised, the adversary is liable to send irregular messages to the devices as well as the sensors. The devices will not be able to identify them. The devices or the sensors respond to these messages. Hence the amount of energy consumption will rise. The proposed work reduces energy by 7.8% compared to (Riaz et al., 2019). The same is represented in Figure 5.2.

e. *Confidentiality*

Information secrecy (Papadopoulos, Kiayias, & Papadias, 2012) is the insurance of transmitted information from inactive assaults, for example, listening in. Delicate data, for example, strategic military data transmitted over a combat zone, require secrecy. The spillage of such data to adversaries could cause pulverizing results. The remote divert in a powerless domain is threatening especially for those being listened to stealthily. Messages transmitted over the air can be listened stealthily from anyplace without having physical access to the system parts. Traditionally, secrecy is accomplished by

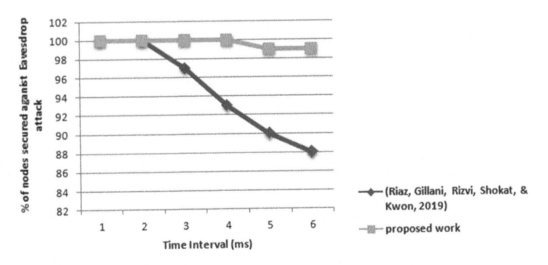

FIGURE 5.1
Comparison of proposal with (Riaz et al., 2019) against eavesdropping attack.

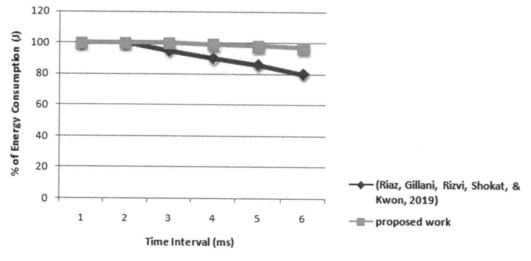

FIGURE 5.2
Comparison of energy consumption.

cryptography. In any case, the constrained assets, for example, the restricted battery force and preparing capacity, confine the utilization of computationally concentrated encryption plots in a system.

The previous system (Riaz et al., 2019) maintains secrecy to some extent by hashing the credentials while transmitting it to the sensors. The identity will be liable to get compromised as it is sent directly to the sink node or the trusted node without any encryption. The proposal aims at maintaining secrecy, as the user encrypts the identity using the public key provided by the sink node before transmitting the same. Hence the work provides 8.14% more confidentiality than (Riaz et al., 2019). The same is represented in Figure 5.3.

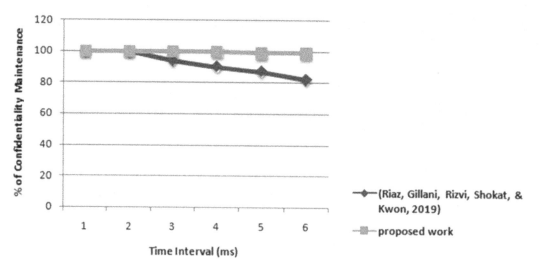

FIGURE 5.3
Comparison of both work w.r.t confidentiality.

5.9 Conclusion

Sensors are tiny devices deployed in any arena. They are devised to track any object of interest. The users can contact sensors and get readings on their devices. To undergo this procedure, the user has to share his credentials. The sensors and users undergo mutual authentication before commencing the actual transmission of data. Every time a request for data is sent, the identity of the sender is suffixed to the message. The procedure aims to increase confidentiality and reliability while reducing different kinds of attacks. The suggestion aims for transmitting secure data. The user is provided with the public key to encrypt credentials and data before transmission. The work also takes care of randomly choosing a packet for evaluation (performed by the sink node). The proposal aims to minimise eavesdropping attacks by 5.28%, improve confidentiality by 8.14% and reduce energy consumption by 7.8% compared to the previous work.

References

Abbate, S., Avvenuti, M., Corsini, P., Light, J., & Vecchio, A. (2010). Monitoring of human movements for fall detection and activities recognition in elderly care using wireless sensor network: a survey. in *Wireless Sensor Networks: Application-Centric Design*, 1.

Akyildiz, F. I., Su, W., Sankarasubramaniam, Y., & Cayirci, E. (2002). A survey on sensor networks. *IEEE Communications Magazine*, 40(8), 102–114.

Ambika, N., & Raju, G. T. (2010). Figment authentication scheme in wireless sensor network. In *Security technology, disaster recovery and business continuity* (pp. 220–223). Jeju Island, Korea: Springer, Berlin, Heidelberg.

Ambika, N. (2020). Diffie-Hellman algorithm pedestal to authenticate nodes in wireless sensor network. In B. K. Bhargava, M. Paprzycki, N. C. Kaushal, P. K. Singh, & W. C. Hong (Eds.), *Handbook of wireless sensor networks: Issues and challenges in current scenario's* (Vol. 1132, pp. 348–363). Cham: Springer Nature.

Arampatzis, T., Lygeros, J., & Manesis, S. (2005). A survey of applications of wireless sensors and wireless sensor networks. In *International Symposium on, Mediterrean Conference on Control and Automation Intelligent Control*, Limassol, Cyprus (pp. 719–724). IEEE.

Arikumar, K. S., & Thirumoorthy, K. (2011). Improved user authentication in wireless sensor networks. In *International Conference on Emerging Trends in Electrical and Computer Technology*, Nagercoil, India (pp. 1010–1015). IEEE.

Benenson, Z., Gärtner, F., & Kesdogan, D. (2004). User Authentication in Sensor Networks. In Z. Benenson, F. Gärtner, & D. Kesdogan (Eds.) *Informatik 2004, Informatik verbindet, Band 2, Beiträge der 34. Jahrestagung der Gesellschaft für Informatik e.V. (GI)* (pp. 385–389). Bonn: Gesellschaft für Informatik e.V.

Benenson, Z., Gedicke, N., & Raivio, O. (2005). Realizing robust user authentication in sensor networks. *Real-World Wireless Sensor Networks (REALWSN)*, 14(52), 1–5.

Boginski, V. L., Commander, C. W., Pardalos, P. M., & Ye, Y. (2011). *Sensors: theory, algorithms, and applications* (Vol. 61). Springer Science & Business Media.

Bohge, M., & Trappe, W. (2003). An authentication framework for hierarchical ad hoc sensor networks. In *2nd ACM workshop on Wireless security* (pp. 79–87). Diego CA USA: ACM.

Callegati, F., Cerroni, W., & Ramilli, M. (2009). Man-in-the-middle attack to HTTPS protocol. *IEEE Security and Privacy*, 7(1), 78–81.

Cao, X., Kou, W., Dang, L., & Zhao, B. (2008). IMBAS: Identity-based multi-user broadcast authentication in wireless sensor networks. *Computer Communications, 31*, 659–667.

Chang, C.-C., & Le, H.-D. (2015). A provably secure, efficient, and flexible authentication scheme for ad hoc wireless sensor networks. *IEEE Transactions on Wireless Communications,* 15(1), 1–20.

Chakraborty, C., Gupta, B., & Ghosh, S. K. (2013). A review on telemedicine-based WBAN framework for patient monitoring. *Mary Ann Libert Inc.: International Journal of Telemedicine and e-Health, 19*(8), 619–626. ISSN: 1530-5627, 10.1089/tmj.2012.0215

Chakraborty, C., & Rodrigues, J. J. P. C. (2020). A comprehensive review on device-to-device communication paradigm: Trends, challenges and applications. *Springer: International Journal of Wireless Personal Communications, 114*, 185–207, doi: 10.1007/s11277-020-07358-3.

Chen, S. L., Lee, H. Y., Chen, C. A., Huang, H. Y., & Luo, C. H. (2009). Wireless body sensor network with adaptive low-power design for biometrics and healthcare applications. *IEEE Systems Journal, 3*(4), 398–409.

Das, A. K., & Bruhadeshwar, B. (2013). A biometric-based user authentication scheme for heterogeneous wireless sensor networks. In *27th International Conference on Advanced Information Networking and Applications Workshops,* Barcelona, Spain (pp. 291–296). IEEE.

Das, A. K., Chatterjee, S., & Sing, J. K. (2015). A new biometric-based remote user authentication scheme in hierarchical wireless body area sensor networks. *Adhoc & Sensor Wireless Networks, 28*(3-4), 221–256.

Das, A. K., Sharma, P., Chatterjee, S., & Sing, J. K. (2012). A dynamic password-based user authentication scheme for hierarchical wireless sensor networks. *Journal of Network and Computer Applications, 35*(5), 1646–1656.

Das, A. K., Sutrala, A. K., Odelu, V., & Goswami, A. (2017). A secure smartcard-based anonymous user authentication scheme for healthcare applications using wireless medical sensor networks. *Wireless Personal Communication, 94*(3), 1899–1933.

El-Basioni, B. M., El-Kader, S. M., & Abdelmonim, M. (2013). Smart home design using wireless sensor network and biometric technologies. *Information Technology, 2*(3), 413–429.

Fan, R., Ping, L.-D., Fu, J.-Q., & Pan, X.-Z. (2010). A secure and efficient user authentication protocol for two-tiered wireless sensor networks. In *Second Pacific-Asia Conference on Circuits, Communications and System (PACCS),* Beijing, China (pp. 425–429). IEEE.

Garg, L., Chukwu, E., Nasser, N., Chakraborty, C., & Garg, G. (2020). Anonymity preserving IoT-based COVID-19 and other infectious disease contact tracing model. *IEEE Access, 8,* 159402–159414. 10.1109/ACCESS.2020.3020513, ISSN: 2169-3536

He, D., Kumar, N., Chen, J., Lee, C. C., Chilamkurti, N., & Yeo, S. S. (2015). Robust anonymous authentication protocol for health-care applications using wireless medical sensor networks. *Multimedia Systems, 21*(1), 49–60.

Jiang, C., Li, B., & Xu, H. (2007). An efficient scheme for user authentication in wireless sensor networks. In *21st International Conference on Advanced Information Networking and Applications Workshops (AINAW'07),* Niagara Falls, Ont., Canada (pp. 438–442). IEEE.

Juang, W.-S. (2006). Efficient user authentication and key agreement in wireless sensor networks. In *7th International Workshop on Information Security Applications,* Jeju Island, Korea (pp. 15–29). Springer-Verlag Berlin Heidelberg.

Ko, L.-C. (2008). A novel dynamic user authentication scheme for wireless sensor networks. In *IEEE International Symposium on Wireless Communication Systems,* Reykjavik, Iceland (pp. 608–612). IEEE.

Kumar, P., Gurtov, A., Ylianttila, M., Lee, S.-G., & Lee, H. (2013). A strong authentication scheme with user privacy for wireless sensor networks. *ETRI Journal, 35*(5), 889–899.

Lee, H. S., Lee, S., Song, H., & Lee, H. S. (2009). Wireless sensor network design for tactical military applications: Remote large-scale environments. In *IEEE Military communications conference (MILCOM),* Boston, MA, USA (pp. 1–7). IEEE.

Lee, T.-H. (2008). Simple dynamic user authentication protocols for wireless sensor networks. In *The Second International Conference on Sensor Technologies and Applications,* Cap Esterel, France (pp. 657–660). IEEE.

Li, X., Niu, J., Kumari, S., Liao, J., Liang, W., & Khan, M. K. (2016). A new authentication protocol for healthcare applications using wireless medical sensor networks with user anonymity. *Security and Communication Networks*, 9(15), 2643–2655.

Liu, C.-H., & Chung, Y.-F. (2017). Secure user authentication scheme for wireless healthcare sensor network. *Computers & Electrical Engineering*, 59, 250–261.

Luk, M., Perrig, A., & Whillock, B. (2006). Seven cardinal properties of sensor network broadcast authentication. In *Proceedings of the fourth ACM workshop on Security of ad hoc and sensor networks*, Alexandria Virginia, USA (pp. 147–156).

Mahendran, R. K., & Velusamy, P. (2020). A secure fuzzy extractor based biometric key authentication scheme for body sensor network in Internet of Medical Things. *Computer Communications*, 153, 545–552.

Miao, F., Bao, S. D., & Li, Y. (2013). Biometric key distribution solution with energy distribution information of physiological signals for body sensor network security. *IET Information Security*, 7(2), 87–96.

Miao, F., Jiang, L. L., & Zhang, Y. T. (2009). Biometrics based novel key distribution solution for body sensor networks. In *Annual International Conference of the IEEE Engineering in Medicine and Biology Society*, Minneapolis, MN, USA (pp. 2458–2461). IEEE.

Miao, F., Jiang, L., Li, Y., & Zhang, Y. T. (2009). A novel biometrics based security solution for body sensor networks. In *2nd International Conference on Biomedical Engineering and Informatics*. Tianjin, China (pp. 1–5). IEEE.

Mukhopadhyay, S. (2012). *Smart sensing technology for agriculture and environmental monitoring*. Berlin Heidelberg: Springer.

Nagaraj, A. (2021). *Introduction to sensors in IoT and cloud computing applications*. Bentham Science Publishers.

Nyang, D., & Lee, M.-K. (2009). Improvement of Das's two-factor authentication protocol in wireless sensor networks. In *IACR Cryptology* (p. 631). Santa Barbara, California, USA: IEEE.

Papadopoulos, S., Kiayias, A., & Papadias, D. (2012). Exact in-network aggregation with integrity and confidentiality. *IEEE Transactions on Knowledge and Data Engineering*, 24(10), 1760–1773.

Poon, C. C., Zhang, Y. T., & Bao, S. D. (2006). A novel biometrics method to secure wireless body area sensor networks for telemedicine and m-health. *IEEE Communications Magazine*, 44(4), 73–81.

Quan, Z., Chunming, T., Xianghan, Z., & Chunming, R. (2015). A secure user authentication protocol for sensor network in data capturing. *Journal of Cloud Computing: Advances, Systems and applications*, 4, 1–12.

Riaz, R., Gillani, N.-u.-A., Rizvi, S., Shokat, S., & Kwon, S. J. (2019). SUBBASE: An authentication scheme for wireless sensor networks based on user biometrics. *Wireless Communications and Mobile Computing*, 2019.

Sanjukta, B., & Chinmay, C. (2020). Machine learning for biomedical and health informatics, CRC: Big data, IoT, and machine learning tools and applications, Ch. 4, 353–373, ISBN 9780429322990, https://doi.org/10.1201/9780429322990

Tseng, H.-R., Jan, R.-H., & Yang, W. (2007). An improved dynamic user authentication scheme for wireless sensor networks. In *IEEE GLOBECOM 2007-IEEE Global Telecommunications Conference*, Washington, DC, USA (pp. 986–990). IEEE.

Turkanovic, M., Brumen, B., & Hölbl, M. (2014). A novel user authentication and key agreement scheme for heterogeneous ad hoc wireless sensor networks, based on the Internet of Things notion. *Adhoc Networks*, 20, 1–17.

Vaidya, B., Rodrigues, J. J., & Park, J. H. (2010). User authentication schemes with pseudonymity for ubiquitous sensor network in NGN. *International Journal of Communication Systems*, 23(9-10), 1201–1222.

Vaidya, B., Silva, J. S., & Rodrigues, J. J. (2009). Robust dynamic user authentication scheme for wireless sensor networks. *Q2SWinet*. Tenerife, Canary Islands, Spain: IEEE.

Wong, K. H., Zheng, Y., Cao, J., & Wang, S. (2006). A dynamic user authentication scheme for wireless sensor networks. In *IEEE International Conference on Sensor Networks, Ubiquitous, and Trustworthy Computing (SUTC'06).1*, Taichung, Taiwan (pp. 8–15). IEEE.

Yasmin, R., Ritter, E., & Wang, G. (2010). An authentication framework for Wireless Sensor Networks using identity-based signatures. In *10th IEEE International Conference on Computer and Information Technology (CIT 2010)*, Piscataway, New Jersey, USA (pp. 882–889). IEEE.

Yussoff, Y. M., Hashim, H., & Baba, M. D. (2012). Identity-based trusted authentication in wireless sensor network. *International Journal of Computer Science(IJCSI)*, *0*(3), 1–10.

6

Obfuscation to Mitigate Hardware Attacks in Edge Nodes of IoT System

K. Rajalakshmi[1], N. Susithra[2], V. Gowtham[1], and G. Saravanaraj[1]
[1]*PSG College of Technology*
[2]*PSG College of Technology and Applied Research*

6.1 Introduction to Hardware Security in IoT Systems

The Internet-of-Things (IoT) system is a paradigm for interconnecting numerous physical components through the internet to process input and obtain the required output in a regulated environment. The primary aim of IoT is to connect any device at any time with any other device in any place through any network or service. The main components of the system are sensor actuators, computing nodes and communicating networks. The world in 2020 has become dependent on IoT for carrying out almost every daily personal and professional activity online. IoT infrastructure is utilised in the development of smart home, smart city, smart transport, smart agriculture, smart healthcare, smart wearables and smart industries. With technological advancement each day, IoT provides sophisticated lifestyles for end users. The IoT revolution has in fact resulted in the convergence of consumer internet, industrial internet and business internet, which in turn opens the global network that connects people and things through information or data exchange (Fantana et al., 2013).

6.2 Chapter Organization

The chapter is organised as follows with the first section giving details about the origin of hardware security, types of security attacks in general, classification of physical or hardware attacks and the challenges to mitigating such attacks. The second section deals with major contributions.

Major contributions include folding transformation for biquad filter with and without register-minimization techniques, filter design and implementation, methodology adopted for implementing hardware obfuscation through this folding transformation of the filter circuit, varying the security level in this method, salient features of implemented obfuscation, and finally, the comparison of area, power and speed of the different types of hardware implementation. The next section discusses the application of Artificial Intelligence Techniques and Machine learning Algorithms to mitigate hardware attacks on edge nodes of an IoT system. The last section offers the conclusion and future scope.

6.2.1 The Origin of Hardware Security

IoT systems consist of nodes which are used for sensing, computing and storing information as well as connecting with other nodes. IoT nodes are integrated with cloud computing to devise connected architectures that provide complex operations. While the world stays connected with billions of IoT nodes, there is a need to scale down the size of the nodes, to manage power for remotely-located nodes through energy harvesting or to obtain power from miniaturised batteries. It becomes essential to scale down the nodes at all levels of connected architecture to reduce the manufacturing costs of IoT systems, creating affordable benefits to billions of end users. This is facilitated by fabricating these nodes with silicon or other materials, like gallium arsenide and germanium. When communication among these billions of nodes is established, the sensing, computing, storage and communicating nodes become vulnerable to intentional attacks, and face the threat of hardware piracy. Attackers intend to gain access to personal data or hardware architecture. They may demand money or bring manufacturing industries to a halt with ransomware attacks that might continuously obstruct access unless a ransom is paid. They may also tamper in order to reverse engineer the design. Though data communication between these nodes is secured by software routines, physical attacks on the node by intruders cannot be prevented by software routines. This necessitates enhancement of Hardware Security using hardware such as encryption engines to speed up software security, but designers have proceeded still further to create systems that secure connecting metal wires (https://www.arrow.com/en/research-and-events/articles/understanding-the-importance-of-hardware-security). Computing nodes of IoT systems are classified as cloud, fog, edge and mist. Cloud infrastructure includes hardware components such as data storage and servers together with software components such as virtualization software, among others. In fog infrastructure, intelligence and computing power are located at the local area network (LAN) and data is transmitted from endpoints to a fog gateway. In edge computing, intelligence and power can be located in either the endpoint or a gateway. In mist computing, the intelligence and power are located at the sensor network. Each computing node has its own architecture which has to be designed for hardware security.

The security vulnerability in hardware components of the IoT system thus might lead to the loss of confidential personal account details, permanent disruption in the service of IoT applications and IoT hardware piracy.

6.2.2 Types of Security Attacks in IoT

IoT nodes at the user end are more vulnerable to attack by intruders. These attacks are classified as "physical attack, network attack, software attack and encryption attack" (Deogirikar & Vidhate, 2017). Malicious node injection attack is a most dangerous physical attack that causes denial of services and makes changes to the data (Garg, Mittal, & Diksha, 2020). Sink hole attack is a network attack causing congestion of data packets at the base station (Rani, Maheswar, Kanagachidambaresan, & Jayarajan, 2020). The intruder can also launch other threats, like forwarding selected packets, making changes or relinquishing data packets. A worm attack is a software attack which originates when a self-cloning program intrudes through security holes in networking software and hardware, causing attacks by deleting the files in system, stealing data like passwords, modifying passwords without the knowledge of users and bringing the system to lockdown (Deogirikar & Vidhate, 2017). The side-channel attack is the most common encryption attack, because it is difficult to detect. An adversary can intrude through the side channel to access secured data by monitoring the electromagnetic (EM) radiation emitted by the device and analyzing the effect of cipher on the system (Andrea, Chrysostomou, & Hadjichristofi, 2015; Alkhudhayr, Alfarraj, Aljameeli, & Elkhdiri, 2019).

6.2.2.1 Physical Attacks

A direct or indirect attack on physical components of IoT systems is called a physical attack (Huang, Wang, Chen, & Jiang, 2020). The attacker is probably in physical proximity to the system under attack or is in the IoT system via any network or medium to perform an attack. The following are the types of Physical attacks commonly occurring in the IoT nodes.

Node Tampering: In node tampering, the attacker physically makes changes to the entire node or a part of the **node, intrudes into the nodes to access and modify secured data such as public cryptographic keys or routing tables, or effects the functioning of ascending data communication layers.**

RF Interference: The attacker adds noise to the Radio Frequency (RF) signal used for communication between the nodes with Radio Frequency Identification (RFID), tags causing Denial of Service (DoS) between the nodes.

Node Jamming: The attacker prevents wireless communication across a Wireless Sensor Network (WSN), resulting in DoS by deploying a jammer circuit.

Malicious Node Injection: This attack involves injection of newly fabricated malicious node in between two or **more nodes; the injected new node regulates the dataflow through it. The malicious node also causes DoS.**

Physical-Damage: This attack appears in the form damage to the hardware devices of the IoT network to cause DoS.

Social-Engineering: The adversary intrudes into an IoT system and physically interacts with the users of the IoT **network in order to steal personal information from the victim for monetary benefits.**

Sleep Deprivation Attack: For power management in the sensor network, most of the sensors are programmed **to be in sleep mode when they are not in use. The attacker intervenes in the sensor network to keep the sensor active to drain battery power and bring the operation of nodes to halt.**

Malicious Code Injection: The intruder, as the name indicates, physically intrudes into the operation of a node using fabricated malicious code to obtain access to the IoT system in order to regulate that particular node or the whole IoT system.

This chapter throws more light on the physical or hardware attacks in IoT systems.

6.2.3 Classification of Physical or Hardware Attacks in IoT Systems

These physical or hardware attacks can be further classified as invasive, semi-invasive, and noninvasive attacks (Bhunia & Tehranipoor, 2019).

Invasive Attack: When an attacker gains direct access to a chip or device and launches an invasive attack, it may be reversible or irreversible. The device may get damaged or tampered with during the attack, leaving a proof of evidence. The cost and the skill required to perform this attack vary from low to high, usually high. The different means of physical attack that fall under this category are as follows

a. Reverse Engineering (RE): The physical components of the IoT system in the form of chips are de-packaged and de-processed to achieve thorough understanding of the internal circuitry and its functioning in order to clone or duplicate the system for monetary benefits.

b. Micro Probing: This technique is used by the attacker to directly access the internal wire of the security-critical module to extract sensitive information. It is usually performed together with reverse engineering.

c. Fault-Injection Attack: The attacker injects faults into the device, in the form of laser or focused-ion beam in order to modify circuits, reconstruct buses, disconnect existing wires and penetrate layers by depositing or removing material on the die in which circuit is present.

d. Hardware Trojans: With the proliferation of IoT nodes and scaling down of device size to accommodate interconnection for exchange of information, reliability must also be enhanced. Instead, these devices become more vulnerable to hardware attacks. The nodes consist of microchips which are designed with third-party Intellectual Property (IP) cores for short-span manufacturing, enabling insertion of malicious circuits inside the node. These can be triggered to cause abnormal behavior of the node, leading to less reliability, system failure, remote access into hardware and extraction of sensitive information (Sidhu, Mohd, & Hayajneh, 2019).

These are the most common hardware attacks on IoT nodes. Interconnection of IoT becomes the center of attraction for adversaries due to easy access, vulnerabilities and secured data in the memory of these devices. Even if only one device is attacked, other devices in the IoT system become vulnerable to severe security attack, causing the overall system to shut down. This causes failure in system reliability. Hence it becomes mandatory to secure these devices from adversaries. The increase in the number of interconnected devices in IoT systems creates more challenges in securing them (Chakraborty & Kumari, 2016).

Semi-invasive Attack: In this attack the adversary, after de-packaging the chip, performs infrared imaging, fault injection using ultraviolet rays and side-channel analysis to extract sensitive information. This causes malfunctioning or breaks IoT system operation.

a. Imaging Attack: The adversary performs infrared imaging of the device layout to determine security-critical modules and extracts secured information or disrupts the normal functioning of the system.

b. Fault-Injection Attack using ultraviolet rays: The adversary attacks the de-packaged device or chip by focusing UV rays to create or break security fuse links on the chip, causing DoS attack. The laser rays change the memory content of the chip.

c. Side Channel Attack: The adversary utilises a laser pointer to trace the power absorbed by the CMOS transistor before and after UV radiation. Power tracing helps the attacker to determine the operation of the micro chip.

Non-invasive Attack: The adversary performs a physical attack without damaging the device; this is called a noninvasive attack.

a. Side-Channel Attack: The attacker utilises the measurement and analysis of the device's physical parameters to extract the security key of the cryptographic system. The physical parameters are execution time, operating power, operating current, electromagnetic emission and acoustic signals. This passive attack is one of the most common and a major threat to cryptographic devices especially available in the IoT systems.

b. Brute-Force Attack: This active attack searches for secured information, such as key or password, on the chip of the IoT system in an organised manner. The strategies adopted are recovering the design of the device from the truth table of the complex computation using the input and output relation, known as a black box attack, and doubling the power supply voltage or injecting random signals or commands to access factory testing or programming modes. This is known as a backdoor attack

c. Data Remanence: This active attack occurs in the SRAM by retrieving data from memory after power down, recovering data from long-term memory after power up or recovering data from frozen memory at $-20°C$. The recovered data helps to extract secured information. Data remanence can occur in EEPROM or Flash memory, as the threshold voltage of memory changes during each write and erase operation. This enables the extraction of secured data during multiple read and erase operations.

d. Fault-Injection Attack: In this active attack the adversary injects a fault or unexpected command at the input or output interface of the IoT system and observes the chip or device execution. Faults are generated by introducing glitches in clock or power signals to the chip, increasing the temperature of the chip, exposing the chip to white light, laser, x-ray beam, ion beam or electromagnetic flux. These injected faults cause the transistors or flip-flops to change states, which can be observed by the attacker in a systematic approach to extract sensitive information.

6.2.4 The Consequences of Security Attacks

Nodes of end users are more vulnerable to both software and hardware attack. IoT nodes find applications in smart healthcare, traffic system, avionics, home, manufacturing and supply chain systems. In the health care sector, when patients are remotely monitored by doctors, attackers may intrude through any level of IoT architecture to confiscate patient details and blackmail both the patient and the doctor for a ransom. The attack might be fatal for a patient with an implanted device. In smart traffic systems, the Vehicle To Vehicle infrastructure (V2X) system may receive a fake message from an advanced driver assistance system (ADAS); false data about the speed and direction of ongoing traffic can lead to accidents (Banerjee, Chakraborty, & Chatterjee, 2018). These false messages invokes traffic chaos, leading to non-regulatory flow of traffic (Andrea et al., 2015). In smart manufacturing ecosystems, the user-accessible equipment which is most likely prone to harmful attacks, comprises smart power-grid controllers and utility monitoring systems (Andrea et al., 2015). In communication ecosystems, small wireless nodes of 4G/LTE networks are installed at street level in outdoor deployments; these are similarly vulnerable and their operation is through a third-party network-service provider equipped with less-stringent security compared with a smart power grid. These nodes become victims of hackers and vandals, who gain control of networks causing vulnerable attacks such as Global Position System (GPS) tampering, deceiving and other time-related security issues (https://www.arrow.com/en/research-and-events/articles/understanding-the-importance-of-hardware-security; Andrea et al., 2015).

Importantly, apart from the above mentioned threats, any node that is open to access by users is vulnerable to RE and IP theft of the design. In these situations, it becomes mandatory to employ layered-security in a top-down or bottom-up approach to protect these devices from such vulnerabilities and to prevent network attacks. Recent FPGA's support this strategy by combining chip-level design security, board-level hardware security, and data security for all communication among layers of the IoT system (Andrea et al., 2015).

6.2.5 The Challenges of Securing the IoT nodes

The main objective of hardware security is to employ different cryptographic algorithms to obtain security characteristics such as integrity, non-repudiation, confidentiality and authentication. It is a major task to implement these cryptographic algorithms in IoT devices. Based on their extremely small size, low power and limited support from the central processing unit (CPU) and memory, these devices are broadly classified into two types: low-resource devices (LRD) and constrained-resource devices (CRD). Hence inclusion of additional layers of security becomes hard (Sidhu et al., 2019). The physical limitations of IoT devices also restrict information processing and complex algorithm implementation capabilities, leading to serious trade-off among confidentiality, integrity and resource management in the IoT systems.

Though incorporation of cryptographic processes with a public key in the higher layer of IoT devices provides confidentiality and integrity, it demands high computational and memory resources that cannot be met due to physical limitations.

Certificate expiry and updating is another challenge in such public-key infrastructure systems because it is hard to update IoT nodes employed in a remote, inaccessible location. This needs human intervention.

Proper authentication and recognition of the identity of users is another important concern in IoT devices. When authentication is compromised by an attacker, accessing legitimate user resources and launching an attack becomes unnoticeable and easy.

Verifying the identity of a device in the dynamic interconnection of numerous IoT devices is another challenge in IoT systems (Sidhu et al., 2019). IoT devices used in different applications share their own architectures and infrastructures. These are provided by individual service providers as per the requirement of each application. When an anonymous attack occurs, it is difficult to determine accountability, making access control of the IoT system ineffective. It is yet another challenge in the IoT domain. Because IoT devices are resource-constrained, attacks can happen in the form of power drain or exhaustion, leading to DoS. IoT devices may be deployed in easily-accessible and less-supervised areas, allowing adversaries to capture the device and tamper with it physically. Such vulnerabilities are hard to detect. With the advancement of technology, most IoT applications use Artificial Intelligence or Machine learning algorithms to build intelligent systems, leading to further challenges in the future.

6.3 Major Contribution

This section deals with a countermeasure for hardware vulnerabilities, namely obfuscation, in which the hardware design functionality can be modified to be difficult for intruders or attackers to understand. Obfuscation may be attained by means of a secure key or structural modifications. DSP is used in numerous applications like telephony, radio, satellite communication, speech, image and video processing and also in advanced IoT applications like smart connectivity and smart healthcare systems, among others. Digital Signal Processing is the heart of data processing in IoT. Real-time smart applications like voice and facial recognition require advanced DSP processing. Also, real time IoT applications require huge amounts of time-critical data to be processed; the speed at which these data are processed plays a vital role. Increasing performance versus power paradox is the center of IoT design choices. This means more storage, more signal processing and more power consumption and consequently less battery life. So, in order to lengthen battery life, power consumption should be reduced without compromising performance.

There are numerous High-level transformation techniques available for Digital Signal Processing (DSP) which can be applied at the algorithmic level or architectural level to improve performance of DSP architectures and circuits implemented using Very Large Scale Integration (VLSI) technology. The various transformation techniques are pipelining (Parhi, 1995), parallel processing (Parhi, 1995), retiming (Parhi, 1995), folding, unfolding (Parhi, 1995), look-ahead, relaxed look-ahead and numeric strength reduction (Parhi, 1995). The following section explains two such techniques, namely, folding and register minimization algorithms and their implementation using VLSI technology. By using folding transformation, area and power consumption can be reduced to a greater extent without compromising speed. Since IoT nodes are power-hungry, such algorithms help in optimizing power. On the other hand, register minimization reduces the number of registers used in the folded architecture, which in turn reduces the silicon area to a considerable amount. With these algorithms we can reduce both silicon area and power consumption without affecting the speed of operation of the IoT nodes. Moreover, they can be used as a countermeasure for hardware attacks. The following section gives a detailed explanation of these techniques.

A biquad filter is considered a secure design example. The biquad filter is a second-order recursive linear filter. It is often used as the basic building block for more complex filters because it has a high enough order to be useful on its own and also a biquad's coefficient precision is very good at high frequencies. This is the reason the biquad IIR filter is most widely used in audio processing applications. In an IoT ecosystem, filters play a major role in Smart Audio Sensors (SAS) (Antonini, Vecchio, Antonelli, Ducange, & Perera, 2018), biomedical applications (Bagavathiyammal & Muruganantham, 2020), Audio IoT analytics for Home Automation Safety (Shah, Tariq, & Lee, 2018). while it is essential to optimise these filters for better performance and low power consumption, secure design is another factor to be ensured. Optimizing filters will make the IoT ecosystem more efficient with less area and hardware secured.

6.3.1 Folding Transformation

Folding is a technique in which a single functional unit is used to perform two or more algorithmic operations based on time multiplexing. Folding is a methodology to minimise silicon area. While folding reduces the area occupied in the chip, it leads to an increase in the number of registers used in the architecture. To minimise the number of registers, a register-minimization technique is carried out by using a lifetime chart. For performing the folding operation, the folding sets and folding order are selected first.

A folding set is an ordered collection of operations executed by the same functional unit. Each folding set contains N entries, some of which may be null entries, where N is the folding order (Parhi & Chen, 2010). Let's consider the biquad filter shown in Figure 6.1 with folding order N = 4 and folding sets S1 = {4, 2, 3, 1} and S2 = {5, 8, 6, 7}.

The folding equation is given by Parhi and Chen (2010),

$$D_F(U \rightarrow V) = NW(e) - P_u + v - u$$

Where $W(e)$ represents the weight of the edge, P_u represents the pipelined stages, v and u represent the folding order of V and U node, respectively.

The folding equation is applied to all the 11 edges in the filter to find the number of delays present in the folded structure.

$$D_F(1 \rightarrow 2) = 4(0) - 1 + 1 - 3 = -3$$

$$D_F(1 \rightarrow 6) = 4(1) - 1 + 2 - 3 = 2$$

$$D_F(1 \rightarrow 5) = 4(1) - 1 + 0 - 3 = 0$$

$$D_F(1 \rightarrow 8) = 4(2) - 1 + 1 - 3 = 5$$

$$D_F(1 \rightarrow 7) = 4(2) - 1 + 3 - 3 = 7$$

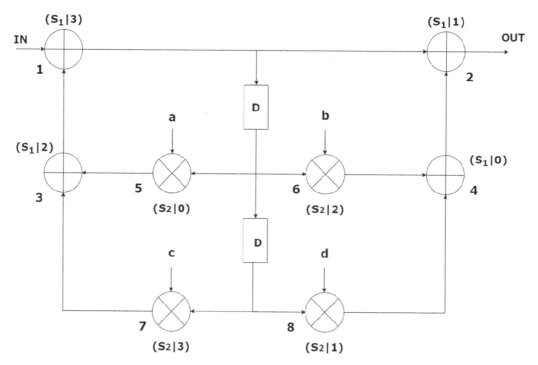

FIGURE 6.1
Biquad filter structure.

$$D_F(5 \rightarrow 3) = 4(0) - 2 + 2 - 0 = 0$$

$$D_F(7 \rightarrow 3) = 4(0) + +2 + 2 - 3 = -3$$

$$D_F(3 \rightarrow 1) = 4(0) - 1 + 3 - 2 = 0$$

$$D_F(6 \rightarrow 4) = 4(0) - 2 + 0 - 2 = -4$$

$$D_F(8 \rightarrow 4) = 4(0) - 2 + 0 - 1 = -3$$

$$D_F(4 \rightarrow 2) = 4(0) - 1 + 1 - 0 = 0$$

Since the folding equations consist of negative values for edge $1 \rightarrow 2$, $7 \rightarrow 3$, $6 \rightarrow 4$ and $8 \rightarrow 4$ cutsets are applied in these paths to perform retiming. Again the Folding equations are applied to the retimed filter to construct the folded filter.

$$D_F(1 \rightarrow 2) = 4(1) - 1 + 1 - 3 = 1$$

$$D_F(1 \rightarrow 6) = 4(1) - 1 + 2 - 3 = 2$$

$$D_F(1 \rightarrow 5) = 4(1) - 1 + 0 - 3 = 0$$

$$D_F(1 \rightarrow 8) = 4(2) - 1 + 1 - 3 = 5$$

$$D_F(1 \rightarrow 7) = 4(1) - 1 + 3 - 3 = 3$$

$$D_F(5 \rightarrow 3) = 4(0) - 2 + 2 - 0 = 0$$

$$D_F(7 \rightarrow 3) = 4(1) - 2 + 2 - 3 = 1$$

$$D_F(3 \rightarrow 1) = 4(0) - 1 + 3 - 2 = 0$$

$$D_F(6 \rightarrow 4) = 4(1) - 2 + 0 - 2 = 0$$

$$D_F(8 \rightarrow 4) = 4(1) - 2 + 0 - 1 = 1$$

$$D_F(4 \rightarrow 2) = 4(0) - 1 + 1 - 0 = 0$$

As there are no negative values in the folding equations, the folded architecture can be constructed from the folding sets and the folding equations. Figure 6.2 shows the folded biquad filter structure.

6.3.2 Register Minimization Technique

To reduce the number of registers used in the folded filter architecture, register minimization technique is carried out by constructing the lifetime table. The lifetime of a node is given by,

$$T_{input} \rightarrow T_{output}$$

$$U + P_U \rightarrow U + P_U + max_V \{D_F(U \rightarrow V)\}$$

Where T_{input} and T_{output} represents the time instance of input and output variables. Lifetime for the folded biquad filter is constructed using the above equations.

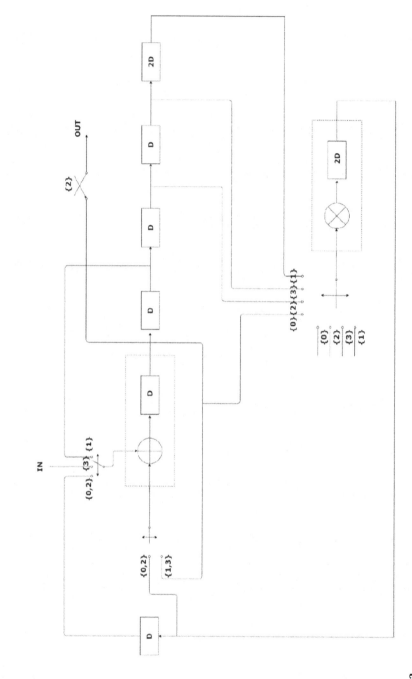

FIGURE 6.2
Folded biquad filter structure.

TABLE 6.1

Lifetime of Retimed Biquad Filter

Node	$T_{input} \rightarrow T_{output}$	
1	$3 + 1 \rightarrow 3 + 1 + 5$	$4 \rightarrow 9$
2	–	–
3	$2 + 1 \rightarrow 2 + 1 + 0$	$3 \rightarrow 3$
4	$0 + 1 \rightarrow 0 + 1 + 0$	$1 \rightarrow 1$
5	$0 + 2 \rightarrow 0 + 2 + 0$	$2 \rightarrow 2$
6	$2 + 2 \rightarrow 2 + 2 + 0$	$4 \rightarrow 4$
7	$3 + 2 \rightarrow 3 + 2 + 1$	$5 \rightarrow 6$
8	$1 + 2 \rightarrow 1 + 2 + 1$	$3 \rightarrow 4$

From the lifetime values shown in Table 6.1, a life chart for the retimed biquad filter is constructed as shown in Figure 6.3.

From the life chart it is observed that maximum lives during a particular instance are 2. Therefore, the maximum number of registers required is 2. By using only two registers, a data allocation table is constructed as shown in Figure 6.4. Using the data allocation table and folding equations, the folded biquad filter with minimum number of registers (the number of registers has been reduced from 5 to 2) is constructed as shown in Figure 6.5.

Methodologies and implementation of the transformed filter structures for secure design will be discussed in the following section.

6.3.3 Obfuscation through High-level Transformation

Obfuscation in general means scrambling information or making it unintelligible. In hardware, obfuscation can be achieved through various techniques. Some popular obfuscation methodologies are functional obfuscation and structural or physical obfuscation. There are also various system-level techniques like component obfuscation, obfuscation with programmable logic, and FSM modification etc.

In this section, a key-based functional obfuscation technique and structural obfuscation are discussed. Structural obfuscation improves security by modifying the structure of the

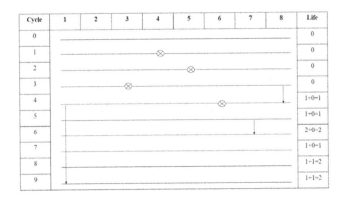

FIGURE 6.3
Life chart of retimed biquad filter.

Cycle	Input	R1	R2	Output
0				
1				
2				
3	8(M)			
4	1(A)	8(M)		8(M)
5	7(M)	1(A)		
6		7(M)	1(A)	7(M)
7			1(A)	
8			1(A)	
9			1(A)	1(A)

FIGURE 6.4
Data allocation table.

design, so that it remains unidentifiable for the attacker. It also helps in optimizing design objectives like area, power and speed. These methodologies can be incorporated while designing the IPs, which ensures the security of the end consumer.

Structural obfuscation in the work is achieved through high-level transformations like pipelining, interleaving, folding and unfolding, applied either in the algorithmic or architectural level to achieve trade-off among different metrics such as area, speed and power (Sunumol & Shanu, 2015).

High-level transformations allow design of circuits using the same datapath but a different control path (Lao & Parhi, 2015). Inserting different modes in the DSP circuits will further secure the design. While some modes generate functionally-incorrect outputs, these may represent correct outputs under a different situation. These meaningful modes can be accommodated in the design with the help of reconfigurable switches. Detailed implementation of the reconfigurable switch is shown in Figure 6.6. The counter in the design will be given as input to the multiplexers in the filter design. This will select the appropriate filter order and coefficients for generating the correct output.

Lao and Parhi (2015) proposed a reconfigurable switch design to achieve obfuscation via High-level transformation. In this method, an activation sequence is required before configuration. This can be accomplished by inserting an obfuscation FSM in the hardware design. From their work it is observed that the above-explained implementation ofa reconfigurable switch is not the only way to achieve obfuscation. We can modify the switch design depending on our application. From Figure 6.6 we can observe that only the input key and multiplexers are used as a switch to achieve design obfuscation. This method of obfuscation is simple to design and powerful enough to prevent attacks like reverse engineering from adversaries.

Security of the system can be further improved by encoding the input key. Several encoding techniques like hashing, Linear Feedback Shift Register (LFSR) or even a Physically Unclonable Functions (PUFs) can be used for encoding the key. The following section points out how various modes of operation can be achieved and gives the security perspective of using high-level transformation for obfuscation.

6.3.4 Variation of Modes to Increase Security Level

As discussed earlier, iinput keys consist of an initialization key of L-bits and data configuration of K-bits. The initialization key and data configuration should be known for proper functioning of the DSP circuit. The correct sequence of the initialization key and

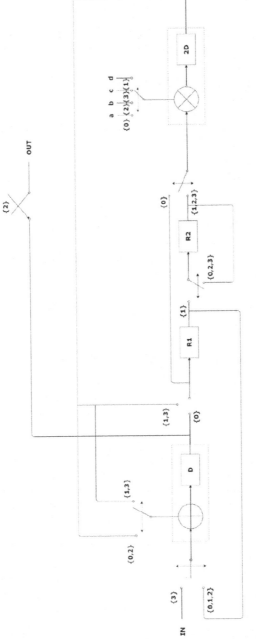

FIGURE 6.5
Folded biquad filter with minimum registers.

data configuration should be applied for proper operation of the DSP circuit. From Figure 6.6 we can see that only after applying the correct initialization key the switch will be enabled for configuration and the output will be generated. So it is impossible to crack the data configuration without applying the correct initialization key, as the switch will not be enabled at the moment.

In case the adversary finds the correct initialization key, then we can induct various modes in our design to improve security. The number of modes depends on the length of the input key (L + K bits). These modes will generate both meaningful and non-meaningful outputs. If the number of non-meaningful modes are more, then it will be easier for the attacker to crack the key with fewer input vectors. It is always better to have more meaningful modes than non-meaningful modes, so the chance of finding the correct key will be lessened. If a large portion of invalid values are generated by a single mode, the chance of predicting non-meaningful modes by an adversary with relatively smaller input vector increases. Finding the correct key depends not onlyon the length of the key but also on analyzing the functionality of the DSP circuit for various input vectors. So we can make this analyzing harder for the adversary by masking the correct functionality with various meaningful modes. An example of mapping the configuration key with its associated modes is illustrated in Table 6.2.

From the above table we can observe that total number of modes depends on the length of the key. Assigning different functionality to each mode lies in the hands of the designer. It can be assigned based on hardware and security requirements depending on the application. Incorporating these meaningful modes in the design stage itself will reduce overall design effort compared to generating these modes by modifying switch instances after performing high-level transformations. A complete flow and the architecture for obfuscating DSP circuit is depicted below.

6.3.5 Methodology Adapted for Obfuscating DSP Circuit

Steps involved in obfuscating DSP circuit is described below:

> *Step-1: Algorithm:* At first, DSP algorithm should be selected based on the required application under study.
> *Step-2: High-Level Transformation:* According to performance requirements (i.e. area, power, speed) high-level transformation technique is chosen.
> *Step-3: Obfuscation through High-Level Transformation:* Variation modes and different configuration of switches are designed and applied simultaneously to secure DSP circuit for the chosen high-level transformation technique.
> *Step-4: Secure Switch Design:* Variations of high-level transformations pave the way for secure switch design. Note that, simple combination-logic synthesis is enough to map various data configuration into the same mode.
> *Step-5: Design of control unit:* In this step, a configuration key is generated and the control unit for obfuscation will be embedded in the design.
> *Step-6: Design specification:* Generation of netlist with hardware description language and hardware synthesis will be done at this stage.

After completing all these steps, the design should be verified for bugs. The bug-free design will be sent to the foundry for manufacturing. With this proposed obfuscated technique incorporated in the design, functionality of the design will remain hidden so the manufacturer will not gain any information either by looking at the structure or by analyzing its functionality. The work flow for implementing this technique is described below.

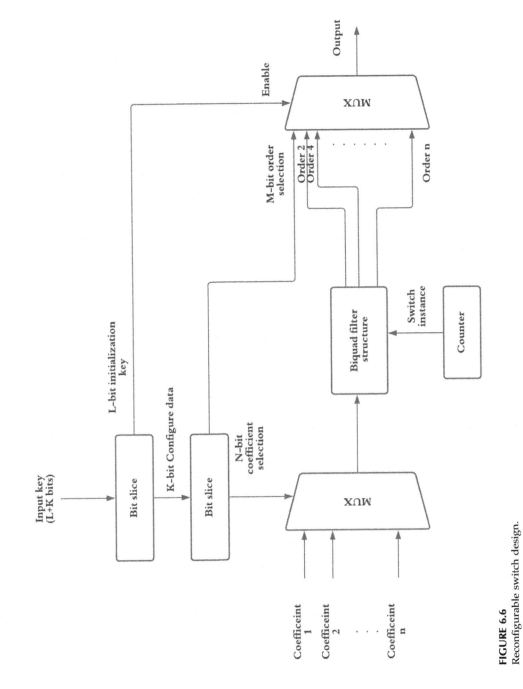

FIGURE 6.6
Reconfigurable switch design.

TABLE 6.2

Switch Configurations

Mode	Filter Operation
1	Second order low pass filter
2	Fourth order low pass filter
.	.
.	.
.	.
$2^{\wedge}(L+K)$	nth order low pass filter

The complete system of the obfuscated DSP circuit is shown in Figure 6.7. After applying the correct initialization key and data configuration, the desired design will be enabled. Applying the wrong data configuration will result in scrambled output. The K-bit data configuration will be again sliced into two parts for generating meaningful outputs, just by varying the order of the filter. M-bits of the data configuration will be used for selecting the order of the filter and N-bits of the data configuration will be used for selecting correct coefficients for the filter. When there is a mismatch in the data configuration, the circuit will produce meaningful DSP output, which will increase ambiguity and enhance security. There is no point in cracking the initialization key alone, as it will generate only scrambled output.

6.3.6 Salient Features of Hardware Security via Obfuscation

Consider this scenario: the adversary has the chip in the hand. The attacker can physically tamper with the chip, analyze the side-channel parameters or try various physical attacks to extract useful information from the chip. Therefore, it is the main duty of the IC designer to prevent these attacks and Obfuscation is the ray of hope in this situation. Salient features of obfuscation are listed below.

- Masking the functionality of the original design with obfuscation will not affect regular functionality, since the design of obfuscation is done at the architectural design level itself.

- The Chance of finding the correct key is meager. The probability for the attacker to find the correct key by random guessing is $1/2^{L+K}$, which is very low. As the length of the key increases, the chance of random guessing will be negligible.

- Since obfuscation is embedded at the design level itself, it will be difficult to remove.

- It can be observed from the implementation reports that the power won't increase significantly. Thus the security is improved without affecting the side-channel parameters.

- Security can be further enhanced by encoding the configuration key using Physically-Unclonable Functions (PUFs) (Enamul Quadir & Chandy, 2019). Currently PUFs are being used to provide unique keys for obfuscation.

- Obfuscated and non-obfuscated circuits differ only by the control unit associated with them, so the original datapath of the design remains unaltered; this would not affect the critical path for the obfuscated design.

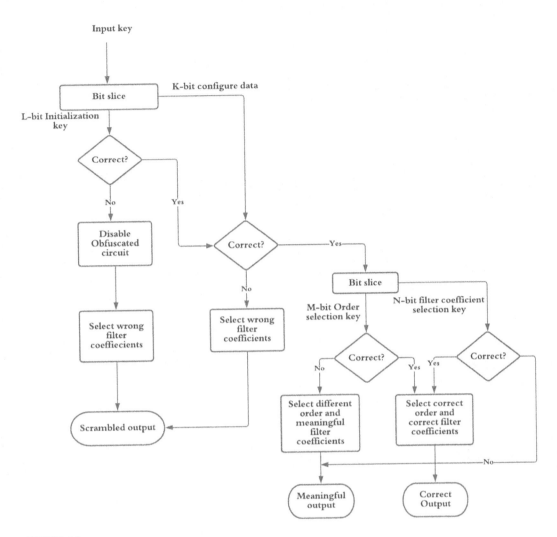

FIGURE 6.7
Workflow of obfuscated DSP circuit.

The implementation of obfuscation as a hardware-attack countermeasure via high-level transformation in Field Programmable Gate Array (FPGA) is explained in the following section.

6.3.7 Hardware Implementation of Obfuscated DSP Circuit

The workflow, as Figure 6.7 explained in the above section, is implemented in this section, from designing a filter to incorporating obfuscation via high-level transformation to the designed filter, implemented in ZyboZ7 AP SoC. High-level transformation techniques discussed in the section on folding were also implemented in hardware and their side-channel parameters were analyzed. Power and area overhead should be minimised in order to create a feasible IoT ecosystem, because higher power consumption will affect the lifetime of an IoT node to a greater extend. So, from these experimental results we can understand how side-channel parameters like power and area will be affected because of

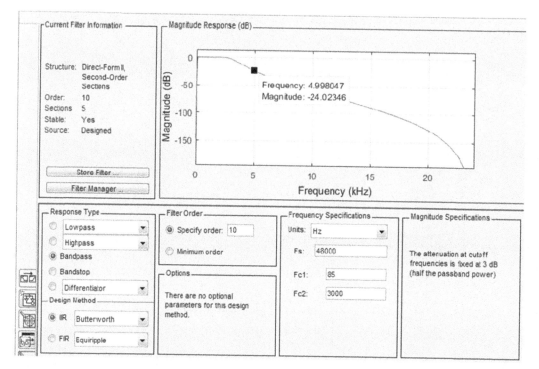

FIGURE 6.8
IIR biquad filter design specifications.

incorporating security in the system. The following section has a Simulink model and Vivado utilization reports for the respective architectures implemented. Results related to hardware resources and power consumption were analyzed for different architectural implementations; comparisons to the state-of-the-art are also presented.

6.3.8 Filter Design using FDA Tool

A Biquad IIR filter for an IoT node that senses speech signals is created using the Filter Design and Analysis Tool (FDA Tool). The sampling frequency (Fs) of the filter is selected as 48 kHZ, which is capable of capturing all audio frequency ranges. The fundamental frequency range for speech is from 85 to 255 Hz. The usable voice-frequency band ranges from approximately 300 Hz to 3400 Hz. So the cutoff frequency of the band pass filter is selected as 85 Hz and 3000 Hz. Upper-midrange frequency (2 kHz to 4 kHz) presence is necessary for clarity of voice. However, too much boost around the 3 kHz range can cause listening fatigue. So, the filter order is selected as 10 which causes gradual attenuation to the upper frequency range without any boosting. Figure 6.8 shows the setting of FDA tool.

6.3.9 Biquad Filter Implementation using System Generator

The biquad filter is also implemented in the system generator as shown in Figure 6.9, using the specifications shown in Figure 6.8. The filter coefficients are taken from the FDA tool. Using the Biquad filter, a tenth-order filter as shown in Figure 6.9 is designed by cascading 5 stages of Biquad filter. The filter is tested withan audio signal added to a noise signal in the form of sine wave of frequency 30 Hz. The input and output of the

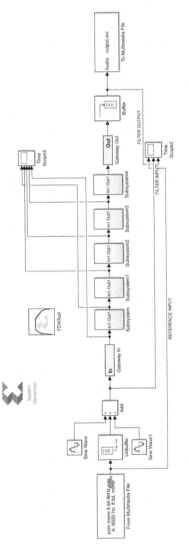

FIGURE 6.9
Simulink model of the tenth-order biquad filter.

filter is 16 bits digital value without any binary point. Internally, the operations are performed with 64 bits having 20 bits floating precision. The output of the filter is shown in Figures 6.10 and 6.11.

Figure 6.12 shows the Simulink model of a biquad filter of order 2. Cascading this biquad filter results in a tenth-order filter.

6.3.10 Folded Biquad Filter Implementation with and without Register Minimization using System Generator

The folded biquad filter without register minimization is implemented in the system generator as shown in Figure 6.13 using the same specifications. The tenth-order filter is designed by cascading 5 stages of the folded biquad filter as shown in Figure 6.14. The output of the filter is shown in Figure 6.15. The total chip power and resource utilisation when the design is implemented in Zybo z7 AP SoC is shown in Figure 6.16.

Similarly, the folded biquad filter with register minimization is implemented in the system generator as shown in Figure 6.17 using the above specifications. The tenth-order filter is designed by cascading 5 stages of the folded biquad filter as shown in Figure 6.18. The output waveform of the filter, total chip power and resource utilization when the design is implemented in Zybo z7 AP SoC are shown in Figures 6.19 and 6.20, respectively.

6.3.11 Verilog HDL Implementation of Folded Biquad Filter Implementation with Register Minimization

The folded biquad filter is implemented in Verilog Hardware Description Language (HDL) and the tenth-order folded biquad filter are constructed by instantiating five-fold biquad filter. The filter is tested by using a sine wave of 20 Hz and 100 Hz input. The input and output of the filter is 16 bits digital value without any binary point. Internally, operations are performed with 64 bits having 20bits floating precision. Figure 6.21 shows the RTL schematic of the Folded Biquad filter Implementation and Figure 6.22 shows the RTL schematic of the tenth-order Folded Biquad filter with register minimization.

Figure 6.23 shows the output of the tenth-order Folded Biquad filter with register minimization using Vivado simulator. It is observed that the 20Hz signal is attenuated and 100 Hz signal is passed through the filter. Figure 6.24 shows power consumption and resource utilization of the tenth-order Folded biquad filter designed using Verilog.

6.3.12 Comparison of Various Methods of Implementation

Table 6.3 shows the results obtained for various implementations. It is observed that folded biquad filter after register minimization utilises fewer LUTs when compared to conventional biquad filter. There is also reduction in power for the folded structure. It is also observed that the folded structure has a critical path delay less than that of a conventional biquad filter. So using this the power consumption of IoT nodes can be reduced and thus battery life of the wireless nodes can be increased.

6.3.13 Implementation of Obfuscated Design via High Level Transformation

Digital Signal processors facilitate various applications in today's smart consumer electronics, like de-noising, filtering, and attenuation. However, in the supply chain

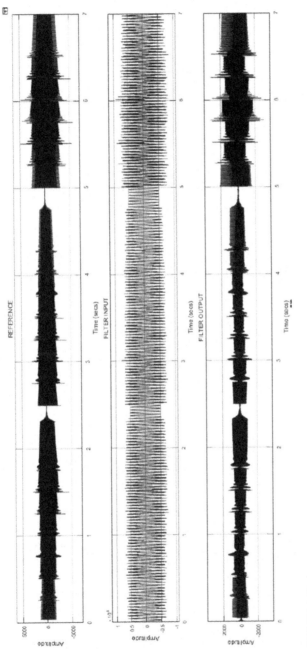

FIGURE 6.10
Output of the tenth-order biquad filter.

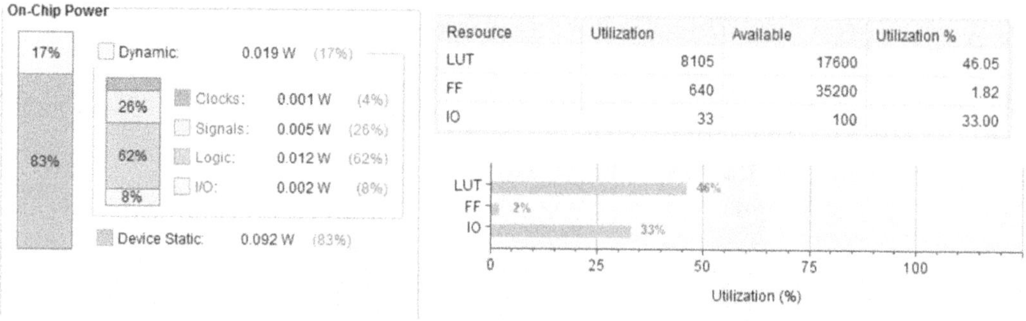

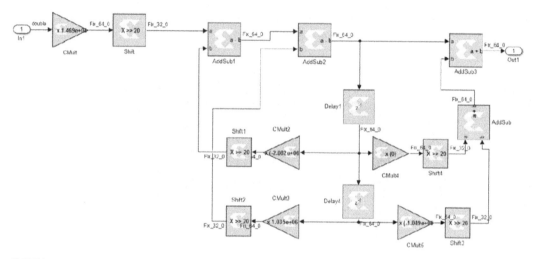

FIGURE 6.11
Power consumption and resource utilization report of tenth-order biquad filter.

FIGURE 6.12
Implementation of biquad filter of order 2.

these DSP processors are vulnerable to distinct hardware attacks. Securing these processors in the IoT ecosystem is the next great challenge. Therefore, these DSP Intellectual Properties (IPs) should be secured to ensure the safety of end customers. Obfuscation is a technique through which hardware can be secured from various attacks. Obfuscation is a countermeasure for preventing hardware attacks and is applied to the above implemented tenth-order folded biquad filter. Figure 6.25 shows the Simulink model of an obfuscated folded biquad filter. Implementation of the obfuscated folded biquad filter is shown in Figure 6.26. Here a control unit is used to achieve the obfuscation. The control unit in the obfuscated structure is built based on a reconfigurable switch design shown in Figure 6.6. The Simulink model of the implemented control unit is shown in Figure 6.27. The input key will be bit-sliced based on needs, and it can be used for generating meaningful outputs to increase ambiguity for the adversary (Figure 6.28).

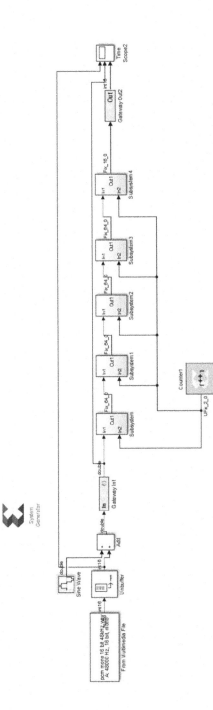

FIGURE 6.13

Simulink model of tenth-order folded biquad filter without register minimization.

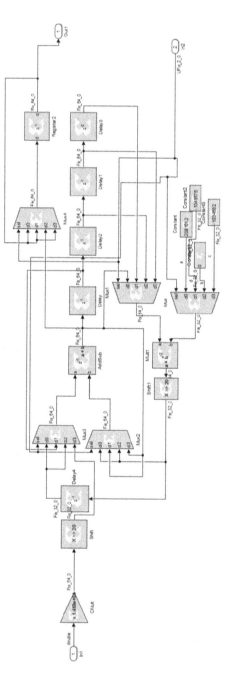

FIGURE 6.14
Folded biquad filter implementation.

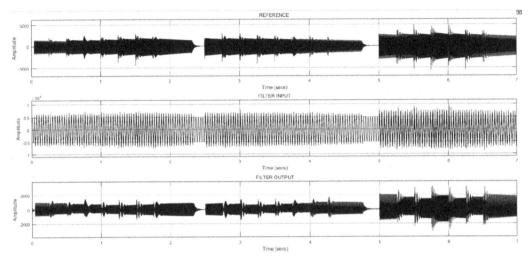

FIGURE 6.15
Output of folded biquad filter.

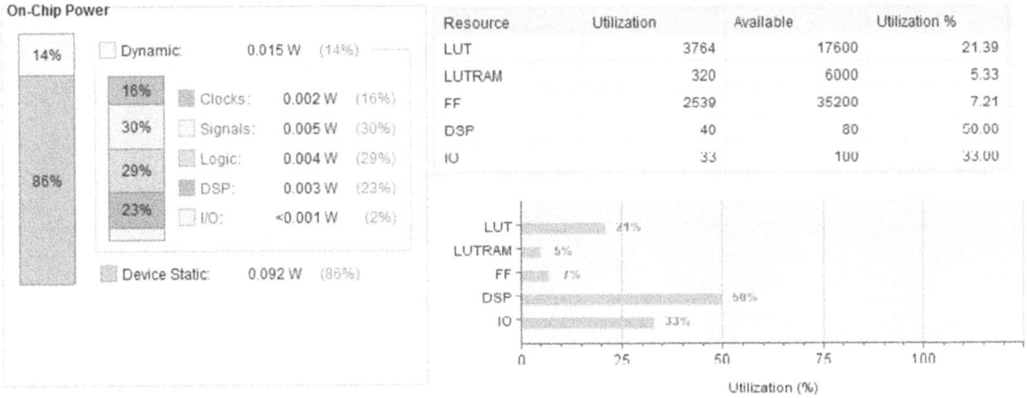

FIGURE 6.16
Power consumption and resource utilization of tenth-order folded biquad filter without register minimization.

6.3.14 Xilinx Vivado Implementation of Obfuscated Folded Biquad Filter

Results related to processing time (or throughput), FPGA area occupancy (or hardware resources) and power consumption were analyzed for different parameters. Figures 6.29 and 6.30 show the Vivado-synthesised structure of an obfuscated folded structure. Figure 6.31 shows the power and hardware resource utilization summary. Table 6.4 shows the device parameter settings of FPGA.

Compared to the folded biquad filter with register minimization, the power consumption of an obfuscated structure doesn't differ very much. A secured system is achieved with very little area and power overhead. From the above implementations, it is inferred that with minimum penalty of area and power, a secure design of a filter for IoT applications is successfully built.

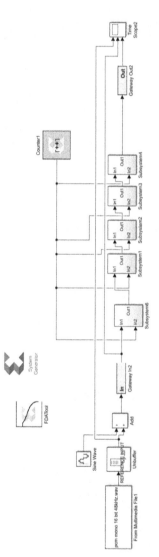

FIGURE 6.17
Simulink model of the tenth-order folded biquad filter with register minimization.

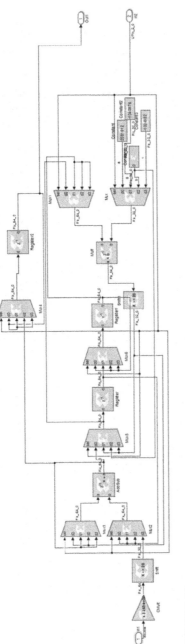

FIGURE 6.18
Folded biquadfilter with register minimization implementation.

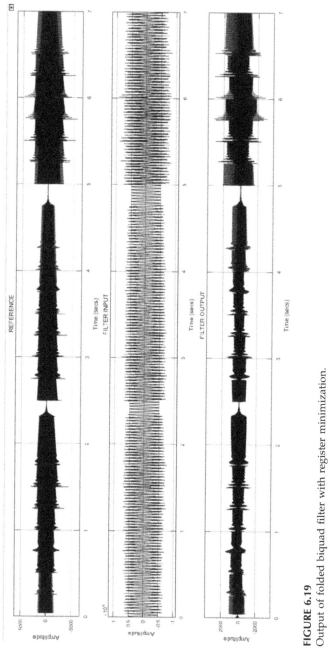

FIGURE 6.19
Output of folded biquad filter with register minimization.

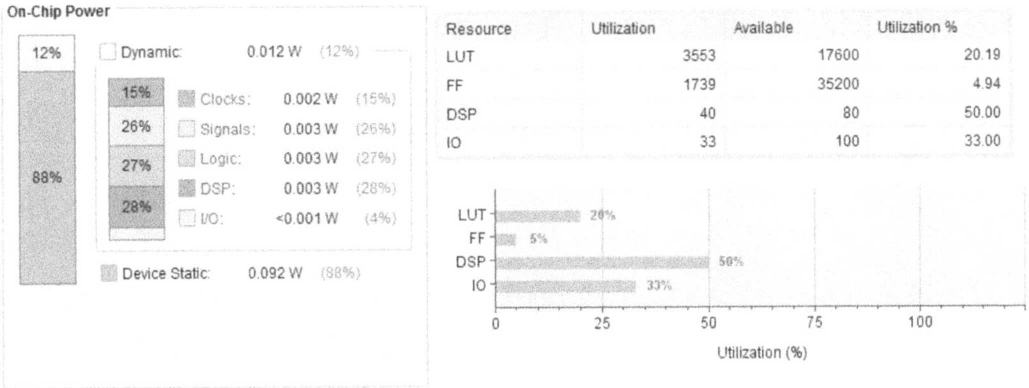

FIGURE 6.20
Power consumption and resource utilization report of folded biquad filter with register minimization.

6.4 Leveraging New Technologies to Mitigate Hardware Attack in IoT Nodes

The billions of connected IoT nodes must be protected from vulnerability to attackers if a reliable, secured, authenticated and resilient IoT application is to be developed. As technology is getting updated frequently, it becomes mandatory to adapt recent commonly-used technologies like Artificial Intelligence (AI) and Machine Learning (ML) in IoT systems to mitigate software and hardware attacks in IoT nodes.

In order to secure data and other high-value assets, security systems incorporating AI and ML based technologies are also required for detecting and stopping attacks on users and IoT devices.

6.4.1 Artificial Intelligence (AI) Technology

AI deals with designing IoT systems to be intelligent and respond like the human brain to security attacks. This objective can be achieved when the system components are trained by learning algorithms. Generally, learning algorithms for implementing a task are done through learning and training from experience. These processing steps, taken together, are called the ML algorithm. There are three types of learning algorithm to train IoT systems.

Supervised learning: It is a process involving training with a large representative data set obtained from previous learning experiences. Pattern classification or regression mechanisms utilise this learning algorithm frequently.

Unsupervised learning: It involves training with newly-defined training datasets. These learning algorithms are utilised to classify, decrease dimensions and estimate density of dataset.

Reinforcement learning: It involves learning based on reward or punishment and is suitable for conditions where the data is insufficient (Talib, Majzoub, Nasir, & Jamal, 2020).

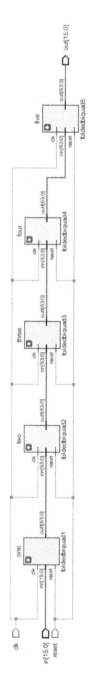

FIGURE 6.21
RTL Schematic of the tenth-order folded biquad filter without register minimization.

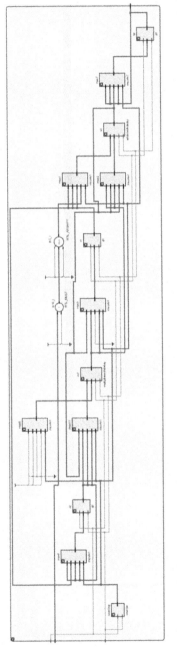

FIGURE 6.22
RTL Schematic of the tenth-order folded biquad implementation filter with register minimization.

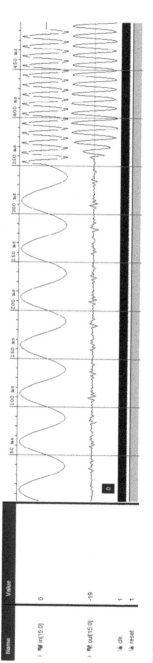

FIGURE 6.23
Output of tenth-order folded biquad filter with register minimization obtained using Vivado simulator.

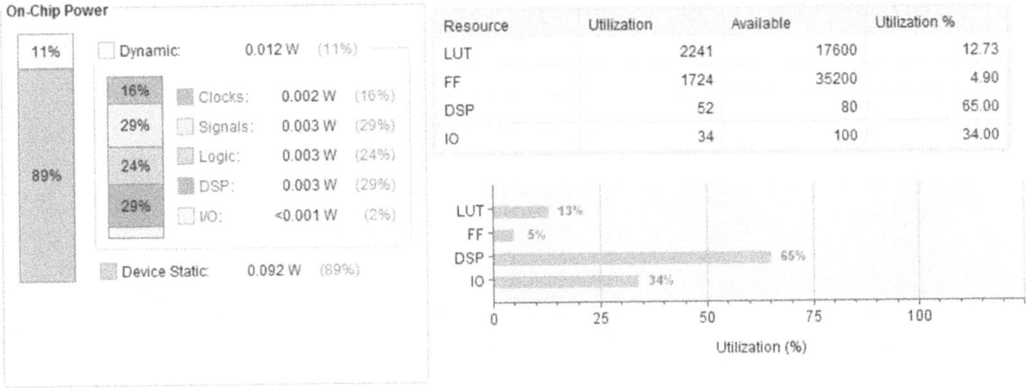

FIGURE 6.24
Power consumption and resource utilization report of the tenth-order folded biquad filter with register minimization.

TABLE 6.3

Comparison of Various Implementation Methods

Implemented Architecture	Implementation Method	Slice LUTs (17600)	Slice Registers (35200)	DSPs (80)	Power (W)	Maximum Delay (ns)	Maximum Clock Frequency (MHz)
Biquad filter	Xilinx System Generator	8105	640	0	0.11	87.412	12
Folded structure without register minimization	Xilinx System Generator	3764	2539	40	0.107	18.197	55
Folded structure with register minimization	Xilinx System Generator	3553	1739	40	0.104	16.970	59
Folded structure with register minimization	Verilog implementation	2241	1724	52	0.103	12.414	81

6.4.2 ML based Hardware Security for IoT Devices

Machine Learning (ML) is employed by both system defenders and attackers; while the former uses it to secure the hardware, the latter launches attacks on hardware using Machine Learning. It provides effective countermeasures to hardware attack as well as the means to cause an attack. ML algorithms help to (1) perform side-channel analysis, (2) launch an attack model on physically-unclonable functions (PUFs), (3) detect Trojans, and (4) take countermeasures against IC overbuilding. By IC overbuilding, the IC manufacturer fabricates more ICs than required for monetary benefits by selling the additional IC without the knowledge of IC designer.

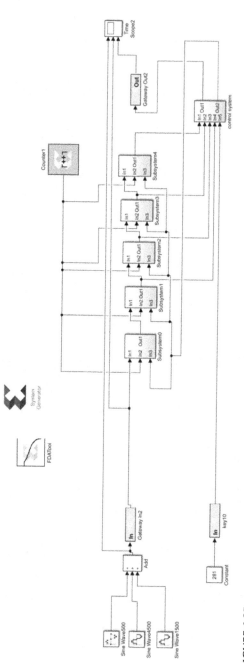

FIGURE 6.25
Simulink model of obfuscated folded biquad filter.

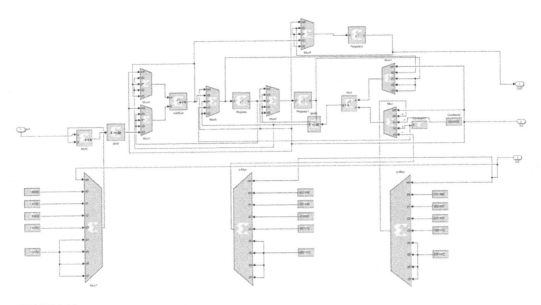

FIGURE 6.26
Obfuscated implementation of second-order folded biquad subsystem.

6.4.2.1 ML based Hardware Trojan Detection

A Hardware trojan activates under special conditions in the circuit as per the requirement of an adversary. These Trojans are hard to detect with traditional tests. Thus,
circuit features like functional or structural models are extracted from the netlist and the
suspicious nets present are analysed. Two quantitative measures for Trojan detection
are switching activity and netlist features. Trojan-infected nets are differentiated from
normal ones by features extracted from the netlist. The SVM (Kulkarni, Pino, &
Mohsenin, 2016) or Artificial Neural Networks (ANN) classifiers (Madden, Harkin,
Mcdaid, & Nugent, 2018) are trained to classify features from an unlabeled netlist of the
circuit (Huang et al., 2020). These classifier methods detect the Trojan circuit by increasing the True Positive Rate (TPR), but there are limitations in determining True
Negative Rate (TNR). TPR is an accurate prediction rate for nets of Trojan and TNR is an
accurate prediction rate of normal nets. The Random Forest(RF) classifiers (Hasegawa,
Yanagisawa, & Togawa, 2017) are applied to select Trojan features and to detect Trojans
from the extracted nets.

6.4.2.2 ML based Side-Channel Analysis (SCA)

Trojans can be detected through side-channel analysis (Karri, Rajendran, Rosenfeld, &
Tehranipoor, 2010) by finding affected process variations and noise, which depend on
Signal-to-Noise ratio (SNR) and Trojan-to-Circuit Ratio (TCR) (Huang et al., 2020). ANN
is used in SCA to determine if Trojans are present in the circuit by sampling extracted
features (Wang et al., 2016). The Back-Propagation Neural Network (BPNN) detection
model is used to detect Trojans from the extracted non-linear features by evaluating the
power consumption of the circuit. To avoid inaccuracy in manual modeling, Extreme
Learning Machine and BPNN (Li, Ni, Chen, & Zhou, 2016) are used for extracting the
features (Yang & Zhang, 2015); it is inferred that these algorithms extract the features in

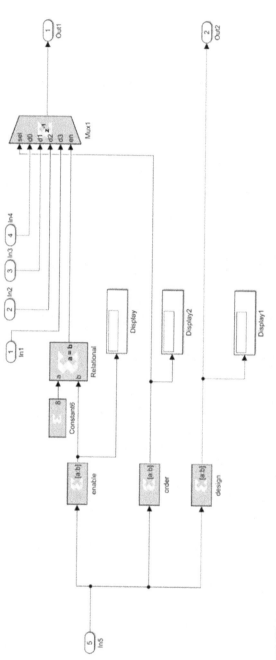

FIGURE 6.27
Implementation of control unit.

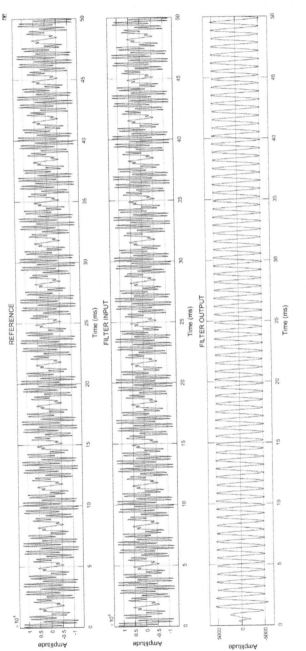

FIGURE 6.28
Output of obfuscated folded biquad filter.

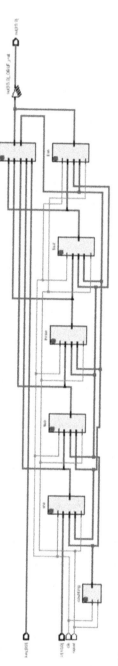

FIGURE 6.29
RTL schematic of obfuscated folded structure.

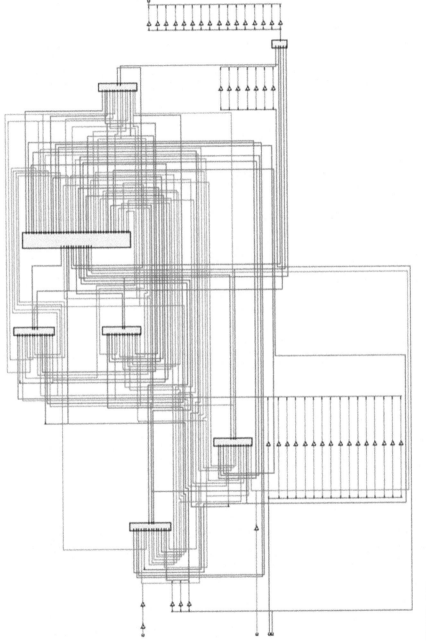

FIGURE 6.30
RTL schematic of folded biquad filter subsystem.

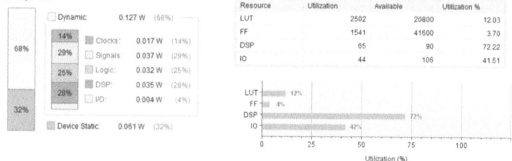

FIGURE 6.31
Power consumption and hardware utilization summary.

TABLE 6.4

Device Parameters

Xilinx Device Utilization Report:	
Maximum frequency = 1 / (T–WNS)	
WNS = –3.441 ns	
T = 10 ns	
Maximum Frequency = 1 / (10 – (–3.441)) ns = 74.4 MHz	
Maximum Clock Frequency (MHz)	74.4
Maximum Delay (ns)	13.322
Power (W)	0.187

better way. However, they lack in sampling of features and therefore the nets used for training are volatile in nature.

To overcome the instability of the ANN method and to train features in order to improve detection ability and accuracy, a more unspecific classification algorithm, like a Supporting Vector Machine (SVM), is used. It outlines detection problems and the effects caused by Trojans. But this Trojan-detection methodology has some disadvantages, e.g., there is poor performance in improving the SNR ratio. The other ML methodologies are the combination of SVM (Daoud, 2018; Sustek, 2011) with other methodologies, like Principal Component Analysis(PCA) or Discrete Fourier Transform (DFT). PCA with SVM uses power transmission waveforms to detect communication-type Trojans in the circuit. DFT in the time domain is used to convert the waveform data, then Trojan detection can be done by SVM. Deep-learning algorithms are also used to detect Trojans.

Thus, these machine learning techniques are very helpful in Trojan detection as they require less computation time even if the circuit is complex.

6.4.2.3 ML in System on Chip (SoC) Architecture

Generally the most commonly used IoT nodes like Edge Computing (EC) are designed as SoC with multicore architecture. As the design for various applications are getting saturated and vulnerabilities in the architecture level of edge nodes (Daoud, 2018) are inserted by adversaries, it is necessary to secure hardware devices; thus, hardware security (Guha, Saha, & Chakrabarti, 2015) is a more necessary part of research in recent days. In the architectural level, as design complexity increases, more readymade IP cores from different vendors are used, and there is a possibility of Trojan insertion in any block of the core during any stage of a design. ML algorithms are used to create attacks on hardware and to develop countermeasures to hardware attacks (Elnaggar & Chakrabarty, 2018). The features of ML-based attacks should be extracted in an organised manner when designing countermeasures against vulnerabilities.

The security of the SoC at the architectural level (Fern, San, Koc, & Cheng, 2017; Lao & Parhi, 2015) can be improved by applying ML to the on-chip modules, consisting of cryptographic algorithms, to secure the device. The analog ANN classifier analyzes the parameters required and classifies the device as Trojan-free or Trojan-infected. The Trojans and their attacks can cause aging through unsupervised strategy. Confidentiality is maintained by the Runtime Trust Neural Architecture, based on adaptive resonance theory. This method utilises the on-chip clock on the SoC, eliminating the requirement of a golden IC.

6.5 Conclusion and Future Scope

This chapter discussed different types of hardware attacks on edge nodes of IoT systems and elaborated on one method of mitigating these attacks, using obfuscation via high-level transformation techniques in the filter design of DSP-processor IPs in edge nodes. The significance of implementing countermeasures which have less area and power overheads against hardware attack for IoT nodes have been discussed. Though there are challenges to securing the hardware of IoT nodes, leveraging new technologies, such as Artificial Intelligence (AI) and Machine Learning (ML) for smart IoT systems, helps in a big way. Also, there is a need for hardware-secured DSPonSOC. To protect such complex hardware from adversaries, hardware obfuscation is employed to prevent reverse engineering. IoT application devices are distributed in a wide geographical area; the main concerns are both hardware design and the programming paradigm (Banerjee, Chakraborty, & Paul, 2019), including heterogeneity caused by hardware, software and communication platforms, the volume of generated data of various forms and sustainability. For a sustainable design and efficient use of IoT nodes, concern for security at all levels of design is indispensable. Use of AI and ML can introduce new strategies for hardware security and help in enhancing the level of obfuscation through dynamic algorithms.

References

Alkhudhayr, F., Alfarraj, S., Aljameeli, B., & Elkhdiri, S. (2019). Information security: A review of information security issues and techniques. In *2019 2nd International Conference on Computer Applications & Information Security (ICCAIS)*.

Andrea, I., Chrysostomou, C., & Hadjichristofi, G. (2015). Internet of Things: Security vulnerabilities and challenges. In*IEEE Symposium on Computers and Communication (ISCC)* (pp. 180–187). Larnaca, 2015. doi: 10.1109/ISCC.2015.7405513

Antonini, M., Vecchio, M., Antonelli, F., Ducange, P., & Perera, C. (2018). Smart audio sensors in the Internet of Things edge for anomaly detection. *IEEE Access, 4*, 1–7, doi: 10.1109/ACCESS.2018.2877523.

Banerjee, S., Chakraborty, C., & Chatterjee, S. (2018). A survey on IoT based traffic control and prediction mechanism. *Springer: Internet of Things and Big data Analytics for Smart Generation, Intelligent Systems Reference Library*, Ch. 4, 154, 53–75. ISBN: 978-3-030-04203-5.

Banerjee, S., Chakraborty, C., & Paul, S. (2019). *Programming paradigm and Internet of Things, CRC: A handbook of Internet of Things & big data*, 148–164. ISBN 9781138584204.

Bao, D. F., & Srivastava, A. (2014, Mar). On application of one-class SVM to reverse engineering-based hardware trojan detection. In *Proceedings of the 15th International Symposium Quality Electronic Design* (pp. 47–54). doi: 10.1109/isqed.2014.6783305.

Bhunia, B., & Tehranipoor, M. (2019). Hardware security: A hands-on learning approach. Morgan Kaufmann, Page iv, ISBN 9780128124772 .

Bagavathiyammal, R., & Muruganantham, T. (2020). A design of Cmos based 4th order low pass biquad filter for biomedical applications. In *2020 Second International Conference on Inventive Research in Computing Applications (ICIRCA)*, Coimbatore, India (pp. 1152–1155). doi: 10.1109/ICIRCA48905.2020.9182851.

Chakraborty, C., & Kumari, S. (2016). Bio-metric identification using automated iris detection technique. In *IEEE: 3rd International Conference on Microelectronics, Circuits, and Systems* (pp. 113–117).

Daoud, L. (2018, Aug). Secure network-on-chip architectures for MPSoC: Overview and challenges. In *Proceedings of the IEEE 61st International Midwest Symposium on Circuits and Systems (MWSCAS)* (pp. 542–543). doi: 10.1109/mwscas.2018.8623831.

Deogirikar, J., & Vidhate, A.. (2017). Security attacks in IoT: A survey. In *2017 International Conference on I-SMAC (IoT in Social, Mobile, Analytics and Cloud) (I-SMAC)* (pp. 32–37).

Elnaggar, R., & Chakrabarty, K. (2018). Machine learning for hardware security: Opportunities and risks. *Journal of Electronic Testing, 34*, 183–201. doi: 10.1007/s10836-018-5726-9.

Enamul Quadir, M. S., & Chandy, J. A. (2019). Key generation for hardware obfuscation using strong PUFs. *Cryptography, 3*, 17.

Fantana, N., Riedel, T., Schlick, J., Ferber, S., Hupp, J., Miles, S., ... Svensson, S.(2013). Internet of things – Converging environment for smart environment and integrated ecosystems (pp 152–204). Denmark: Alborg.

Fern, N., San, I., Koc, C. K., & Cheng, K.-T. T. (2017, Sep). Hiding hardware Trojan communication channels in partially specified SoC bus functionality. *IEEE Transactions on Computer-Aided Design of Integrated Circuits and Systems, 36*(9), 1435–1444. doi: 10.1109/tcad.2016.2638439.

Garg, A, Mittal, N., & Diksha. (2020). A security and confidentiality survey in wireless Internet of Things (IoT). In: V. Balas, V. Solanki, R. Kumar (Eds.), *Internet of Things and big data applications, intelligent systems reference library* (Vol. 180). Cham: Springer.

Guha, K., Saha, D., & Chakrabarti, A. (2015, Jun). RTNA: Securing SOC architectures from confidentiality attacks at runtime using ART1 neural networks. In *Proceedings of the 19th International Symposium on VLSI Design Test* (pp. 1–6). doi: 10.1109/isvdat.2015.7208048.

Hasegawa, K., Yanagisawa, M., & Togawa, N. (2017, May). Trojan-feature extraction at gate-level netlists and its application to hardware-Trojan detection using random forest classifier. In *Proceedings of the IEEE International Symposium Circuits and Systems (ISCAS)* (pp. 1–14).

Huang, Z., Wang, Q., Chen, Y., & Jiang, X. (2020). A survey on machine learning against hardware Trojan attacks: Recent advances and challenges. *IEEE Access, 8*, 10796–10826.

Karri, R., Rajendran, J., Rosenfeld, K., & Tehranipoor, M. (2010, Oct). Trustworthy hardware: Identifying and classifying hardware Trojans. *Computer, 43* (10), 39–46. doi: 10.1109/mc.2010.299.

Kulkarni, A., Pino, Y., & Mohsenin, T. (2016, Mar) SVM-based real-time hardware Trojan detection for many-core platform. In *Proceedings of the 17th International Symposium on Quality Electronic Design (ISQED)* (pp. 362–367). doi: 10.1109/isqed.2016.7479228.

Lao, Y., & Parhi, K. K. (2015, May). Obfuscating DSP circuits via high-level transformations. *IEEE Transactions on Very Large Scale Integration (VLSI) Systems, 23* (5), 819–830. doi: 10.1109/TVLSI.2014.2323976.

Li, J., Ni, L., Chen, J., & Zhou, E. (2016, Oct). A novel hardware Trojan detection based on BP neural network. In *Proceedings of the 2nd IEEE International Conference on Computer and Communications (ICCC)* (pp. 2790–2794). doi: 10.1109/compcomm.2016.7925206.

Madden, K., Harkin, J., Mcdaid, L., & Nugent, C. (2018, Nov). Adding security to networks-on-chip using neural networks. In *Proceedings of the IEEE Symposium Series on Computational Intelligence (SSCI)* (pp. 1299–1306).

Parhi, K.K. (1995). High-level algorithm and architecture transformations for DSP synthesis. *Journal of VLSI Signal Processing, 9,* 121–143. doi: 10.1007/BF02406474, .

Parhi, K. K., & Chen, Y. (2010). Signal flow graphs and data flow graphs. In Bhattacharyya, S., Deprettere, E., Leupers, R., Takala, J. (Eds.), *Handbook of Signal Processing Systems.* Boston, MA: Springer. doi: 10.1007/978-1-4419-6345-1_28.

Rani, S., Maheswar, R., Kanagachidambaresan, G. R., & Jayarajan, P. (Eds.). (2020). Integration of WSN and IoT for smart cities. In *EAI/Springer Innovations in Communication and Computing,* ISBN: 978-3-030-38515-6.

Sengupta, A., Bhadauria, S., & Mohanty, S. P. (2017, Apr). TL-HLS: Methodology for low cost hardware Trojan security aware scheduling with optimal loop unrolling factor during high level synthesis. In *IEEE Transactions on Computer-Aided Design of Integrated Circuits and Systems, 36* (4), 655–668. doi: 10.1109/tcad.2016.2597232.

Shah, S., Tariq, Z., & Lee, Y. (2018). Audio IoT analytics for home automation safety. *IEEE International Conference on Big Data (Big Data)* (pp. 5181–5186). doi: 10.1109/BigData.2018.8622587

Sidhu, S., Mohd, B. J., & Hayajneh, T. (2019). Hardware security in IoT devices with emphasis on hardware trojans. *Journal of Sensor and Actuator Network, 8,* 42, https://doi.org/10.3390/jsan8030042.

Sunumol, K. S., & Shanu, N. (2015). Obfuscation in DSP algorithms using high level transformations for hardware protection. In *2015 IEEE Recent Advances in Intelligent Computational Systems (RAICS),* Trivandrum (pp. 27–32). doi: 10.1109/RAICS.2015.7488383.

Sustek, L. (2011). Hardware security module. In *Encyclopedia of cryptography and security* (pp. 535–537). Boston, MA, USA: Springer.

Talib, M. A., Majzoub, S., Nasir, Q., & Jamal, D. (2020). A systematic literature review on hardware implementation of artificial intelligence algorithms. *Journal of Supercomputing, 77,* 1897–1938.

Understanding the Increased Importance of Hardware Security in IoT Technologies, May 2020. https://www.arrow.com/en/research-and-events/articles/understanding-the-importance-of-hardware-security

Wang, S., Dong, X., Sun, K., Cui, Q., Li, D., & He, C. (2016, Oct). Hardware Trojan detection based on ELM neural network. In *Proceedings of the 1st IEEE International Conference on Computer Communicaton and the Internet (ICCCI)* (pp. 400–403). doi: 10.1109/cci.2016.7778952.

Yang, S. Y., & Zhang, H. (2015). Feature selection and optimization. In *Pattern Recognization and Intelligent Computing* (3rd ed., ch 2.1, sec. 2, pp. 27–28). Beijing, China: PHEI.

7

Lightweight Security Solutions for IoT using Physical-Layer Key Generation

R. Upadhyay and A. Soni
Devi Ahilya Vishvavidyalaya

7.1 Introduction

With the spur of wireless communication, smart networks like Internet of Things (IoT) and wireless sensor networks (WSN) have become some of the most common topics of discussion. These modern networks alter our prospects for observing intelligent devices, and the sensors and services associated with them. IoT refers to the assemblage of sensor nodes and embedded systems, collectively called "things", coupled through the Internet (Wang et al., 2019). One example is smart home automation systems, which include monitoring and automation of facilities like heating, ventilation, and security access control via internet. Such systems improve the quality of our lives. A major service area of IoT includes infrastructure services, like collection of electricity and water bills and traffic in commercial and residential areas. One of the applications of advanced WSN is smart agriculture, which monitors parameters like humidity, heat in the fields and soil characteristics. Smart advertisement and purchasing preferences can also be monitored by IoT (Zhang, Zhao, Xiang, Huang, & Chen, 2019).

In wired networks, the communicating parties are connected through physical media such as cables, and are resilient to malicious attacks by the intruder. By contrast, in networks such as WSN or IoT, communication is broadcast and is thereby susceptible to hazardous attacks, including the denial of service (DoS) attack, spoofing, and eavesdropping. It introduces new challenges for the security of wireless systems along with the privacy of individuals. Some of the important attributes of wireless security design include complexity, authentication, and encryption, as shown in Figure 7.1 (Soni, Upadhyay, & Jain, 2016).

Wireless networks, in general, adopt an open system-interconnection (OSI) model, while traditional security approaches are based on the network and upper layers. Existing systems typically employ cryptographic techniques for achieving communication security between legitimate pairs of users. They assume that the illegitimate user has limited computational power and cannot decode the encrypted data within its lifetime, i.e., they rely on computational security (Aldaghri & Mahdavifar, 2020; Mazin, Davaslioglu, & Gitlin, 2017). Physical-layer security (PLS) is an emerging

Wireless Security Design Attributes

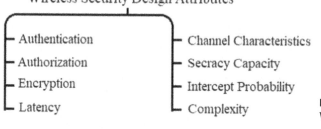

- Authentication
- Authorization
- Encryption
- Latency

- Channel Characteristics
- Secrecy Capacity
- Intercept Probability
- Complexity

FIGURE 7.1
Wireless security design attributes.

substitute to attain information-theoretic security in wireless networks. Physical-layer key generation (PLKG) is one of the PLS techniques employing parameters such as distance, channel state information (CSI), received signal strength (RSS) and angle of arrival (AoA) (Badawy, Elfouly, Khattab, Mohamad, & Guizani, 2016) to achieve security by generating secure keys. This method is elaborated on in this chapter. Performance of the PLKG system can further be enhanced by incorporating the suggestions discussed in the last section of the chapter. This chapter presents a physical-layer-based security system for IoT networks. The traditional security approach is discussed in brief in section 7.3, preceded by the motivation for the study in section 7.2. Wireless physical-layer key generation is discussed in detail in section 7.4 with experimental findings. Applications and future scope is presented in section 7.5, followed by the conclusion in section 7.6.

7.2 Motivation

Security of modern wireless networks like WSN and IoT is crucial, as these modern networks play a vital role in interconnecting various smart sensor nodes around us. These smart nodes are restricted in power and computational resources and require low-power algorithms for their functioning. PLKG is one such solution and is less computationally-complex as compared to higher layer approach; it has been less-often explored in previous works, and is the motivation for this study.

7.3 Wireless Security

Wireless security is an open issue and is an area of research interest. Traditionally, classical encryption is employed to achieve data security (Menezes, van Oorschot, & Vanstone, 1996), which works by assuming that the time required by an illegitimate user to crack the complex cryptographic system is much longer than the lifetime of the information packet; this guarantees backward secrecy. Depending on the keys used by the communicating nodes, classical encryption schemes are further categorised as symmetric encryption schemes and asymmetric encryption schemes as shown in Figure 7.2. Symmetric encryption schemes employ a similar key and are usually used for data protection, while asymmetric encryption schemes use a common public key

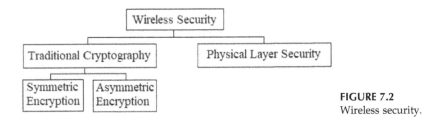

FIGURE 7.2
Wireless security.

and different private keys. This is called public-key cryptography and usually requires a key-distribution system.

Classical encryption schemes have several vulnerabilities, which is evident from the example of public-key cryptography. Firstly, they depend on the computational difficulty of some mathematical problems, e.g., discrete algorithms. This computationally difficult nature of security algorithms may not hold in future due to rapid enhancement in low power applications. Besides, it requires a key-management infrastructure that should be secured as well. This approach is, therefore, less attractive for many WSNs and ad hoc networks applications, because sensor nodes have limited computational capacity, while ad hoc networks are decentralised. Issues like privacy, authentication, and confidentiality are taken care of in the higher layers of the protocol stack by the use of key-based cryptographic systems (Stallings, 2013). Even though classical encryption schemes are applicable in the upper layers of the communication protocols, the physical layer can also be explored to improve wireless security, which is the primary concern of this chapter.

PLS schemes leverage unpredictable and random characteristics of wireless channels to achieve information-theoretic security. In PLS, confidential information is shared over a wireless medium in the presence of an adversary. A high degree of security can be achieved in wireless networks by utilizing the inherent characteristics of the physical layer (Mucchi, Nizzi, Pecorella, Fantacci, & Esposito, 2019) without relying on higher-layer encryption. The PLS techniques are usually quantified in terms of secrecy rate, complexity, mean square error (MSE), energy efficiency, signal-to-interference noise ratio (SINR), Channel State Information (CSI) requirements and bit disagreement (BDR). Commonly-utilised PLS techniques are: Algebraic Channel Decomposition Multiplexing (ACDM), Code Division Multiple Access (CDMA), Multiple Input Multiple Output (MIMO), Orthogonal Frequency Division Multiplexing (OFDM), and physical-layer key generation (PLKG) (Soni et al., 2016) as listed in Figure 7.3. Of these techniques, PLKG is focused on in this chapter.

7.4 Physical-layer Key Generation

PLKG is another security technique and is an unexplored field of research. PLKG is straightforward, as the keys are generated at the communicating nodes themselves, so a public-key infrastructure is not required. Applications like wireless sensor networks, smart medicine, etc., can utilise PLKG to achieve security. The idea of attaining security by the PLKG has been explored theoretically and experimentally (Ye et al., 2010;

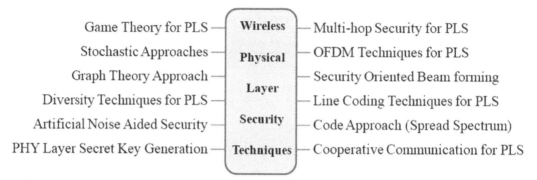

Game Theory for PLS — **Wireless** — Multi-hop Security for PLS

Stochastic Approaches — **Physical** — OFDM Techniques for PLS

Graph Theory Approach — **Layer** — Security Oriented Beam forming

Diversity Techniques for PLS — **Layer** — Line Coding Techniques for PLS

Artificial Noise Aided Security — **Security** — Code Approach (Spread Spectrum)

PHY Layer Secret Key Generation — **Techniques** — Cooperative Communication for PLS

FIGURE 7.3
Different wireless physical-layer security techniques.

Ren, Su, & Wang, 2011; Premnath et al., 2013; Chen, Jiang, & Zou, 2015; Saad, Mohamed, Elfouly, Khattab, & Guizani, 2015). A general review of key-generation principles, procedure, and performance metrics, with practical design guidelines, is presented in Zhang, Duong, Marshall, and Woods (2016), Zhang, He, Duong, and Woods (2017). Performance of wireless PLKG systems can be improved by implementing power-efficient key- management approaches along with preprocessing the input RSS values (Moara-Nkwe, Shi, Lee, & Eiza, 2018; Li et al., 2018). Discrete cosine transform (DCT) and discrete wavelet transform (DWT) preprocessing is applied over the RSS samples to improve system performance (Margelis et al., 2017; Zhan & Yao, 2017). Further, moving-window-based preprocessing is also applied to improve the BDR of the key-generation system in the presence of colored noise (Soni, Upadhyay, & Kumar, 2019a; 2019b). The primary requirement of the PLKG system is the generation of similar keys at both communicating ends with sufficient randomness. The key-generation system, in general, is considered a wiretap model with two legitimate communicating nodes, Alice and Bob, and an illegitimate node, Eve, over the wireless channel, as depicted in Figure 7.4. Alice and Bob communicate by encrypting messages using generated secret keys in this model. Eve, as a burglar, seeks to listen to the messages. It is assumed that all the nodes are single-antenna systems with the static channel, and exchange the probe signals within one coherence period. Eve is a passive adversary and is placed several wavelengths away from Alice and Bob, with an assumption that the channels h_{AE}, h_{BE} are uncorrelated respectively with h_{AB}, h_{BA}. Eve is not capable of mining any significant information regarding Alice and Bob's measurements.

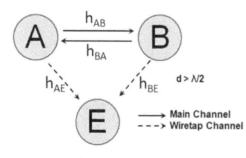

→ Main Channel
- - -> Wiretap Channel

FIGURE 7.4
Wiretap model with Alice and Bob as legitimate nodes, and Eve as an illegitimate node.

7.4.1 Wiretap Channel Model

Taking into account the Gaussian wiretap model (Wyner, 1975), let $x_A(t_1)$ and $x_B(t_2)$ be the pilot signals exchanged between Alice and Bob, $h_{AB}(t_1)$ and $h_{BA}(t_2)$ be the time-varying channel-impulse response between them at the time t_1 and t_2. Both nodes, along with the known probe signal and the received signals $y_A(t_1)$ and $y_B(t_2)$, estimate the channel. The signals received at Alice and Bob may be written as:

$$y_B(t_1) = h_{AB}(t_1) * x_A(t_1) + \eta_A(t_1)$$

$$y_A(t_2) = h_{BA}(t_1) * x_B(t_2) + \eta_B(t_2) \tag{7.1}$$

Where $\eta_A(t_1)$ and $\eta_B(t_2)$ are the identically-distributed and independent zero-mean Gaussian noise with variance σ_η^2, it is assumed that Eve can listen to the public channel and eavesdrop on signals between Alice and Bob given as:

$$y_E(t_1) = h_{AE}(t_1) * x_A(t_1) + \eta_E(t_1)$$

$$y_E(t_2) = h_{BE}(t_2) * x_B(t_2) + \eta_E(t_2) \tag{7.2}$$

Where $h_{AE}(t_1)$ and $h_{BE}(t_2)$ are the impulse response of Alice-Eve and Bob-Eve channel and $\eta_E(t_1)$ and $\eta_E(t_2)$ are the noise terms, due to channel reciprocity, a high correlation is maintained between the estimates of Alice-Bob channel made at Alice and Bob. The estimates of the channel between the Alice-Eve and Bob-Eve are highly uncorrelated due to the assumption that Eve is positioned much further away than $\lambda/2$ distance from Alice and Bob. Therefore, regardless of having the probe signals $x_A(t_1)$ and $x_B(t_2)$, Eve is incapable of working out any significant estimates of the Alice-Bob channel from its received signals.

7.4.2 Principles of Key Generation

Data packets transmitting over wireless channels are affected by various physical phenomena like reflection and scattering. Channel variations are also affected by mobility. Such physical phenomena and mobility produce random fluctuations in the RSS. Fortunately, wireless key generation exploits these random fluctuations for The generated keys must be the same with sufficient randomness. PLKG utilises the following principles for key generation:

7.4.2.1 Temporal Variation

This principle accounts for the variations arising due to the reflection or scattering of received signals from the movement of communicating nodes or other objects in the surroundings. These temporal variations introduce sufficient randomness in the channel parameters, which is necessary for key generation (Liu, Wang, Yang, & Chen, 2013). However, these temporal variations introduce disagreement of keys between both ends. Hence, to maintain agreement between the keys, samples are taken within one coherence time. However, randomness is limited in the static environment. It is specified in the form of the autocorrelation function (ACF). If μ is the mean, σ is the standard deviation of a random variable X_t, then the ACF is represented as:

$$R(\tau) = \frac{E[(X_t - \mu)(X_{t+\tau} - \mu)]}{\sigma^2} \qquad (7.3)$$

7.4.2.2 Channel Reciprocity

This principle is very important and represents the similar nature of channels on both communicating links at same carrier, i.e., $h_{AB} = h_{BA}$. This is because of the same propagation path from both directions of the channel. Channel reciprocity facilitates identical key generation at both ends. As the mode of communication is half-duplex for almost all the practical radio modules, it results in non-simultaneous measurement at both ends, which causes unevenness in the channel measurement and discrepancies in the generated keys (Chen et al., 2011). A possible solution to this issue is to apply suitable signal processing like DCT, DWT or MWA. Channel reciprocity can be quantified in terms of the cross-correlation of collected signals at legitimate pair of nodes. If μ is the mean, σ is the standard deviation of a random variable X_t; thus, the cross-correlation is represented as:

$$\rho_{AB} = \frac{E[AB] - E[A]E[B]}{\sigma_A \sigma_B} \qquad (7.4)$$

7.4.2.3 Spatial Decorrelation

Fading caused by a transmitter is independent and different for receivers at diverse locations. As per the theory of communication, any entity placed farther than half a wavelength from any of the communication nodes experiences uncorrelated multipath fading. This is an essential attribute for security purposes, as it is assumed that Eve is situated farther than one-half wavelength from legitimate users in real-time systems and is experimentally verified (Liu et al., 2013). Spatial decorrelation states, $h_{AB} \neq h_{AE}$ and $h_{BA} \neq h_{BE}$. It is clear in the literature that the signals perceived by Eve are not correlated to signals of authorised users. Spatial decorrelation can be quantified by finding cross-correlation among the signals of Alice, Bob and Eve.

7.4.3 Performance Metrics

Effectiveness of generated keys depends on two aspects; firstly, the agreement between the keys generated at both ends and secondly, their randomness. Different performance matrices are used to evaluate different aspects of the key-generation system. Some of the metrics are based on information theory like mutual information, while others may be categorised under statistical metrics, which includes tests like frequency test, cumulative sum test and mono bit test, which reflects randomness. Different randomness tests such as NIST, Diehard (Göhring & Schmitz, 2015) are available to access key randomness, while the concept of bit-disagreement rate is used to measure the similarity of keys. BDR and NIST randomness test are discussed in this chapter.

7.4.3.1 Bit Disagreement Rate (BDR)

Bit-disagreement rate depicts the strength of the PLKG. It is the ratio of the number of bits that do not agree between the legitimate users to the total number of bits.

It is calculated after quantization and can also be represented in terms of the key-disagreement rate (KDR) which is measured after privacy amplification and is given as

$$KDR = \frac{\sum_{i=1}^{N_k} | K^A(i) - K^B(i) |}{N_k} \tag{7.5}$$

where K^A and K^B are the keys generated at Alice and Bob, respectively. This disagreement is corrected in the information reconciliation stage. BDR can be decreased by improving the signal cross-correlation and using signal-preprocessing techniques.

7.4.3.2 Key Randomness

Randomness is the most important feature for the key sequences, which should be distributed uniformly. A less random key will result in a smaller search space for brute force attackers, thus compromising the security of the cryptographic system. There are various tools for randomness checks, but we used the NIST suite for this purpose. The NIST test suit is a group of 15 statistical tests, among which seven are used commonly to test the randomness in generated keys (Rukhin et al., 2001). These are Frequency Test, Block Frequency Test, Approximate Entropy Test, Serial Test, Runs Test, Longest Runs Test, and Cumulative Sums Test. The frequency test confirms that the bit stream is uniformly distributed (Ambekar, Hassan, & Schotten, 2012), i.e., having approximately the same number of zeros and ones for a bit stream to be truly random. The serial test certifies the uniform distribution of the number of occurrences of two-bit patterns in a bit sequence. The testing procedure and test report are discussed later in this chapter.

7.4.4 Key Generation Procedure

In general, a secret-key-generation scheme mainly consists of four stages: channel probing; quantization; information reconciliation; and privacy amplification, as shown in Figure 7.5. In this section the algorithm followed for secure key generation is discussed along with the functioning of constituent blocks.

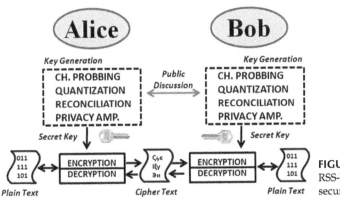

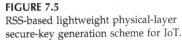

FIGURE 7.5
RSS-based lightweight physical-layer secure-key generation scheme for IoT.

Algorithm 1   RSSI based PHY Layer Key Generation.

INPUT: Probed RSSI values R_x

$\quad\quad (x = a(Alice), b(Bob))$

Step 1: $\hat{h}_x \leftarrow R_x$ // Channel Estimation

Step 2: $\hat{h}_x^{MWA} \leftarrow \hat{h}_x$

$\quad\quad\quad\quad$ **for** $(z = 1, z < x - m, z = z + 1)$

$\quad\quad\quad\quad\quad\quad \hat{h}_x^{MWA} = \frac{1}{m} \cdot \Sigma_{q=0}^{m-1} \hat{h}_{x-q}$

$\quad\quad\quad\quad$ **end**   // Moving Window Averaging

Step 3: $[\hat{H}_x] \leftarrow \hat{h}_x^{MWA}$ // Constructing Input Vector

Step 4: $[\hat{Y}_x]: \{0, 1\} \leftarrow Q([\hat{H}_x])$ // Quantization// Q = { Linear-I L-I,// Linear-II L-II,// Lloyd Max LMQ }

Step 5: $K_a: \{0, 1\} \leftarrow lbc([\hat{Y}_x]: \{0, 1\})$ // Information Reconciliation

Step 6: $K_{Alice}^{160}: \{0, 1\}^{L=160} \leftarrow SHA(K_a: \{0, 1\})$ // Privacy Amplification

OUTPUT: $\{ K_{Alice}^{160} \}$ // 160 Bit Cryptographic Key

7.4.4.1 *Channel Probing*

This is the first step of key generation in which RSSI samples are collected along with the effect of channel on the probed signals. These probed signals become the foundation of key generation. There are various ways to probe signals and extract the inherent randomness from them. It is extracted from parameters such as AoA, RSSI, distance and CSI. In the case of CSI, it is extracted in the form of channel amplitude, phase, and frequency response. In this chapter, RSSI from the received packets is considered to extract channel randomness for generating secret keys. RSSI is acquired from the network interface card (NIC). It is not the actual power transmitted from an access point. It is a relative quantity. The energy in the preamble of the received beacons gives the RSSI values. Each of the received beacons gives a single RSSI value. The calculated RSSI value and physical parameters do not possess a constant relationship, i.e., RSSI ranging from 0–100 dBm on one hardware can be mapped on another hardware for a range of 0–255 dBm. Table 7.1 shows various wireless standards and related test beds for RSSI extraction.

TABLE 7.1

RSSI-Extracting Test Beds for Various Wireless Standards (Zhang et al., 2016)

Standard	Testbed
IEEE 802.11	All of NIC's
IEEE 802.15.4	MICAz, TelosB
Bluetooth	Smart Phones/Blue Tooth Dongle
LTE	Smart Phones

The time difference between the sampling of signals should be such that a sufficient number of samples will be recorded within the channel coherence time. System performance, like the BDR of the PLKG system, can be enhanced by supporting the channel-probing phase with suitable preprocessing techniques. Signal preprocessing, like the DCT (Margelis et al., 2017), DWT (Zhan & Yao, 2017), and moving averaging (Soni et al., 2019b), should be applied before quantization. It is essential to make key generation appropriate for resource-constrained networks like WSN or IoT. The foremost benefit of incorporating these preprocessing techniques is to dispose of high variations in RSS patterns, which lowers BDR. This lowering in BDR is required in key-generation systems to match the keys at both ends. In this chapter, MWA averaging is applied as a preprocessing technique. If X_n represents the nth RSS sample, then the nth MWA for a window size of m is given by:

$$MWA_n = \frac{X_n + X_{n-1} + \ldots + X_{n-(m-1)}}{m}$$

$$= \frac{1}{m} \cdot \sum_{a=0}^{m-1} X_{m-a} \tag{7.6}$$

MWA-based preprocessing is considered here in order to discuss its effect on scattering and standard deviation of the collected RSSI sequences. It is expected that the MWA is significantly effective at lower SNR range, which makes PLKG suitable at low SNR environments like WSN or IoT. MWA not only improves BDR, but also makes it suitable for power-constrained applications.

Figure 7.6 shows the effect of applying MWA with different values of window size (m = 1, 10, 20, 30) for various SNR (24dB (bottom row), 16dB (middle row), 8dB (top row)) on the scattering of RSSI values probed at Alice and Bob. From the plot it is apparent that the raw RSSI values logged at both ends have very large variations. It is shown in the scatterplot for m = 1. Variations in the scatter plots can be minimised by using the MWA preprocessing. This reduced scattering will facilitate the bit agreement between the two communicating ends.

The reduction in the standard deviation of the collected RSSI sequences as a function of window sizes (m = 1, 2, 5, 10, 15, 20, 25) at different SNR (24 dB, 16 dB, 8 dB) is shown in Figure 7.7. IoT environments are noisy and are characterised by an SNR ranging from 16 dB to 8 dB. It is clear from Figure 7.7 that standard deviation is improved significantly at lower SNR. This reduction in the standard deviation increases the correlation among the measured RSSI sequence, which in due course reduces the BDR and contributes to overall performance improvement.

7.4.4.2 Quantization

In this step, continuous or discrete-valued RSS samples are converted into binary bits. A wide range of quantization schemes is available which can be useful for key generation (Graur, Islam, & Henkel, 2016; Guillaume, Zenger, Mueller, Paar, & Czylwik, 2014; Sheng, Xu-Jian, & Li-Wei, 2007a). One such classification is linear or non-linear quantization. The first of three quantization schemes in this classification is (A) Linear Simple (Linear-I); in this scheme, the entire range of the acquired RSS sequence is divided into 2^n number of quantization levels with equal step size, where n represents the number of quantization bits. In scheme (B) Linear Based on Mean and Variance (Linear-II), the decision thresholds and quantization levels are determined on the bases

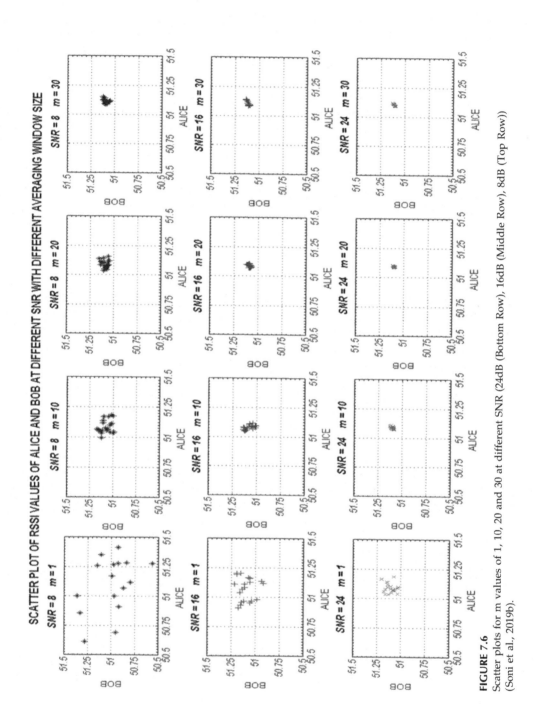

FIGURE 7.6
Scatter plots for m values of 1, 10, 20 and 30 at different SNR (24dB (Bottom Row), 16dB (Middle Row), 8dB (Top Row)) (Soni et al., 2019b).

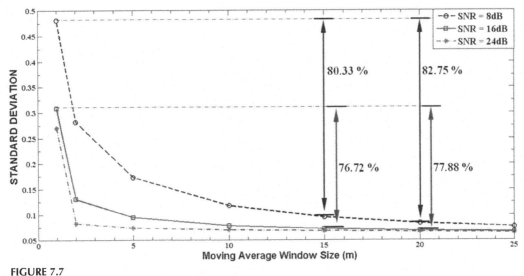

FIGURE 7.7
Standard deviation improvement at various SNR (Soni et al., 2019b).

of mean and variance of the input signal (Ambekar et al., 2012) In scheme (C) Non-Linear Lloyd Max Quantization, the parameters, like decision thresholds and quantization levels, are given by the Lloyd Max algorithm, which is based on the probability density function of the RSS samples (Sheng, Xu-Jian, & Li-Wei, 2007b). It gives the quantization interval and quantization level by iteratively solving the equations alternately such that the quantization error is minimal.

The BDR is calculated after quantization, and it should be low so that a low-complexity error-correcting code can be utilised in the information-reconciliation block. Further, it is interesting to compare the effect of the above-mentioned quantization schemes on the BDR performance of the key-generation system, as shown in Figure 7.8. It is observed that the non-linear adaptive quantization scheme, i.e., the Llyod max Quantization, is the most efficient as compared to the linear quantization in the desired SNR range.

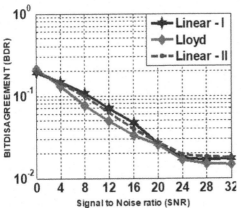

FIGURE 7.8
BDR performance with different quantization schemes.

7.4.4.3 Information Reconciliation

Reconciliation theoretically means to make two sets of data agree with each other. The binary bit sequences generated at Alice and Bob constitute some degree of bit disagreement, particularly at lower SNR values, due to various reasons such as interference, noise, and hardware constraints. Information reconciliation in key generation is applied to improve further bit disagreement in the binary bits generated by the quantizer from pre-processed RSS signals. In information reconciliation, binary bits generated at both Alice and Bob are further treated with a set of pre-defined rules at both nodes with some minimum mutual information exchange to reduce bit disagreement (Etesami & Henkel, 2012). There are various error-correcting codes in the literature, like Turbo code, Golay code, BCH code, low-density parity-check (LDPC), etc., which are applicable for the information reconciliation block of key generation. In this chapter linear block coding is used in the information reconciliation block.

7.4.4.4 Privacy Amplification

Sufficient randomness, reproducibility, and uniform distribution are the desired attributes in the generated key sequences. Since in the information-reconciliation block some mutual information is exchanged publicly in the wireless channel, the eavesdropper can make use of this information to estimate secure keys. To attain sufficient randomness and to avoid possible key prediction, privacy amplification is used. This can be performed by extraction of a universal hashing function (e.g. SHA-1, SHA-2, SHA-3, etc.) (Dang, 2009). Secure Hashing Algorithm (SHA-1) is used in this chapter to a generate 160-bit key. Enough randomness is guaranteed in the generated keys by testing the randomness by NIST statistical test suit.

We generated the keys from the simulation model of the key-generation system with MWA-based preprocessing incorporated in the channel-probing phase, followed by the Lloyd max quantizer. The BDR is calculated at different SNR after the quantization. The linear-block coding reconciles the quantized RSSI samples and sufficient randomness is introduced by secure Hash coding. Figure 7.9 shows the improved BDR performance of the system on applying moving-window averaging. Furthermore, the NIST test suit verifies key randomness, and the keys are found sufficiently random as the p-values for all tests which are far greater than the threshold commonly considered 0.01, as shown in Table 7.2

7.5 Applications and Future Scope

RSS-based PLKG, which exploits inherent channel randomness, is a promising field in wireless node security. Due to easy acquisition of RSS, it can easily be deployed over the network of emerging technologies like the Internet of Things, where mass numbers of wireless nodes are interconnecting day by day. PLKG is comparatively less computationally-complex than existing cryptographic techniques and may be deployed for resource-constrained devices. Some of the major application areas include smart sensors, wireless sensor networks, programmable and wearable medical devices (Chakraborty, Gupta, & Ghosh, 2013; Banerjee, Chakraborty, & Paul, 2019), smart agriculture and metering. Although

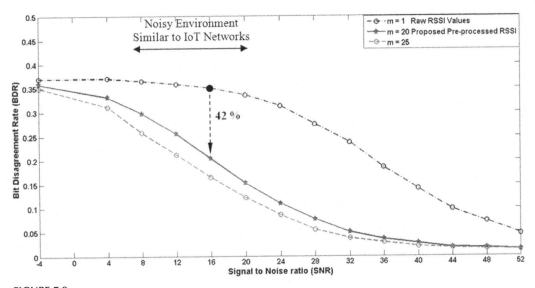

FIGURE 7.9
Effect of window size on BDR performance (Soni et al., 2019b).

TABLE 7.2

Report of Randomness Testing

S. No.	Test	P-Value	
		Alice	Bob
1	Frequency	0.5846	0.3241
2	Block Frequency	0.5229	0.9185
3	Cumulative Sums	0. 1435	0.3241
4	Runs	0.1263	0.0271
5	Longest Runs	0.1626	0.3972
6	Approximate Entropy	0.5680	0.2327
7	Serial	0.2071	0.5009

RSS-based PLKG has a wide range of applications, it has some unresolved challenges to overcome. Some of the open fields for future research are discussed as follows:

- *Active and Passive Attacks*: Key generation is susceptible to active as well as passive attacks by adversaries. Most of the research in this field covers only passive adversaries, which can only hear communication between a legitimate pair of nodes. Active adversaries, where the Eve can transmit signals to Alice and Bob, still needs to be addressed.

- *Multi-User Key Generation*: Most of the work in this domain is based on the wiretap-channel model of Wyner where a legitimate pair of nodes is considered for secure key generation along with an intruder, although group key generation is common in a practical scenario. Group Key generation is extensively applicable in ad-hoc networks and needs attention.

- *RSS Preprocessing*: BDR performance of secret key-generation systems will be improved by preprocessing the RSS signals before quantization. RSS acquired at the Alice and Bob is a time series data. Many signal preprocessing techniques, like discrete cosine transform (DCT), discrete wavelet transform (DWT), differential averaging and moving-window filtering can be applied to reduce the mismatch in the RSS pattern at Alice and Bob, which further improves the BDR performance of the system.
- The effects of parameters like channel fading, environmental noise, electronic noise and correlated colored noise components on the performance of key-generation systems may also be analyzed and is an open perspective.

A number of unresolved issues, like implementation in 5G networks, cross-layer design issues, joint optimization of parameters like security, throughput and reliability, millimeter waves and MIMO for physical layer security, among others, are also less explored in the literature and have an open perspective for research.

7.6 Conclusion

PLS is one of the promising solutions for securing power-constrained networks like WSN or IoT, being less complex than traditional cryptographic techniques. In this chapter, we presented an overview of the physical-layer key-generation system for achieving security in power constraint networks. Multiple-channel indicators like RSSI, CSI and AoA are available for PLKG, due to ease of acquisition of RSSI. PLKG based on RSS samples seems to be the most appropriate choice for networks like IoT. PLKG involves channel probing followed by quantization and information-reconciliation steps. The last step is privacy amplification. The key sequences generated at Alice and Bob should be similar and possess sufficient randomness. To ensure these attributes, key disagreement is quantified in terms of the BDR and the randomness is ensured by the NIST test suit. This chapter attempts to further improve the PLKG by improving the BDR performance, along with making a system suitable for low-power applications. MWA isone of the low-complexity preprocessing techniques applied in the PLKG. Results show that with proper window size for MWA, substantial improvement in BDR is achieved in the desired range of SNR for low-power applications. Moreover, other constituent blocks of the system, like quantization and channel reconciliation, can also be modified to achieve a better response. In this way, PLKG can be considered a lightweight security technique for power-constrained networks.

Acknowledgment

This research work is supported by the Visvesvaraya Ph.D. Scheme for Electronics and IT, Government of India, sanctioned to Devi Ahilya University under grant number PhD-MLA-4(37)/2015-16, Dated 11-09-2015.

References

Aldaghri, N., & Mahdavifar, H. (2020). Physical layer secret key generation in static environments. *IEEE Transactions on Information Forensics and Security*, 15, 2692–2705.

Ambekar, A., Hassan, M., & Schotten, H. D. (2012). Improving channel reciprocity for effective key management systems. In *International Symposium on Signals, Systems, and Electronics (ISSSE)*, IEEE. Potsdam, Germany.

Badawy, A., Elfouly, T., Khattab, T., Mohamad, A., & Guizani, M. (2016). Unleashing the secure potential of the wireless physical layer: Secret key generation methods. *Physical Communication*, 19, 1–10

Banerjee, S., Chakraborty, C., & Paul, S. (2019). *Programming Paradigm and Internet of Things* (pp. 148–164). CRC: A Handbook of Internet of Things and Big Data. ISBN 9781138584204.

Chakraborty, C., Gupta, B., & Ghosh, S. K. (2013). A review on telemedicine-based WBAN framework for patient monitiring. *Mary Ann Libert Inc.: International Journal of Telemedicine and e-Health*, 19(8), 619–626. ISSN: 1530-5627, 10.1089/tmj.2012.0215

Chen, C., Member, S., & Jensen, M. A. (2011). Secret key establishment using temporally and spatially correlated wireless channel coefficients. *IEEE Transactions on Mobile Computing*, 10, 205–215.

Chen, M., Jiang, T., & Zou, W. (2015). Differential physical layer secret key generation based on weighted exponential moving average. In *2015 9th International Conference on Signal Processing and Communication Systems (ICSPCS)*, Cairns, QLD (pp. 1–4).

Dang, Q. (2009). NIST SP 800-106: 'Randomized hashing for digital signatures', February 2009.

Etesami, J., & Henkel, W. (2012). LDPC code construction for wireless physical-layer key reconciliation. In *IEEE International Conference on Communications in China (ICCC)*, Beijing, China (pp. 208–213).

Göhring, M., & Schmitz, R. (2015). On randomness testing in physical layer key agreement. In *2015 IEEE 2nd World Forum on Internet of Things (WF-IoT)*, Milan (pp. 733–738). doi: 10.1109/WF-IoT.2015.7389145NIST.

Graur, O., Islam, N., & Henkel, W. (2016). Quantization for physical layer security. *IEEE Globecom Workshops*, 1–7. doi: 10.1109/GLOCOMW.2016.7849013.

Guillaume, R., Zenger, C., Mueller, A., Paar, C., & Czylwik, A. (2014). Fair comparison and evaluation of quantization schemes for PHY-based key generation, OFDM 2014. In *Proceedings of the 19th International OFDM Work. 2014 (InOWo'14)* (pp. 1–5).

Li, G., Hu, A., Zhang, J., Peng, L., Sun, C., & Cao, D. (2018, July). High-agreement uncorrelated secret key generation based on principal component analysis preprocessing. *IEEE Transactions on Communications*, 66(7), 3022–3034.

Liu, H. Wang, Y., Yang, J., & Chen, Y. (2013). Fast and practical secret key extraction by exploiting channel response. *Proceedings of the 32nd IEEE International Conference on Computer Communications (INFOCOM)*, Turin, Italy, Apr. 2013, 3048–3056.

Margelis, G., Fafoutis, X., Oikonomou, G., Piechocki, R., Tryfonas, T., & Thomas, P. (2017). Physical layer secret-key generation with discreet cosine transform for the Internet of Things. In *2017 IEEE International Conference on Communications (ICC)*, Paris (pp. 1–6).

Mazin, A., Davaslioglu, K., & Gitlin, R. D. (2017). Secure key management for 5G physical layer security. In *2017 IEEE 18th Wireless and Microwave Technology Conference (WAMICON)*, Cocoa Beach, FL (pp. 1–5). doi: 10.1109/WAMICON.2017.7930246.

Menezes, A. J., van Oorschot, P. C., & Vanstone, S. A. (1996). *Handbook of Applied Cryptography*. Boca Raton, FL, USA: CRC Press.

Moara-Nkwe, K., Shi, Q., Lee, G. M., & Eiza, M. H. (2018). A novel physical layer secure key generation and refreshment scheme for wireless sensor networks. *IEEE Access*, 6, 11374–11387. doi: 10.1109/ACCESS.2018.2806423.

Mucchi, L., Nizzi, F., Pecorella, T., Fantacci, R., & Esposito, F. (2019). Benefits of physical layer security to cryptography: Tradeoff and applications. In *2019 IEEE International Black Sea*

Conference on Communications and Networking (BlackSeaCom), Sochi, Russia (pp. 1–3). doi: 10. 1109/BlackSeaCom.2019.8812778.

Premnath, S. N., Jana, S., Croft, J., Gowda, P. L., Clark, M., Kasera, S. K. ... Krishnamurthy, S. V. (2013, May). Secret key extraction from wireless signal strength in real environment. *IEEE Transactions on Mobile Computing, 12*(5), 917–930. doi: 10.1109/TMC.2012.63.

Ren, K., Su, H., & Wang, Q. (2011, August). Secret key generation exploiting channel characteristics in wireless communications. In *IEEE Wireless Communications, 18*(4), 6–12.

Rukhin, A., Soto, J., Nechvatal, J., Smid, M., Barker, E., Leigh, S. ... Vo, S. (2010). A statistical test suite for random and pseudorandom number generators for cryptographic applications. *National Institute of Standards and Technology*, Gaithersburg, MD, USA, Tech. Rep. Special Publication 800-22 Revision 1a, Apr. 2010.

Saad, A., Mohamed, A., Elfouly, T. M., Khattab, T., & Guizani, M. (2015). Comparative simulation for physical-layer key generation methods. In *2015 International Wireless Communications and Mobile Computing Conference (IWCMC)*, Dubrovnik (pp. 120–125). doi: 10.1109/IWCMC.2015. 7289068.

Sheng, F., Xu-Jian, L., & Li-Wei, Z. (2007a). A Lloyd-Max-based non-uniform quantization scheme for distributed video coding. *ACIS- SNPD, 1*, 848–853.

Sheng, F., Jian, L. X., & Wei, Z. L. (2007b). A Lloyd-Max-based non-uniform quantization scheme for distributed video coding. In *International Conference on Software Engineering, Artificial Intelligence, Networking and Parallel/Distributed Computing, IEEE Computer Society*, Qingdao, China (p. 848).

Soni, A., Upadhyay, R., & Jain, A. (2016). Internet of things and wireless physical layer security: A survey. *Proceedings of the International Conference on Computer and Communication Technology (IC3T 2016)*, Vijaywada, India. doi: https://doi.org/10.1007/978–981-10-3226-4_11.

Soni, A., Upadhyay, R., & Kumar, A. (2019a). Performance improvement of wireless secret key generation with colored noise for IoT. *International Journal of Communication Systems, 32*(16). doi: //doi.org/10.1002/dac.4124.

Soni, A., Upadhyay, R., & Kumar, A. (2019b). Wireless physical-layer key generation with improved bit disagreement for the Internet of Things using moving window averaging. *Elsevier: Physical Communication, 33*, 249–258.

Stallings, W. (2013). *Cryptography and network security: Principles and practice* (6th ed.). Englewood Cliffs, NJ, USA: Prentice-Hall.

Wang, D., Bai, B., Lei, K., Zhao, W., Yang, Y., & Han, Z. (2019). Enhancing information security via physical layer approaches in heterogeneous IoT with multiple access mobile edge computing in smart city. *IEEE Access, 7*, 54508–54521. doi: 10.1109/ACCESS.2019.2913438.

Wyner, A. (1975). The wire-tap channel. *Bell System Technical Journal, 54*, 339–348.

Ye, C., Mathur, S., Reznik, A., Shah, Y., Trappe, W., & Mandayam, N. B. (2010, June). Information-theoretically secret key generation for fading wireless channels. In *IEEE Transactions on Information Forensics and Security, 5*(2), 240–254.

Zhan, F. I., & Yao, N. (2017). On the using of discrete wavelet transform for physica- layer key generation. *Elsevier: Ad Hoc Networks, 64*, 22–31. doi: https://doi.org/10.1016/j.adhoc.2017.06.003.

Zhang, J., Duong, T. Q., Marshall, A., & Woods, R. (2016). Key generation from wireless channels: A review. *IEEE Access, 4*, 614–626. doi: 10.1109/ACCESS.2016.2521718.

Zhang, J., He, B., Duong, T. Q., & Woods, R. (2017, April). On the key generation from correlated wireless channels. *IEEE Communications Letters, 21*(4), 961–964. doi: 10.1109/LCOMM.2017. 2649496.

Zhang, Y., Zhao, H., Xiang, Y., Huang, X., & Chen, X. (2019). A key agreement scheme for smart homes using the secret mismatch problem. *IEEE Internet of Things Journal, 6*, 10251–10260. doi: 10.1109/JIOT.2019.2936884.

8

Threat and Attack Models in IoT Devices

S.T. Naitik, J.V. Gorabal, and K. Chatrapathy
Sahyadri College of Engineering & Management,
India

8.1 Need for Security in IoT Devices

Internet of Things (IoT) is the latest application of Wireless Sensor Networks (WSN); it deals with communication across smart things like devices, hosts, sensors, actuators, etc. In IoT energy harvesting is crucial; remote control and communication among devices are challenging and therefore security among devices participating in communication is essential. To facilitate seamless communication, security is one of the most critical parameters to be implemented and optimised in IoT architecture in Meneghello, Calore, Zucchetto, Polese, and Zanella (2019).

It is predicted that 50 billion IoT devices will be connected by 2021; recently the use of IoT is booming in industrial applications. Due to improper design mechanisms in IoT architecture, protocols, and issues of interoperability, security and privacy flaws can still be explored while using any IoT commercial products. We can observe and analyze risk assessment of services accessing private and sensitive information from video recording in a restrictive environment, patient health monitoring, building access control systems and real-time personal localization issues.

Resiliency in the design of IoT devices is the need of the hour today. Heterogeneity in IoT devices makes security-based design methods difficult to implement. As a result, we have seen cyberattacks even at the hardware level of the device. Limitations in the architecture and operational environments like energy, communication, computation, and storage capacities of IoT devices leads to issues in implementing advanced security mechanisms to protect from cyberattacks.

The real security challenge we face arises when we connect devices to global network like the Internet. Manufacturers are not focused on integrating security mechanisms deeply and completely. Due to an insufficient timeline for the production of IoT devices, vendors have failed to focus on the security and privacy of devices. The majority of cyberattacks occur due to ignorance of the end-user. We are interested in looking for a device's operation or functionality rather than following the instruction manual where they instruct to change the default password. Hence our devices will be prone to cyber espionage to control the entire application remotely by hackers.

8.2 IoT Architecture

Present-generation technologies like IoT are event-driven by sensors and actions initiated by actuators. During the operation of devices connected smartly to sense intelligence in the environment, we can observe many flaws that occur and propagate, along with its architecture which makes protocols less efficient. Flaws are present from the edge layer to the data-analytics layer. Unforeseen scenarios like buggy apps, device/communication failures, bad interactions, and stagnate synchronization lead to unsafe physical states. The mitigation of cyber-attacks on IoT devices needs special skills and thorough analysis of flaws in component devices, interactions, and configuration mechanisms (Song, Nguyen, Qian, & Krishnamurthy, 2018). Nowadays there are various smart things available in the market like Amazon Alexa, Google home mini, Google home, smart watches, etc. These devices are under the control of hackers when accessed remotely. The real-world IoT deployment is still young because developers secure cyberspace rather than physical space (Song et al., 2018).

Misconfiguration is a typical reason for security infringement. When introducing a smart application, a client needs to arrange the application Sensor(s) and actuator(s) with the IoT system, which can be transformed to unsafe physical states by powerless arrangements. For such misconfiguration, there are several periodic explanations: (i) the application's portrayal is hazy; (ii) there is an excessive number of arrangement choices; and (iii) typical clients frequently don't have sufficient information to comprehend the practices of brilliant gadgets. For example, Figure 8.1 shows the customer's necessary information sources that include a temperature determination sensor (lines 2–4), and the electrical plugs that the warmer sensor is attached to. Even though the engineers use the word or and the application just from the understudy assumed that the application controls both a radiator and an AC to maintain the optimum temperature and misarranged the application to control both the AC outlet and the radiator outlet. To intensify the disarray, the application anticipates the design of outlets (capabilityswitch) of the real gadgets that are connected to the outlets Because of misconfiguration, when the temperature is higher than a predefined limit, the Virtual Indoor regulator would turn on both the arranged outlets (i.e. both the radiator and the AC). This disregards the accompanying two rational properties: (I) a radiator is turned on when the temperature is over a predefined limit and (ii) an AC and a radiator are both turned on as in Song et al. (2018) (Figure 8.2).

In this work, we will probably identify IoT system security problems (i.e. weaknesses) that are exploitable by aggressors to advance the system into awful physical states or release delicate data. Instances of faulty physical states are that when no one is at home, (i) the front entrance is opened, and (ii) when the temperature is below a predefined maximum, a radiator is destroyed. Concerning data spillage, we require that: (i) private

FIGURE 8.1
Series of events in IoT system.

```
1 preferences {
2    section("Choose a temperature sensor... "){
3       input "sensor", "capability.temperatureMeasurement", title:
         "Sensor"
4    }
5    section("Select the heater or air conditioner outlet(s)... "){
6       input "outlets", "capability.switch", title: "Outlets",
         multiple: true
7    }
8    section("Set the desired temperature..."){
9       input "setpoint", "decimal", title: "Set Temp"
10   }
11   section("When there's been movement from (optional)"){
12      input "motion", "capability.motionSensor", title: "Motion",
         required: false
13   }
14   section("Within this number of minutes..."){
15      input "minutes", "number", title: "Minutes", required: false
16   }
17   section("But never go below (or above if A/C) this value with
         or without motion..."){
18      input "emergencySetpoint", "decimal", title: "Emer Temp",
         required: false
19   }
20   section("Select 'heat' for a heater and 'cool' for an air
         conditioner..."){
21      input "mode", "enum", title: "Heating or cooling?", options:
         ["heat","cool"]
22   }
23 }
```

FIGURE 8.2
Configuration of virtual thermostat app using user inputs.

data is transmitted via only message interfaces (e.g., in SmartThings, send an SMS message and PushMessage), not system interfaces (e.g., HTTP post in SmartThings); and (ii) the beneficiaries of techniques for sending messages to coordinate the arranged telephone numbers or contacts. In such circumstances, we agree that customers guide whether to permit or prohibit such operations because of their security inclinations (Song et al., 2018).

All gadgets (network center, sensors, and actuators), the cloud, as well as our trusted registration base (TCB) friend application, and try not to consider programming assaults against them. Be that as it may, IoTSan mitigates physical attacks that can infuse event(s) into the system (e.g., by really expanding the temperature or staggering the sensors) or disappointments (e.g., by sticking) with the malevolently actuated gadget or correspondence. From IoTSan in Atlaeenm and Wills (2020) looks to recognise and forestall such

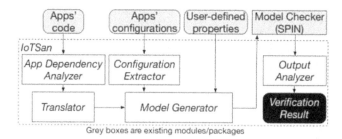

FIGURE 8.3
IoT SAN architecture.

occasions from driving the framework into wellbeing infringement. Be that as it may, directed answers for those assaults (e.g., the infringement of sensor data and its usage out of application are out-of-scope) (Figure 8.3).

In the Gregorian calendar month 2014, the IoT design consortium discharged the related model of reference for vulnerability. This model serves as a standard system to help speed up IoT deployments for trade. This reference model is intended to combine and inspire IoT preparation models to collaborate and evolve.

It is built to be seven layers to provide additional details for each layer to create a common nomenclature, as shown in Figure 8.4. It additionally wherever varied types of process device It worked through The model has entirely distinct layers. This model also helps many manufacturers to supply compatible IoT products, converting the IoT from an abstract model into a functional and usable unit. The physical layer is Layer One. It requires physical devices and controllers for the management of different objects. Within the IoT, these artifacts represent items that require different types of devices that send and receive info; for example, sensors that collect info concerning the encompassing surroundings. The communications layer is used to link completely different IoT objects together with devices such as switches, gateways, routers, and firewalls for interconnection manipulation. Layer three is about computing at the core devices. The process components operate with a large amount of information at this layer and will conduct any transformation of data to reduce information dimensions. Layer four is data aggregation. This layer is concerned with storing the returns of data from completely different IoT devices. This layer is used to interconnect entirely different IoT objects with devices such as switches, gateways, routers, and firewalls for interconnection manipulation. Edge computing filters this layer of information and processes it. Edge computing is at layer three. The data returned from the property layer is taken by this layer and translated into storage data and improved leveling processes (Chanda, Das, Banerjee, & Chakraborty, 2021). The process components operate with a large amount of information at this layer. To reduce the dimensions of information, it will perform some data transformation. The data aggregation layer addresses the storage of returns of data from completely different IoT devices. Edge computing filters and processes this data in Atlaeenm and Wills (2020).

From the edge-computing layer for storing data in different formats from heterogeneous processors, various types of information might come back. During this process, the aggregates and formats of the knowledge-abstraction layer keep information that builds via many manageable and productive methods; they are accessible by applications. The data layer discusses the data interpretation of different IoT apps. Some applications that use IoT devices or machines for management are protected by this layer. It also

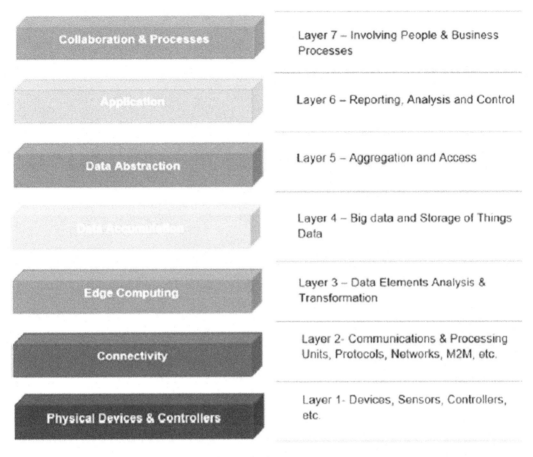

Collaboration & Processes	Layer 7 – Involving People & Business Processes
Application	Layer 6 – Reporting, Analysis and Control
Data Abstraction	Layer 5 – Aggregation and Access
Data Accumulation	Layer 4 – Big data and Storage of Things Data
Edge Computing	Layer 3 – Data Elements Analysis & Transformation
Connectivity	Layer 2- Communications & Processing Units, Protocols, Networks, M2M, etc.
Physical Devices & Controllers	Layer 1- Devices, Sensors, Controllers, etc.

FIGURE 8.4
IoT reference model.

includes various applications for the exchange of information and data management over the internet.

IoT system security is assessed with classical measures for defense and risk analysis. Within the IoT scheme, typical specifications for CIA (Confidentiality, Honesty, and Availability) protection should be used. Confidentiality means that it is necessary to protect the exchange of messages between a sender and a recipient from any malicious or unauthenticated user. Confidentiality must not only be warranted within the communication network for the IoT system, but also jointly once transmission messages are exchanged between different smart things.

Integrity is used to ensure that message content that is changed by the accepted recipient of the appropriate sender is secure from any exploitation by an attacker's manipulation, while not the user can track this exploitation. Availability is used to ensure that it is difficult for a malicious user to disrupt or harmfully affect the communication or service quality enabled by smart devices or the network of devices.

The CIA is vital for smart applications. For each IoT design stage, alternative security criteria need to be implemented, as shown in Figure 8.5. Node authentication, which prevents unauthenticated access to nodes and protects lines between smart nodes from breaches, is the primary security problem for the physical layer. The cryptologic rule

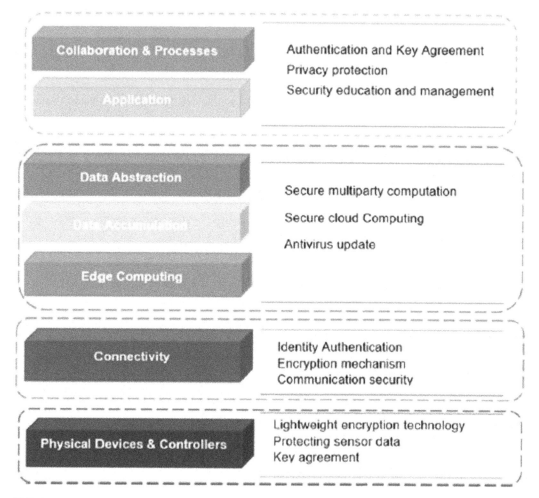

FIGURE 8.5
Requirements of Security at each layer.

and protocol are crucial facets to code-transmitted information, particularly for resource-depleted smart devices. Communication protection measures are necessary for the property or network layer in the same way as identity authentication to avoid unlawful nodes. The Distributed Denial of Service (DDoS) attack is also popular at this stage, so during this layer, there is a desire to protect against DDOS attack in defenseless nodes.

To ensure that knowledge holds on to cloud computing, many mechanisms for application protection are needed. In addition to updated antivirus protection, robust secret writing algorithms are required, while a willingness to protect the user's privacy, key negotiation, and authorization are required for the application and collaboration stage. Also, for data protection at this level, education and key management are essential.

8.2.1 Challenges Facing by IoT Security

The security issues of device, like all emerging technology, is still the main concern that fills the trail of successful innovations of the IoT system several security issues in the

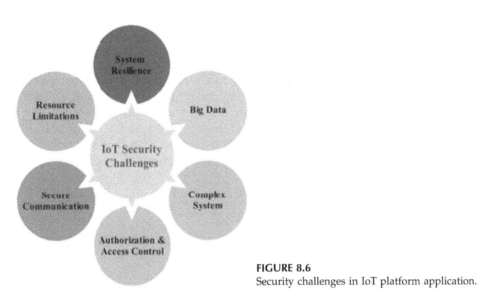

FIGURE 8.6
Security challenges in IoT platform application.

region need self-addressing to increase people's trust in IoT adoption devices (Chanda et al., 2021) (Figure 8.6).

Asset Limitations-Most smart gadgets have confined cycle and capacity limits, because of little and lightweight alternatives for building them that sudden spike in demand for lower vitality. In this manner, refined security calculations don't appear to be adequate for these influenced gadgets as they are not equipped for performing complex cycle activities progressively.

The IoT platform contains numerous devices that evolve large volumes of knowledge, as mentioned earlier. In terms of structure, such information is variable and also arrives in real-time. The number, pace, and selection generate a very complex challenge for storing and analyzing methods that are used to obtain significant data. Distributed computing will make it simpler to store this tremendous amount of data for a significant period. However, working with this enormous data can be a major challenge, because the information management service is highly passionate about the entire performance of various applications. Also, one of the essential elements of enormous knowledge is the integrity of data sources. Expertise ensuring the safety of this enormous amount of information is significantly inflated in a critical approach that must be taken in additional security measures in Atlaeenm and Wills (2020).

Authorization and access control—provision of a relevant degree for IoT-system authorization and access-management mechanism is one of the foremost basics for producing a secure system. System authentication has many concerns: the use of weak or default passwords, for example, that exploit computer information or perhaps physically harm it by giving access to attackers from the World Health Organization. Adopting IoT device security by choice, allowing two-factor to authenticate these problems, and imposing the use of secure passwords will be easier to deal with in Atlaeenm and Wills (2020).

It is not enough to secure IoT devices to ensure that security within the entire IoT system has been achieved. Instead, the interaction between different devices is necessary to shield communication nodes, such as smart devices and cloud app service, from any attack. A suitable encryption technology should also be used. The mistreatment of different networks would also decrease security through analytical equipment and private communication channels.

One of the most challenging problems that the IoT architecture has to answer is system resilience. Framework versatility alludes to the capacity of the framework to react to unannounced attacks or other situations. The system should therefore be ready to shield alternative network nodes from any attack if some IoT devices are hacked.

There are billions of heterogeneous devices in complex IoT system that create realtime achievable management of this large-scale network, especially with limits on memory, resources, and time. There are many computers, entities, communications and interfaces, and opportunities for security breaches.

8.3 IoT Attacks Taxonomy

Reliance on distant transmission is the IoT guideline factor that makes it defenseless against ambushes. Due to the natural safety weakness of the sensor centers, IoT are extraordinarily exposed to a wide extent of security threats. Also, since sensor center points are usually open, it is slanted to ambushes at the physical level since they are not guaranteed. Security ambushes over different layers of the TCP/IP demonstration stack are referred to in Figure 8.7. The attacks referenced there are ambush dynamics and idle

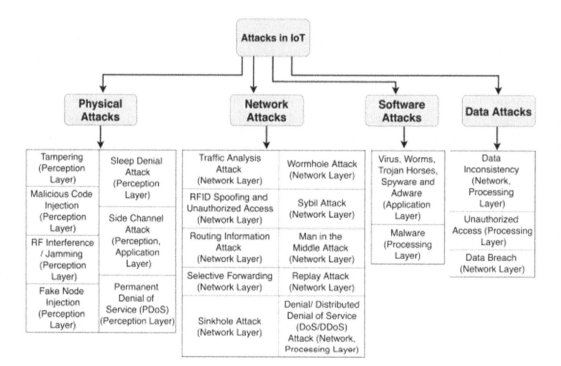

FIGURE 8.7
Layers-based IoT attacks.

attacks. The following are the different forms of dynamic attacks and impacts on layers in Krishnan and Mallya.

Active Attacks: In dynamic assaults, there is an interruption of administration and change or defilement of information. A portion of significant dynamic attacks that are influencing the IoT condition is depicted. In this attack, the messages are captured in the system then diverted to another region to trigger one layer-based IoT hub attacks to show up nearer than the other. These sorts of assaults are ordinarily coordinated towards retail location gadgets that work on close to field correspondence.

Sybil Attacks: In Sybil assaults, an assailant makes various counterfeit personalities, called Sybil characters. These are common in recommendation-based frameworks in which an assailant makes numerous characters to make more impact. Such types of assaults may impact a ballot system in which Sybil hubs may project counterfeit votes to negotiate the structure's reliability. Misleading Sybil hubs in the IoT setting may affect consumer security and trust.

Particular Forwarding Attacks: In a particular sending attack, messages can be directly forwarded by malignant hubs; numerous messages are just lost. Specifically, the attacker may drop traffic from specific hubs and forward the hubs from distinct hubs. The malicious hub can postpone its unsafe position exercises because of its irregular hurtful nature. It is one of the most hazardous assaults, as it can influence the greater part of the directing conventions.

Sinkhole Attack: A Sinkhole attack is one in which the aggressor bogusly directs data and publishes that the assailant hub has a place with the most limited course to the goal. The malicious hub adjusts the data in the directing bundles to lower esteem and pulls in all rush hour gridlock to itself. When all hubs begin sending traffic to the adversary hub, the aggressor hub can adjust bundles, rebroadcast the parcels, or just drop the parcels. Correspondingly, for additional assaults, sinkhole assaults are taken off platforms, as in Krishnan and Mallya.

Hub replication: The aggressor builds its minimum-effort sensor hubs with the ID of a current hub in the system in this form of attack. As a component of the current system, the aggressor can make a new false hub.

Hi Flooding: To overwhelm the data transmission capability, an aggressor rebroadcasts each overhead parcel to its neighbors. This may clog traffice in genuine parcel drops.

Passive Attacks: These are latent assaults when unapproved customers listen in or screen the transmission. Aloof attacks are otherwise referred to as the Attack on Privacy. Some are reported below in Krishnan and Mallya:

Traffic Analysis: Traffic investigation is an aloof assault that is central to data privacy. Any of the sniffing devices such as Wireshark built with port reflection can be used by intruders to do a parcel capture of system traffic. To acquire imperative data about the content of the message, this traffic will later be analyzed. The sniffing of scrambled traffic can give the intruder crucial details, such as information about individuals, the existence of communications, and so on. The release of message substance or disrupting the confidentiality and security of the data is another classification of latent assaults. This is enormous in the context in which IoT is now widely used in the human services industry. The protection of patient data is important in the social insurance industry. Moreover, confidentiality of payment details in the banking industry is also extremely basic. Thus, the confidentiality and secrecy of data are essential in the IoT environment.

8.3.1 Software Attacks

The most significant factor behind security attacks in software package systems are software attack devices. They target vulnerabilities found in communication interfaces within the implementation of the system. Exploits: As the Internet is connected to the IoT device, the attacker uses this facility to create malicious scripts that attempt to gain access to sensitive information or interrupt the system's handiness. Such malicious scripts assault room, dead-by-wrong, via device users.

- Spoofing Attack: A kind of malicious attack that through compromised emails or phishing websites targets user login credentials and alternative sensitive data.

- Malware: This attack is analogous to the installation of malicious code attacks in which the attacker injects malicious software packages into the system to obtain system access, steal confidential data or interrupt system access.

- Refusal of Service Assault: An assailant may execute an attack denying the requested service on the IoT system over pertinent layer that influences all clients of the IoT organization. This sort of assault mutually obstructs lawful clients and empowers full admittance to delicate data for the aggressor.

- Encryption Attack: Via various communication channels, the IoT framework links all objects. The hidden writing algorithms device was used to safeguard the communication system. Nothing is unbreakable, however. Devices for secret writing attacks violate encryption frameworks for smart applications.

- Timing Attack: This attack targets devices for hidden timeslots that victimie safe techniques inside the IoT system to achieve secret writing and cryptography keys used in the method of secret writing information.

- Key-Determining Attacks: if we need to feel that the aggressor has the ciphertext or plaintext effectively, at that point the assailant will probably locate the key, composing key by breaking the mystery composing system of the gadget.

- MITM Attack: they share secret writing and cryptography keys for 2 nodes to talk to each other on a safe line victimization and secret writing algorithmic law. By intercepting signals sent between 2 nodes, MITM attack attempts to understand access to the current information and try to independently conduct a key trade with every hub, permitting the transgressor to encipher and change any likely signals between correspondence hubs.

8.3.2 Privacy of IoT

Billions of recent web sensors and devices continue to develop in the IoT, disseminating a huge quantity of data concerning people, together with their places, ties, search records, financial transactions, images, voices, and interactions, state of health, etc., with or without their permission. This vast volume of knowledge makes it a daunting job to maintain our privacy. Privacy will take several forms within the IoT system, but first of all, we need what security implies that protection is characterised as the case of people, gatherings, or associations to see with their own eyes once, yet how much information about them is given to others per Westin.

Protection has to do with four primary components: records; correspondence; body; and domain. Data security is dependent upon different types of individual information acquired and handled by a substance; for example, clinical data. Contact protection issues protective knowledge are sent to any communication channel between two nodes of human activity victimization. Privacy of the body is linked to the physical protection of people on board any external damage. Territorial privacy in the IoT context deals with creating restrictions on physical areas such as residences, geographical points, and public places, and the privacy of defensive citizens has become a tremendously awful thing to remember (Atlaeenm & Wills, 2020).

This can be due to the additional passive, pervasive, and less intrusive info assortment method, which creates fewer responsive users to be tracked. A threat to privacy is described as the possible danger of losing ownership of private data. This risk is also one of the customers' main concerns and has a crucial influence on the degree of acceptance of any new technology in the market (Atlaeenm & Wills, 2020).

8.3.3 Privacy Threats

One of the IoT's key features is that objects can comprehend and sense their environment. This ability, however, ends in chasing and observing user activity and activities that compromise user privacy and lead to several problems that cause individuals to lose lives in reality. This chapter discusses common threats to privacy within the IoT system in Atlaeenm and Wills (2020).

Identification—In essence, the IoT method is universal, enabling devices to identify and capture various forms of user information and their environmental interactions. Usually, this data is obtained by service providers, who are positioned outside customer oversight. Identifying is the hazard of linking a partner (e.g., name, address) to a private person's data. In the IoT, emerging technology and the interconnection of different methods extend the danger of detection. Police cameras compare with such strategies in non-security environments, where consumer activity is observed for interpretation and analysis and sale. Generally, attribute-based authentication is recommended to minimise information to deal with this issue that a tool gathers inside the IoT and to retain security over the disclosure of information.

Localization and Monitoring—The threats of determining and documenting The entirely different location of an entity over time and area indicates that mobile phone location, net or GPS data are placed, for example, and subsequent device. The availability of large-scale and full Spatio-temporal information has contributed to a growing interest in geographic information and the integration of spatial information analysis. Many will exacerbate the localization threats with the advancement of IoT. The nature of information assortment technology, and interaction with IoT, such as developing location-aware applications and enhancing their accuracy, Devices that record the identity, position, location of the user and activity for location-aware applications.

Profiling—Profiling is the aggregation and process information tool for long-term classification of human behaviors and behavior following certain features. Usually, the information is collected when not approved to form a great deal of complete profile by users and combined with alternative personal data. Profiling is currently used in an overwhelming number of fields, such as e-commerce, targeted advertising, and credit scoring. Alternative users could also be exposed to one in every risk as alternative users who share the same PC and browser could read their targeted promotional content, this is important to the private data. By the growth of the smart world, because of proliferation of the information bases and smart devices, the information spectrum would improbably increase quantitatively. Also, information may be qualitatively changed when gathered from previously-inaccessible elements of the non-public life of individuals, such as knowledge gathered by wearables and entirely distinct devices from home, according to Chanda et al. (2021).

Lifecycle transitions—This sort of threat to privacy applies to the speech act of personal information wherever a consumer product's owner is altered during its life cycle. Because consumer items containing non-public data for their entire life cycle, smartphones, cameras, and computers are owned by the same owner. New, related products may include non-public information; due to owner change, the risk to private speech may increase.

Data warehouse attacks—Inventory attacks are connected to unlawful collecting of data on the existence and characteristics of goods in an excessively unreasonable manner. Inventory attacks can often be carried out by mistreating the IoT device's fingerprint, such as their contact speed, latent duration, and so on. If the IoT's pledge is fulfilled, all sensible items are accessible on the internet; the ability for unauthorised individuals to take advantage of this and make a list of stuff happy for a target is gaping. For profiling individuals, a list attack can be used, because owning unique items discloses non-public information about the owner.

Due to the combination of separate data sources and the relation of entirely different systems, the linkage threat refers to uncontrolled revelation details. The integration of different kinds of data about the person shows new information that the owner doesn't seem to expect. In the IoT context, the data discovered is taken into account a privacy violation at intervals, the linkage hazard is inflated by the incorporation of entirely different entities that generate a lot of heterogeneous and distributed networks that can increase the quality of the system and make the function of the information assortment less transparent.

8.4 Attacks, Threats, and Vulnerabilities

The technological buzz within the trade of computing is rising out of the desktop territory; it's not hidden. Technology has accelerated steadily within the last decade. Continuous evolvement and adaptation of technology are capable of connecting varied devices through the net. Moreover, a reduction in underlying price permits everybody to procure access to knowledge. To amass data instantly, there's a desire to get web services anytime, anywhere. Therefore, new computing paradigms together with massive knowledge, cloud computing, blockchain, and web of Things (IoT) coined up that alter remote storage and access of knowledge in Garg, Chukwu, Nasser, Chakraborty, and Garg (2020) (Figure 8.8).

Rapidly-growing IoT offers several opportunities, but it's still under development and has privacy and security problems. Every day additional IoT devices square measure being additional to the prevailing networks, the egg-laying additional load on the network that's so much on the far side what has been ascertained within the past (Figure 8.9) (Garg et al., 2020).

According to a recent Symantec report, between 2017 and 2018, there were a large number of IoT device attacks; the total attack range was about 5,200 per month. For these attacks, the top source countries are shown in Figure 8.10 (Table 8.1).

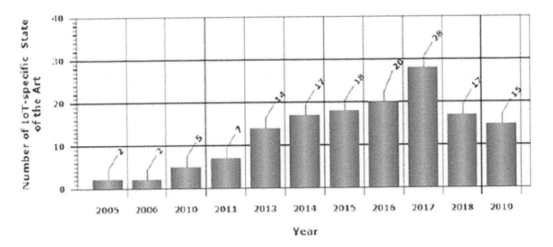

FIGURE 8.8
IoT state of art specific growth.

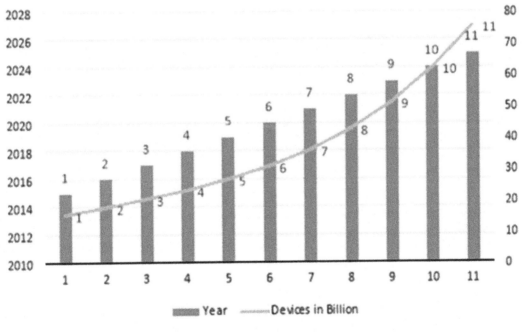

FIGURE 8.9
Estimation of IoT devices demand until 2020.

The biggest problem for IoT is privacy issues. We appear to be unable to quote IoT while failing to answer the privacy considerations relevant to it. The ease of implementing the latest technologies often outweighs the need for security and privacy. Nevertheless, the issue of privacy is just too important to neglect inside the world of IoT. The advantages of large data will end in the premature implementation of IoT software before it is completely developed. Knowledge collection by IoT devices is monumental in magnitude. There are several basic security issues that we have in mind, namely, how

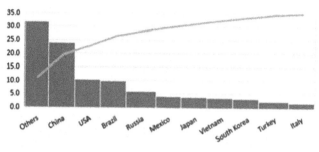

FIGURE 8.10
Top countries for IoT attacks.

TABLE 8.1

Examples of Attacks on IoT Devices (Srivastava, Gupta, Quamara, Chaudhary, & Aski, 2020)

Device Type	Vulnerability Possible Exploits/Attacks
Cars	Chrysler car company had to recall 1.4 million vehicles after researchers proved that attackers were able take control of the vehicle remotely.
Smart home devices	Millions of homes are affected. Multiple vulnerabilities were discovered in smart home devices, such as smart door locks, that could be hacked and opened remotely without using a password.
Medical devices	Multiple vulnerabilities in medical devices such as insulin-pumps, Xray and CT-scanners devices, and implantable sensors have been found.
Smart TVs	Millions of Internet-connected televisions are vulnerable to multiple attacks, such as click fraud, data theft, and ransomware.
Embedded devices	Everyday devices such as routers, watches, cameras, and smart phones using the same hard embedded code SSH and HTTPS server, certificates left by manufactures leaving other millions of devices vulnerable to attacks such as interception and interruption

knowledge is gathered, processed, transported, and retained, as explained in Srivastava et al. (2020).

The security issues in this layer, accoriding to (Obaidat, Obeidat, Holst, Hayajneh, & Brown) stem from the very presence of the IoT framework in the open, unsafe environment of the desired applications. Furthermore, this is due to the existence of IoT nodes and computers that are constrained by resources. Physical-layer problems with IoT systems include physical damage and change of state. Attacks on this layer2 device were directed at the definition of data shaping.

Node attack: this can either physically alter components of the network or system's hardware, or move the whole network to a malicious node. The aggressor's goal is to make the node or IoT system aware of access and management. This may be accomplished by destroying the practicality of the hardware elements or by sacrificing the device's confidential data, such as communication keys. Injection is mistreatment of the interface to insert malicious code that spreads and affects the IoT node to the rest of the network or computer physically to obstruct the system's supply and proper practicality. Perhaps it will be reprogrammed by the aggressor with a physical node or computer control, manipulating software system components and reconfiguring or extracting cryptological data.

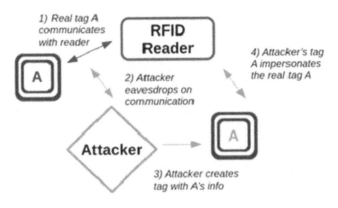

FIGURE 8.11
RFID-tag-spoofing attack scenario.

Distribution of data: relevant in Computer Driver Hacker can steal the encoding keys while accessing the driver.

Cloning of tag research in Obaidat et al. or spoofing RFID tag attacks: the attacker copies into another RFID tag the intended victim's RFID tag information that replicates another real tag (Garg et al., 2020). Interaction between the RFID tag and its reader or physical tampering is accomplished by capturing this. As shown in Figure 8.11 below, the perpetrator can copy information from the compromised RFID tag into another RFID tag. The symbol (ID) or electronic product code (EPC) is also used for this detail that can be the transmitted serial range that any reader or operation key within the range can browse. The goal is to confuse the reader, which provides the perpetrator with RFID access to confidential information impersonation in keeping with, The reader does not understand the difference between a real RFID tag and a corrupted RFID tag.

RFID tag attack tracking: Because the device of these tags is often unprotected, everyone can search for them. This provides the wrongdoer with a wealth of items or people with the following knowledge. Once this tag is added to confidential personal data, this becomes very risky. Tracking data about individuals may well be connected by connected readers who read everything passing through RFID tags, including their movements, monetary transactions, and social communications. To come back up with a trend, this date can then be relevant. This is also a significant issue and a threat to the privacy of individuals.

8.4.1 Attacks on Layer—Network

One of this layer's most important tasks is to relay information. Keeping the network practical and on the market is the greatest challenge. Furthermore, wireless links are subject to completely different security threats.

Denial of service attacks: This attack would drain resources essential for driving IoT apps for a system to be inaccessible and unable to provide services. At different layers of the IoT architecture, this attack will take different forms. By creating a large amount of traffic or targeting IoT network protocols, as shown in Figure 8.12 below, it may overwhelm the network at the network layer, leading to the inconvenience of related IoT devices or systems. This includes numerous attacks, such as floods by SYN, floods by UDP, death pings, etc. Leaky, unencrypted user information is one of the most dangerous threats.

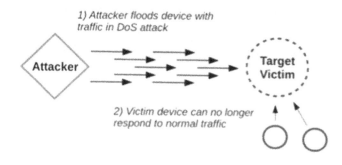

FIGURE 8.12
DOS attack scene.

8.4.2 Attacks on Layer—Application Use

By creating a large amount of traffic or targeting IoT network protocols, as shown in Figure 8.12 below. In Obaidat et al., data from the network layer is also processed by the layer. Code attacks (i.e. the manipulation of bugs in programs or application-layer protocols) and lifetime permissions are a particular threat to this layer. These attacks are aimed at accessing sensitive IoT user information, which ends up compromising the privacy of users and the confidentiality of information.

8.4.3 Spoofing—Phish Attack

To obtain non-public user information such as ID and password, i.e., authentication credentials, an infected email or phishing website is used by the attacker, as shown in Figure 8.13 below. When the victim accesses their email address, the intruder gets access to private information such as login credentials or email accounts.

8.4.4 Injection of Malware

IoT apps suffer from malware vulnerabilities that can propagate and circulate on their own, which is considered to be one of the IoT system's most difficult attacks. The attacker

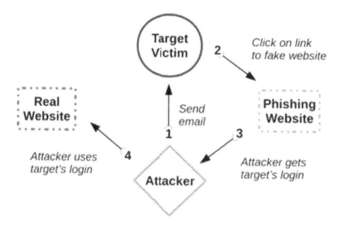

FIGURE 8.13
Attacker masquerades as the victim.

can infect the IoT application, then interpose the system and gain access to sensitive information. Also, the device may be corrupted by malicious computer code, which can lead to DoS, meddling with, or stealing information.

8.4.5 Malicious Scripting Code

To damage the IoT system, the program is corrupted by these scripts by inserting or changing machine code and its practicality in a planned manner. As IoT applications square measure mainly all internet-based, an attacker achieves its target whenever the victim attempts to access a service on the net. The intruder can access sensitive information or cause the machine to crash.

8.5 Design of Malware Attacks

8.5.1 Structure of Testbed

Our testbed in Waraga, Bettayeb, Nasir, and Talib (2019) uses a standard design, whereby modules that can be expanded or may be completely replaced are created by each part of the testbed. Also, the framework provides for the simple insertion of more security checks. The architecture of the structure, as shown in Figure 8.14, consists of five modules.

8.5.2 Module Interface

This module serves as an interface connected to an I/O. It consists of two units: the graphical user interface (GUI)-related unit and the output unit. The graphical user interface delivers (if required) user input to the Testing Module for analysis. During the

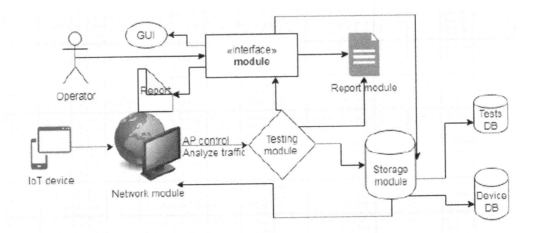

FIGURE 8.14
An automated testbed for IoT security.

testing process, this technique decreases user interference. If the review is complete, the Report Module produces a summarised report and sends it to the customer. The esting Module handles and launches cases to take a look at them as well. All look at the case and script devices stored in the Storage Module when the testbed is working, data will be accessed by the modular scripting to look at the IoT device's final network characteristics. For example, dedicated cases are launched to check the vulnerability of these ports when any open ports within the IoT framework are recognised. Outdated services, setup or authentication for low-security can embody such vulnerabilities. Also, IoT system responses are tested to see if all test suites have succeeded or not.

8.5.3 Computer Networking

This module performs communication operations for networks and IoT computers. The Network Access (AP) function is created and monitored and can be defined in any manner. The report module provides a final report on the results of the system's safety evaluation. It is built at results and logs from a glance.

The secondary memory module stores all events triggered by different modules for later retrieval and final report generation once the evaluation is complete. This retains all knowledge about the devices checked and stores scripts.

Testbed main components:

In terms of operating, our projected automatic testbed depends on five parts that use the structure modules of the testbed. Figure 8.15 summarises the components of the testbed and their purposes. The testbed has the following key operational components:

1. There is also an IoT gadget to be tested, such as a smart socket, a good wireless camera, etc.—DUI (IoT Under Investigation Computer) It will be connected with the wireless network of the test bed.

2. System Admin—This is also the bulk of the protection. A testbed running the Kali Linux OS. The computer networking module is malicious and tests and analyses network packets of all network traffic. It also sends Vulnerability Relevant warnings if maltreatment python scripts and Tshark are detected in some form of attack. Any DUI-requested computer address is reviewed against computer Blacklists to decide whether it is harmful or not. Blocked and reportable are malicious calls. Inside the testbed, the storage module stores information about all registered devices and excludes unregistered devices from the network.

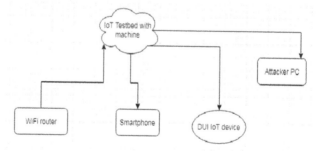

FIGURE 8.15
Testbed devices for IoT security.

3. The Non-public Cellular Network—In model network conditions, testbeds typically use wireless routers. To gather data that is transmitted over the native network, Sniffing is then implemented as a software kit. Interception instruments such as spoofing the Address Resolution Protocol (ARP) to obtain additional data or packets within the network system between two nodes custom launch applicable to vulnerability attack are used. New IoT devices, assisted by our studies, identify spoofing attacks and immediately disconnect from the network once they are detected. A wireless hotspot that builds a Wireless Native Access Network (WLAN) using a virtual AP tool is mistreated by the admin machine (Chakraborty, Joel, & Rodrigues, 2020). This strategy is preferred over the use of a physical router because It provides privileged access to the Dynamic Host Configuration Protocol (DHCP) server tools and to the testbed automation system for various functionalities within the AP. In order to address the Artist Spoofing requirement, this enables the testbed to inspect network traffic with administrative privileges. The network module's WPA2-PSK Wi-Fi key is fitted with a virtual AP tool to encrypt communications and breach security. Also, the outboard DUI relationship is tracked. The computer would also be detected and removed from the network if the system is already malware-infected. This move meets the safety needs of the applications that dominate the testbed. Some IoT devices are regulated solely by the mobile application relevant to them. For example, the testbed in Waraga et al. (2019) includes a smartphone system fitted with approved mobile apps for IoT system management under research. It creates traffic and packets with dominant network commands and can be used to verify whether or not the commands and messages of the controlling device are transmitted in plain text; that is, decipherable by attackers. Additionally, to create replay attacks, it replicates packets, exposing IoT system vulnerabilities.

4. Offensive computer—To discover vulnerabilities, the offensive The (Kali Linux) computer will be used to initiate DUI attacks. It can, for example, cause replay attacks or identification attacks by brute-force. It also executes DoS attacks against the testbed and also the DUI to check their resistance and capacity to dam such attacks. The state of affairs for a DUI linked to the testing network is summarised in Figure 8.16 to provide a more comprehensive plan of how the testbed modules get together. As shown in Figure 8.14, the testbed will verify its identity whenever a brand new computer attempts to connect to the network by wanting it up in the native information records of the testbed. Before beginning to take a look, Inside the software details, the test bed operator records the devices to be tested. If the network module fails to notice the device inside the data, the test bed will reject the device from the network.

8.5.4 Methodology: Automated Testbed Process

1. Ensure that the DUI is equipped with a testbed wireless network.

2. Once the operator chooses one or more devices to be tested, the testbed assesses each one individually.

3. To test each IoT device, the testbed first excludes all IoT devices from the network other than the DUI.

4. The testing module then launches some initial test cases, including an extensive Nmap test to check network activity and report open ports.

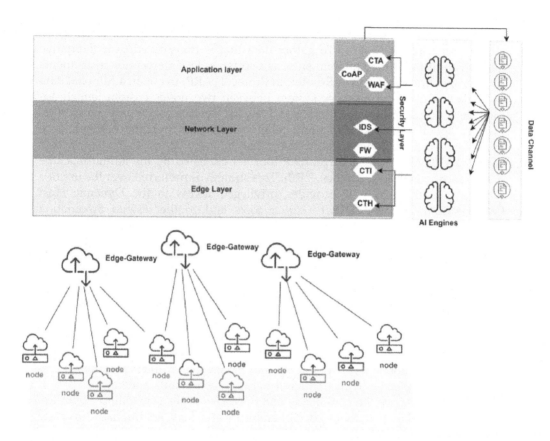

FIGURE 8.16
AI4SAFE IoT architecture model.

5. Based on the results of these preliminary tests, the testing module will determine whether or not there is a web server hosted in the IoT device.

6. Based on this data, the testing module will launch the corresponding tests of its Web server, as follows:

  • Gather information and scan for vulnerabilities
  • Perform an extensive scan using Nmap
  • Inspect SSL Certificate
  • Check Firewall and exploits
  • Check SQL injection and XSS attacks
  • Bypass basic HTTP authentication
  • Analyze communication between IoT device and Host
  • Inspect that there are no other Open ports
  • Perform brute force attack on credentials and directories
  • Perform brute force attack on SSH/FTP/Telnet port(if found)
  • Check asynchronous connection with a Time server
  • Check requested IP addresses

If its not Web server then:

- Gather information and scan for vulnerabilities
- Perform an extensive scan using Nmap
- Check Exploits
- Analyze communication between IoT device and Host
- Inspect that there are no other Open ports
- Perform brute force attack on SSH/FTP/Telnet port(if found)
- Check asynchronous connection with a Time server
- Check requested IP addresses

8.6 Impact of Attacks on Security Objectives

Confidentiality, honesty, and comfort (the triad of the Central Intelligence Agency) security targets may be compromised by attacks. The possible effect of the absence of one of those three safety goals is illustrated in NIST's FIPS 199 publication:

- Low: small outcome on activities, properties, or individuals
- Moderate (Mod): extreme impact on behaviors, properties, or people
- High: extreme or ruinous impact on practices, properties, or individuals

Not applicable: refers solely to confidentiality due to the context within which the Related in Vulnerability attack occurs, the potential impact can vary. We prefer to think about the possible impact of selected attacks on the CIA triad in Table 8.2 in

TABLE 8.2

Comparison between Two Research Phases for IoT Security Testbed Analysis in (Waraga et al., 2019)

Research phases	Set up	Expected result	Methodology
Phase 1: Extensive Analysis	Each device is tested individually using a list of tools conducting different hacking attempts to find the IoT device's vulnerabilities.	Each test will generate different results/outcomes.	Tests are done manually. Results are obtained manually through analysis.
Phase 2: Automated Testing	The testbed system's software is based on a modular structure. Each testing module will automatically run a list of tests to check the IoT device's vulnerabilities.	The test results will be expressed as Vulnerable or Not Vulnerable. If the device passes a given test, it is not vulnerable in that area; if it fails, the device is vulnerable to attack.	The module's test cases are in Python code. The module analyzes the IoT device's responses to each test script to determine if it passed or failed the security test.

Obaidat et al. The overall type of computer on which the attacks are directed is determined by user details. In one case, the attacks concentrate on a smart lightweight light bulb, a smart health monitor with different scenarios, within the intensity of the effect, the difference in applications would produce a difference (Tables 8.3 and 8.4).

TABLE 8.3

Summary of Test Cases Results (Obaidat et al.)

Tests	Used tools	Expected results
Gather information and scan for vulnerabilities	Snitch, OWASP ZAP, Wascan, Skipfish	Gathers information about the DUI's vulnerabilities or about previous attack attempts recorded in the CVE database. Wascam and Skipfish are security penetration testing tools that recursively crawl web pages hosted in webservers. They assess security and look for vulnerabilities such as flaws, links, email addresses and any other information that could lead to social engineering, malware injections, etc.
Nmap scanning	Nmap	Lists all open ports along with their services and DUI
Firmware check	Binwalk	Outdated firmware is usually vulnerable. The firmware is therefore checked to ensure that the device is using the latest version.
Hardware analysis	UART	This test attempts to dump firmware from the hardware using UART in order to obtain root shell and access sensitive information

TABLE 8.4

Important Attacks on CIA Triad of Information Security (Obaidat et al.)

Sample Attacks	Potential Impact on Confidentiality of User Information		Potential Impact on Integrity of User Information		Potential Impact on Availability of User Information	
	Smart Home Heating Control	Smart Health Monitor	Smart Home Heating Control	Smart Health Monitor	Smart Home Heating Control	Smart Health Monitor
RFID tag tracking	Low/Mod	Low/Mod	Low	Low/Mod	Low	Low/Mod
Denial of Service (DoS)	Low	Low/Mod	Low	Low	Low	Mod/High
Man in the Middle	Low/Mod	Low/Mod	Low	Mod/High	Low	Mod/High
Traffic analysis	Low/Mod	Low/Mod	Low	Low	Low	Low/Mod
Phishing	Low/Mod	Low/Mod	Low	Low/Mod	Low	Low
Malicious virus/worm	Low/Mod	Low/Mod	Low	Low/Mod	Low	Mod/High

8.6.1 IoT Network Privacy Preservation Solutions

The protection of IoT interface privacy in Obaidat et al. must be the most critical goal for efficient adoption and deployment of the IoT system. The device has many techniques that are recommended to protect privacy. Within the IoT scheme, this section offers a brief discussion of methods to address the privacy dilemma.

Design Privacy—Intentional privacy is a valuable key to the security of privacy within the IoT environment. IoT customers should have the options they need to process their data and decide how it can be used by an organization. Some businesses are currently using a type of arrangement that allows bound providers to access data as desired. Therefore, to be engineered as a necessary part of any product, resources to protect consumer privacy are inherently important.

Privacy awareness—The lack of public awareness is one of the most controversial problems of privacy infringement. IoT users should be completely tuned in to keep themselves safe from any kind of threat to private data (Obaidat et al.).

Data usage reduction—IoT service providers can take advantage of the reduction in data by reducing the spectrum of just what is relevant to the service they are implementing with personal data. The competence they have for the service should be retained jointly (Banerjee, Chakraborty, Kumar, & Biswas, 2019).

Cryptology— One of the most important methods for protecting privacy on The required cryptological approach to data encryption is IoT equipment. This solution is also achieved, however, with restricted storage and processing resources in IoT devices.

Data anonymization—When knowledge assortment is required, each person who likes the variety of social security and To get rid of the identity of people in databases, from the information records DL numbers should be omitted.

Authentication—a specific access-management model for the IoT system to adjust device access to supply is one of the choices for protecting IoT user's privacy.

8.6.2 Application Layer Security

Usually, edge devices trot out applications for IoT devices via net services in the application layer. To maintain a stable interaction between devices and repair providers, we suggest two separate safe modules:

CoAP-DTLS CoAP serves as a form of limited-device hypertext-transfer protocol that enables devices such as sensors or actuators to speak about the IoT, to be managed also to spend half of a system on their data. The protocol is intended for low information measurement reliability and via its low power extraction and low overhead network, high congestion. wherever TCP-based protocols such as transport calculation message queuing (MQTT) are used, fail to complete an acknowledgment, CoAP will continue to function. CoAP may be a plaintext protocol like hypertext transfer, but it works below UDP TLS, and by default, it can be used. The most widely-performed encoding is datagram transport layer protection (DTLS) in the victimization datagram in Pajouh, Khayami, Dehghantanha, Choo, and Parizi (2020).

The CoAP procedure for establishing a secure affiliation in the application layer to maintain affiliation and integrity between two hosts (edge devices) is shown in Algorithm one. Most edge IoT environment buyers are supported by the REST API hypertext transfer protocol. In the security architecture, Relevant in Vulnerability intelligent WAF operates as algorithmic program two for detective work anomalies in the request for the provided services via hypertext transfer protocol.

Algorithm 1 CoAP-DTLS procedure for securing between hosts communication

---

1: **function** CoAP-DTLS($host_A$, $host_B$)
2:     $host_A \rightarrow$ sendHello($host_B$);
3:     $host_B \rightarrow$ verifyHelo($host_A$);
4:     $host_A \rightarrow$ generateCookie();
5:     $host_B \rightarrow$ serverHello(hostA):keyShare,Certificate;
6:     $host_B \rightarrow$ certificateRequest($host_A$);
7:     $host_A \rightarrow$ sendCertificate($host_B$);
8:     $host_A \rightarrow$ certificateVerify(certificate$_B$);
9:     $host_A \rightarrow$ finishHanshake();
10:     hostB$\rightarrow$ finishHanshake();
11:     **if** certificate$_A$ && certificate$_B$ **then**
12:         $host_A \rightarrow$ sendRequest($host_B$, payload);
13:         $host_B \rightarrow$ sendRespond($host_A$, payload);
14:     **end if**
15: **end function**

FIGURE 8.17
CoAP-DTLS procedure.

Cyber Threat Attribution (CTA): one of the most difficult problems inside the application layer is to solve the root of threats and to select the most suitable action correlates to a threat. Knowing the attacker's ways, tactics, and procedures (TTP) during this layer would make it easier to find the source of attacks and create optimum choices based on the character of the attacker's campaign. Inside the application layer, the CTA module is responsible for attributing the malicious behavior residing in jittery layer devices to its original malicious actor through a profile matching engine and recommends the optimal threat/attack call (course of action) (Figure 8.17).

8.6.3 Protection on IoT

IoT protection is a priority for preventing a smart device's components from producing physical harm or undesirable risk; there is also a need for environmental protection from such harm. Furthermore, safety-critical practices ought to keep up responsibility within the IoT system making certain within the IoT method, protection and reliability are not an easy job. However, it does not only need to create the same application but jointly create new approaches to design. The physical harm of IoT devices and their surroundings affects safety. An attack space is evident than an automatic data processing system is created by the physical system connected to an automatic data processing system. It jointly provides element-channel attacks to locate and exploit PC devices by altering intruders.

Also, safety issues are amplifying the size of sophisticated security attacks. Security is incorporated with either the physical system or automated data processing system in the look portion of the merchandise lifecycle and at runtime search. Since the network-connected IoT system device and the latest threat device are abused daily, To recognise the latest threats and search for the most effective method to control them, a runtime search is necessary in Atlaeenm and Wills (2020). A device must therefore be controlled to discover different threats in the process. Security inside the IoT framework should be considered in normal usage. A machine may go drift secure, But if the system is compromised, the attacker may attempt to exploit the practicality of the system that damages the objects handled by the device or endangers individuals who come into contact with it. To preserve the physical protection and safety of IoT products, this problem should therefore be addressed in future design.

8.7 Conclusion

Today, smart things have become an integral part of individuals' lives. Demand for IoT implementation and integration into various applications has grown exponentially. As a result, exploitation of vulnerabilities has led to a series of cyberattacks on the private data of organizations. The observations on IoT security architecture are examined with several real-world examples. We have focused on how a testbed could be automated to test the IoT security for various devices. We have used the automated testbed to inspect the level of security exhibited by various devices when subjected to the testbed framework, which will help assist the designers to focus on secure coding in both hardware and software of the device and application thereby driving the design towards secure coding arena.

References

Atlaeenm, H. F. & Wills, G. B. (2020). IoT security, privacy, safety and ethics. In *Digital twin technologies and smart cities, Internet of Things*. Springer Nature. https://doi.org/10.1007/978-3-030-18732-3_8

Banerjee, A., Chakraborty, C., Kumar, A., & Biswas, D. (2019). *Emerging trends in IoT and big data analytics for biomedical and health care technologies* (Ch. 5, pp. 121–152). Elsevier: Handbook of Data Science Approaches for Biomedical Engineering. ISBN:9780128183182.

Chakraborty, C., Joel, J. P. C., & Rodrigues, A. (2020). Comprehensive review on device-to-device communication paradigm: Trends, challenges and applications. *Springer: International Journal of Wireless Personal Communications, 114*, 185–207. doi: 10.1007/s11277-020-07358-3.

Chanda, P. B., Das, S., Banerjee, S., & Chakraborty, C. (2021). *Study on edge computing using machine learning approaches in IoT framework* (1st ed., Ch. 9, pp. 159–182). CRC: Green Computing and Predictive Analytics for Healthcare.

Garg, L., Chukwu, E., Nasser, N., Chakraborty, C., & Garg, G. (2020) Anonymity preserving IoT-based COVID-19 and other infectious disease contact tracing model. *IEEE Access, 8*, 159402–159414. 10.1109/ACCESS.2020.3020513, ISSN: 2169-3536.

Krishnan, D., & Mallya, M. A. A security attacks report on the Internet of Things and problems in current countermeasures. In Z. Gajic, H. Vasudevan, & A. Deshmukh (Eds.) *Proceedings of the International Wireless Communication Conference. Lecture Notes on Data Engineering and Technology of Communications* (Vol. 36). Singapore: Springer. https://doi.org/978-981-15-1002-147/10.1007.

Meneghello, F., Calore, M., Zucchetto, D., Polese, M., & Zanella, A. (2019, Oct). IoT: Internet of Threats? A survey of practical security vulnerabilities in real IoT devices. *IEEE Internet of Things Journal, 6*(5), 8182–8201.

Obaidat, M. A., Obeidat, S., Holst, J., Hayajneh, A. A., & Brown, J. A comprehensive and systematic survey on the Internet of Things: Security and privacy challenges, Security frameworks, enabling technologies, threats, vulnerabilities and countermeasures. www.mdpi.com/journal/computers,www.mdpi.com/journal/computers.

Pajouh, H. H., Khayami, R., Dehghantanha, A., Choo, K.-K. R., & Parizi, R. M. (2020). AI4SAFE-IoT: a stable AI-powered Internet of Things edge layer architecture. In *Neural computing & applications*. Springer. https://doi.org/10.1007/s00521-020-04772-3.

Song, C., Nguyen, D. T., Qian, Z., & Krishnamurthy, S. V. (2018, December). IoT san: Fortifying the safety of IoT systems. CoNEXT'18. In *Proceedings of the 14th International Conference on emerging Networking Experiments and Technologies* (pp. 191–203).

Srivastava, A., Gupta, S., Quamara, M., Chaudhary, P., & Aski, V. J. (2020). Future IoT enabled threats and vulnerabilities: State of the art, challenges, and prospects. *Wiley Publishers: International Journal of Communication Systems*, e4443. https://doi.org/10.1002/dac.4443

Waraga, O. A., Bettayeb, M., Nasir, Q., & Talib, M. A. (2019). Design and implementation of automated IoT security testbed. In *Computers & security*. Elsevier Publishers. doi: https://doi.org/10.1016/j.cose.2019.101648.

9

Review on Hardware Attacks and Security Challenges in IoT Edge Nodes

Nagarjuna Telagam[1], D. Ajitha[2], and Nehru Kandasamy[3]

[1]GITAM University, Bangalore, India
[2]Sreenidhi Institute of Science and Technology,
 Hyderabad, India
[3]National University of Singapore, Singapore

9.1 Introduction

The security of Internet-of-Things nodes is of massive interest to researchers around the world. The devices suffer from Trojan attacks on the network, software or database. Methods such as hardware Trojan taxonomy and insertion are discussed (Sidhu, Mohd, & Hayajneh, 2019). In only a few years, the IoT ecosystem involved a significant number of interconnectied devices. Their security issues should be considered during development and updating of the software. Currently, these devices suffer vulnerability to different attacks, leading to immense damage. Various components in IoT devices, such as sensors and actuators, are more vulnerable to tampering, which leads to incorrect data transmission to the nodes from unauthorised sources. Patients in hospitals wear IoT devices which collect health information (Tao et al., 2018). Hardware attacks happened in 2017. These attacks, called ransomware attacks, targeted Windows operating system devices. These ransomware attacks encrypt files on the hard disk, making them inaccessible to users (Maple, 2017).

Security vulnerabilities make a significant impact on economic loss and societal problems. Solutions for hardware security depend on physically unclonable functions [PUF]. These functions have unique characteristics and offer flexibility for cryptographic key generation. The advantage of PUF's and key-generation methods is that they use resistive random access memories based on embedded systems (Bhayani, Patel, & Bhatt, 2016).

Wireless communication also play a crucial role in emerging IoT devices. The sensors communicate with automobile devices. Sensors in mobile phones will connect factories and hospitals. They gather the user's location and medical data and track devices (Galleso, 2016). The best example of sensor data collection is the smartwatch. These watches reduce shopping time for users; they also pay bills using the Near-field communication interface application (Alison, 2018).

Some of the major concerns in the large-scale IoT device networks correspond to data fusion, management and data complexity. Eavesdrop attacks are initiated by hackers, who get network data with tampering techniques, causing the entire system to fail.

Cybercriminals target Internet-of-Things devices; these attacks will increase from 2016 to 2018. There were 32.7 million IoT attacks (Shamsoshoara, Korenda, Afghah, & Zeadally, 2019). Attackers mostly use ransomware; hackers use bitcoins or credit cards to demand money (O'Gorman and McDonald, 2012). The variant called Wanacry has affected 300,000 computers in 2017. Another version of a ransomware attack is Petya, which grows itself in network interconnections (Nick Heath, 2017).

A 2016 cyberattack caused denial of service, due to insufficient security in different IoT nodes (Dyn cyberattack, 2018). The same attacks are penetrating health monitoring systems; these attacks lead to many consequences for patient data and may result in patient death (American College of Cardiology, 2018). Cryptographic keys play a significant role in IoT security nodes. These keys can be private or public. They must contain sensitive information about users, and they are stored in the non-volatile memory of IoT devices. Non-volatile memory is easily vulnerable to physical attacks. The software named white-box cryptography [WBC] is used to protect the cryptographic keys (Joye, 2008). This WBC requires more processing power and higher memory storage, and concern is applicable for symmetric methods explained (Kinney, 2006). Broadly speaking, there are two classifications of software and hardware-based mechanisms for protecting IoT devices from different attacks. The first classification depends on the software. Mostly the algorithms play a role in decryption. But these algorithms are vulnerable to attacks, i.e., the attackers will decode the algorithms for private keys. Here also, the keys are stored in non-volatile memory, which leads to susceptibility. Another classification is a hardware-based mechanism that uses integrated circuits to keep the keys. The man-in-the-middle attack is mostly used in hardware-based security attacks (Barker & Roginsky, 2011). In the year 2002, a function was introduced for hardware-based security, which is called physically unclonable functions [PUF] (Gassend, Clarke, Van Dijk, & Devadas, 2002). These PUF's have the significant advantage of unclonability, offering a low-cost solution for future generations for cryptographic public and private keys. PUF responses are susceptible to environmental variations and will change over time due to noise and device ageing (Chatterjee et al., 2018).

This book chapter discuss security vulnerabilities, hardware attacks, architectures, specifications and impact on threats on edge nodes in IoT. This chapter focuses on vulnerabilities in edge nodes and countermeasure protocols and concludes with many research directions for researchers in IoT security attacks.

Section 9.1 offers a brief introduction about security vulnerabilities of IoT, security attacks and consequences in recent years. Section 9.2 discusses the basic blocks and functions of Edge nodes architecture and specifications used in IoT motes. Section 9.3 discusses security challenges in IoT nodes. Section 9.4 discusses Impact on threats on IoT nodes. Section 9.5 discusses vulnerabilities in IoT three layer architecture. Section 9.6 discusses countermeasures for hardware attacks in Edge nodes, and section 9.7 concludes the chapter with future scope.

9.2 IoT Edge Nodes Architecture

The most common elements present in the architecture are shown in Figure 9.1. The model consists of microcontroller units like STM32, radio communication links, sensors, and a long-life lithium battery. Energy harvesters such as photovoltaic cells, supercapacitor cells, etc., are used for recharging the batteries.

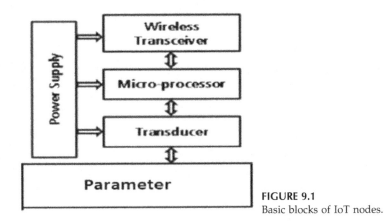

FIGURE 9.1
Basic blocks of IoT nodes.

The IoT nodes architecture is shown in Figure 9.2. The model shows the subsystems, interfaces, sources, assembly, and integration of devices. Many sensors include amplifiers, which are connected and multiplexed to Analog-to-digital converter circuits; battery status can be monitored by circuits present in the power management unit. The battery is recharged from the energy harvesting circuits. Concerning memory in IoT node architecture, RAM is mostly used in the microprocessing/microcontrolling unit for processing data. Another memory is non-volatile memory that has information on programming instructions and settings.

9.2.1 Specifications of IoT Nodes

The present generation IoT nodes are designed according to different system-integration approaches, i.e., systems on board, system on a chip, etc. Table 9.1 shows the survey results from ninety commercially motes (IoT nodes in the form of systems on board) and from thirty microcontroller units and sensor hubs.

The explanation of typical parameters used in IoT motes are given below.

RAM: Chip size is around 512 KB's. The system on chips size is 32 KB.
Flash memory: Microcontroller units have inbuilt memory, and size is around 32 KB.
Battery lifetime: The lifetime of a battery is given in months.
Volume: The System on board's volume numbers are very high and shown in mm^3
ADC resolution: The ADC bit size ranges from 8 bits to 16 bits.
DAC resolution: The DAC bit size ranges from 10 bits to 12 bits.
Wireless receiver: The radios' signals have an operating frequency of 2.4 GHz band and transfer from motes and microcontroller units.
Operating voltage: The typical values are from 1.8 Volts to 3.3 Volts in motes and microcontroller units.

9.3 Challenges in Security IoT Nodes

9.3.1 Security Taxonomy

In this section, the security challenges of IoT devices are explained. Generally, in IoT networks, many devices are connected through the Internet, creating mobility and

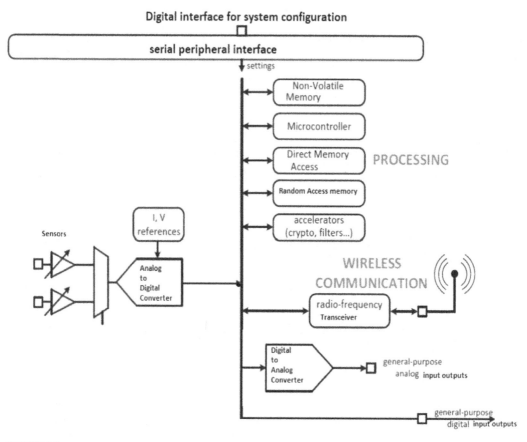

FIGURE 9.2
General architecture of IoT nodes with subsystems.

TABLE 9.1

Parameter Specifications Used in IoT Nodes

Parameters	Motes (SoB)		Sensor Hubs (SoC)	
	Min	Max	Min	Max
RAM in Kilobytes	2	128	4	16
Flash memory in Kilobyte	32	64000	-	-
ADC (resolution bits)	8	11	12	16
DAC (resolution bits)	10	11	-	-
Operating Voltage	1.8	15	1	3.2

different heterogeneity characteristics. The IoT nodes have only limited energy and less computational capacity in the network (Alaba, Othman, Hashem, & Alotaibi, 2017; Mendez, Papapanagiotou, & Yang, 2017). IoT nodes' key challenge is related to the heterogeneous nature of different applications, and every application is related to various security protocols. This heterogeneous behavior degrades

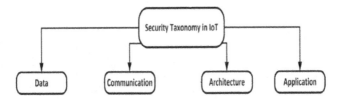

FIGURE 9.3
Classification of security in IoT nodes.

cryptographic algorithms' efficiency, which depends on private and public keys (Zhao & Ge, 2013). Figure 9.3 shows the different classifications of security taxonomy for IoT devices based on four domains, i.e., data, communication, architecture, and application.

Data: Data privacy and confidentiality in IoT devices are fundamental challenges in real-time, and still, they remain concerns for many organizations (Alessio, De Donato, Persico, & Pescapé, 2014). The confidentiality will ensure that user information is accessed by only authorised users and defines the level of access for users. The trust concept applies to all users in every network. It imposes security on three or four layers of IoT devices (Capkun, Buttyán, & Hubaux, 2003; Lin et al., 2017).

Communication: Information exchange between devices will happen in different layers, and it follows many communication protocols. Many attacks, such as man-in-the-middle attacks (Han, Ha, & Kim, 2015) and eavesdropping attacks (Pongle & Chavan, 2015), are possible in the communication medium (Hashem et al., 2016).

Architecture: IoT network architecture is not typical all over the world. The architectures may vary concerning applications. Some of the architectures are software-defined networks (Valdivieso Caraguay, Benito Peral, Barona Lopez, & Garcia Villalba, 2014), object-security architecture (Vučinić et al., 2015) and smart cities (Gaur, Scotney, Parr, & McClean, 2015).

Authentication: This refers to verification of the device's identity. It also assures nodes will exchange information from authorised sources. In the real-time world, multiple IoT networks are present; these networks are authenticated by different vendors and depend on certification authority. These authentication schemes depend on hash functions.

Exhaustion of resources: Resource-exhaustion attacks cause a high rate of energy consumption in IoT networks. The energy will be drained in the nodes (Botta, De Donato, Persico, & Pescapé, 2016).

Trust establishment: Trust can be explained in three ways: trust for security within every layer; trust among layers; and trust between end users and networks (Chou, Chen, Chang, Chih, & Chang, 2017).

Privacy: The data information of users leads to threats in IoT. The conventional system in IoT says the data is collected from all the devices without noticing user privacy. The tough challenge is maintaining data information in a heterogeneous network.

Profiling: Profiling is related to data information in many applications. The major challenge is to disallow the identity threat in IoT devices.

Tracking: The primary threat in IoT nodes is the location of the user with time and coordinates.

9.4 Impact of Threats/Attacks on IoT Architecture

Most of the threats in the architecture are from hardware components. The significant features are integrated circuits; every hardware device has an embedded integrated circuit. The ICs are used in substantial control systems such as weapons, nuclear power plants, and transportation. So, ICs are mandatory for all devices. IC design and distribution are emerging problems for many companies. The companies or factories around the world depend on outsourcing the production of ICs. The threats are generated in ICs, mostly from cheaper factories. The blueprints of ICs can be known to third-party sources, leading to the insertion of hardware Trojans or implantation of malicious codes by adversaries at any stage in the design manufacturing process (Rostami, Koushanfar, & Karri, 2014).

The impact of threats on IoT architectures describes a pyramid of pain (Hu 2003). The least impact is mentioned at the top of the pyramid. The figure shows the sensors in architecture which have the most vulnerable nature. The next most vulnerable part is communication or information exchange between components. The next part is data accumulation, where the hackers get permission to access the sensor data. The next part is hardware abstraction and firmware, where the demand for application programming interface leads to interaction between data and application. The least impacted region is the system on-chip and field-programmable gate arrays. The hardware platform needs to be given more attention because it has a significant impact on the IoT Cyber system. The greatest threat to hardware components is the hardware Trojan.

9.4.1 Hardware Trojan

The hardware Trojan is a modification to a conventional circuit that leads to a change in the circuit's functionality. Third parties will insert hardware Trojan in IC for less reliability (Sidhu et al., 2019). The hardware Trojans have a severe impact on devices such as data leakage, network security bypass, integrated-circuit infection, and delay of the chip's security measures.

A hardware Trojan's behavior can't be changed after insertion. Software Trojans can be eliminated by updating the firmware software. The hardware Trojan can damage IoT networks, leak data, degrade service, etc. It can also infect any integrated circuit, some of which are field-programmable gate arrays and digital-signal processors (Venugopalan & Patterson, 2018).

Third-party sources insert hardware Trojans in IoT devices that impact many sectors such as telecommunications, financial industries, commercial factories, and even the military (Wang, Salmani, Tehranipoor, & Plusquellic, 2008).

The analysis showed that only a few logical gates are enough to trigger an HT without any noticeable variation in circuit parameters. These small variations in circuit parameters after Trojan insertion make the HT very hard to detect during the testing phase.

9.4.2 Hardware Trojan Taxonomy

9.4.2.1 Physical

Trojan's physical qualities are distributed into Trojan size, Trojan type, Trojan distribution, and Trojan structure. Trojan size depends on the number of components in the chip.

If the size of the Trojan is small, there is a higher probability of activation. There are two Trojan types: functional and parametric. Trojan distribution indicates the identification of Trojan location in the physical layout of the chip. Trojan structure can be modified by third-party sources or Adversaries (Wang, Tehranipoor, & Plusquellic, 2008).

9.4.2.2 Insertion Phase

The Trojan can be inserted (Bhunia, Hsiao, Banga, & Narasimhan, 2014) during the design of the integrated circuit. There are various Trojan insertion points: fabrication; testing; and assembly, among others.

9.4.2.3 Activation

Trojans can be activated by using triggers at any time. The stimuli may be internal or external. The internal trigger can sense environmental conditions surrounding the device. The trigger can be activated by sensors such as temperature or humidity. The external trigger can be activated based on a time delay or instructions.

The authors (Yang, Hicks, Dong, Austin, & Sylvester, 2016) implemented a fabrication-time attack, which reduces system security. The authors (Wang, 2014) show how intelligent attacks can reverse engineer.

9.4.2.4 Payload

The payload is activated by a trigger condition with pass transistor logic condition. A digital Trojan modifies the logic values at payload IoT nodes. In some examples, the fault will happen by inserting an additional resistor in the circuit, increasing the capacitance value or creating a delay in the path. Finally, it makes a delay in the lifespan of the integrated circuit.

9.4.2.5 Threats

Based on the hardware Trojan, threats can be classified—the attacks such as the denial of service, modifying the function of an integrated circuit, etc.

9.4.2.6 Location

The hardware components used in IoT architecture are memory elements, regulated power supply unit, input functions, output functions and trigger circuits. All these components are vulnerable to Trojans in the IoT boards. The exact physical location of the Trojan can be identified in these components.

9.5 Internet-of-Things Layer's Security Vulnerabilities

9.5.1 Perception Layer

Figure 9.4 shows the three-layer architecture of IoT. This layer collects and senses information. The collection is done from sensor nodes and RFID authentication tags.

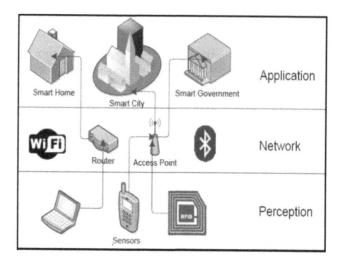

FIGURE 9.4
Three-layer architecture of IoT.

This layer is divided into two sub-layers: the perception network and the node. The node sub-layer sends control signals to gateways. There are three vulnerability issues in this layer. The first one is a physical attack in which the hardware components can be tampered with easily. The second security issue is the heterogeneous character of networks power dissipation, database limitation, etc. This naturally leads to many threats and attacks. The third threat for this layer is confidentiality, which depends on wireless signal strength (Choi, Li, Wang, & Ha, 2012; Gupta, Chakraborty, & Gupta, 2019) (Tables 9.2 and 9.3).

TABLE 9.2

Possible Security Concerns in Three Layers

Security Attacks	Caused by
Replay attack	It was caused by Spoofing the information of IoT devices.
Timing attack	Analyzing the time required to do encryption of IoT devices
Node capture attack	The attacker takes over the IoT nodes and captures the data
Denial of service attack	This attack will arise from the node's energy by preventing them from going into the sleeping node.
Spoofing attack	With the malware node, the attacker successfully masquerades as an IoT end device, end node, or end gateway by falsifying data.
Selfish threat	Some IoT end nodes stop working to save resources or bandwidth to cause the failure of the network
Malicious code attack	Trojans, virus, and junk message that will cause software failure
Routing Attack	Attacks on routing path
Transmission threats	In data transmission, the threats can happen as interrupting blocking, data manipulation, and forgery.
Data breach	The information release of secure information to an untrusted environment
The public and private key	They comprise of keys in networks

TABLE 9.3

Security Threats Which happened in the Perception Layer in Real-Time

Reference	Attack Name	Launched Year	Effects	Countermeasure
(Li et al., 2010)	Wireless Sensor Network Node Jamming	2010	Jam Node Communication	IPSec Security channel
(Burgner & Wahsheh, 2011)	Hardware tampering	2011	Data Leakage	Security for physical designs of hardware components
(Zhao & Ge, 2013)	Fake node injection	2013	Fake Data manipulation	Secure Booting
(Halim & Islam, 2012)	Malicious code injection	2012	Halt Transmission	Intrusion detection Technology(IDT)
(Bhattasali, Chaki, & Sanyal, 2012)	Sleep denial attack	2012	Node shutdown	Authentication
(Li, 2012)	RF interference of RFIDs	2012	Distortion in node Communication	Authentication

9.5.1.1 Security Solutions to Perception Layer

The components in this layer are radio frequency identification readers, sensors, and gateways. The user's data is secure from physical threats, as explained in (Zhang, Zhai, Yang, & Cui, 2014). The proposed cryptographic framework protects the user's privacy with encryption and decryption algorithms (Nagarjuna, Lakshmi, & Nehru, 2019). The algorithms are compared on the different sensor nodes, and the most ensured cryptographic algorithm is the elliptic curve, which is more secured than the RSA algorithm.

9.5.2 Network Layer

The network layer in the IoT architecture will manage both wireless and wired connections. This layer transfers sensor data to all computers across the interconnected network. Routing occurs across IoT hubs; switching, gateways, and routing can be operated with Long Term Evolution technology, Bluetooth, and Wireless Fidelity (Kraijak & Tuwanut, 2015), (Chakraborty, Gupta, & Ghosh, 2015). The network layer suffers leakage of private information in a passive monitoring attack, leading to the confidentiality of this layer being compromised by an eavesdropping attack.

Passive monitoring and eavesdropping attacks occur in the information exchange between IoT nodes and gateway mechanisms (Chakraborty, Gupta, & Ghosh, 2014). Another possible attack in the network layer is the man-in-the-middle attack because the network protocols are not updated. In secure-key exchange, the attacker can get privacy information and perform identity theft (Table 9.4).

9.5.2.1 Security Solutions for the Network Layer

Network-layer security depends on two sub-layers. The first sublayer is wireless security, which is managed by authentication protocols. The second sub-layer is wired security, which depends on channel communication between IoT devices; it is

TABLE 9.4

Security Concerns in the Network Layer

Reference	Attack Name	Launched Year	Effects	Countermeasure
(Thakur & Chaudhary, 2013)	Traffic Analysis attack	2013	Data leakage in the network	Routing Security
(Grover & Berghel, 2011)	RFID spoofing	2011	Intrusion in network Data manipulation	GPS Location System
(Uttarkar & Kulkarni, 2014)	RFID unauthorised access	2014	Node data can be modified (Read, Write & Delete)	Network Authentication
(Soni, Modi, & Chaudhri, 2013)	Sinkhole attack	2013	Data leakage (Data of the Nodes)	Security Aware AdHocRouting
(Padhy, Patra, & Satapathy, 2011)	Man in the middle attack	2011	Data Privacy Violation	Point-to-Point Encryption
(Chen, Guha, Kwon, Lee, & Hsu, 2011)	Routing information attack	2011	Routing loops (Network Destruction)	Encrypting Routing Tables

protected by firewalls (Telagam, Kandasamy, & Nanjundan, 2017; Kandasamy, Telagam, Seshagiri Rao, & Arulananth, 2017; Kandasamy, Ahmad, Reddy, Telagam, & Utlapalli, 2018).

9.5.3 Processing Layer

This layer will interact between the network layer and the application layer. Cloud-computing operations are done in these layers. It also includes data storage from low-level layers' services. It can computate and process data to upper layers simultaneously (Table 9.5).

9.5.3.1 Security Solutions for the Processing Layer

The processing layer involves cloud-computing services and data storage in IoT devices. The Cloud Security Alliance suggests many standards to solve security challenges and has protective measures using strong encryption and decryption algorithms with antivirus protection (Telagam, Kandasamy, Nanjundan, & Arulanandth, 2017).

9.5.4 Application Layer

This layer acts as an interface between end users and applications. It supports different services such as recognizing resource allocation, processing, screening, and computation of data in business applications. Spam data is recognised easily by the application layer and can filter data (Nick Heath, 2017). The process layer is keen to make decisions to allow intelligent processing and controlling of devices. Big data rely upon this application layer; some security concerns are explained below (Table 9.6).

TABLE 9.5

Security Concerns in the Application Layer

Reference	Attack Name	Launched Year	Effects	Countermeasure
Patil, Bhavsar, and Thorve (2012)	Application security	2014	Violation of private data	Web application scanner
Chandramouli and Mell (2010)	Data security	2012	User Data Leakage on the cloud	Homomorphic encryption
Hashizume, Rosado, Fernández-Medina, and Fernandez (2013)	Underlying infrastructure security	2010	Service hijacking	Fragmentation redundancy scattering
Nagaraju and Sridaran (2012)	Third-party relationships	2013	User Data Leakage on the cloud	Encryption
Dahbur, Mohammad, and Tarakji (2011)	Virtualization threats	2012	Resources destruction	Hyper safe
Brakerski and Vaikuntanathan (2014)	Shared resources	2011	Resources theft	Hyper safe

TABLE 9.6

Security Concerns in the Application Layer

Reference	Attack Name	Launched Year	Effects	Countermeasure
Kumar, Singh, Singh, and Ali (2013)	Phishing Attacks	2016	Data leakage in user credentials	Biometrics Authentication
Koo, Hur, and Yoon (2013)	Virus, Worms, Trojan Horse, Spyware	2012	Resource Destruction & Hijacking	Protective Software
Tewari, Jain, and Gupta (2016)	Malicious Scripts	2011	Hijacking	Firewalls
Heer et al. (2011)	Denial of Service(DoS)	2010	Resource Destruction	Access Control Lists
Medaglia and Serbanati (2010)	Data Protection and Recovery	2011	Data loss & Catastrophic Damage	Cryptographic Hash Functions
Merry and Hajeck (2009)	Software Vulnerabilities	2011	Buffer over flow	Awareness of security

9.5.4.1 Security Solutions for the Application Layer

Solutions for the application layer are similar to those for the network layer. It is classified into two sub-layers, one of which depends on encryption algorithms and authentication protocols. The second sub-layer depends on national applications in which the intrusion detections and access the user data can be modified (Ongtang, McLaughlin, Enck, & McDaniel, 2012).

9.6 Countermeasures

9.6.1 Trojan Detection

The goal of adversaries or third-party sources who manufacture the IC design is to launch a successful attack without the user's being notified. Figure 9.5 shows Trojan detection classed into two stages during manufacturing. The first classification is pre-silicon and the second is post-silicon.

9.6.1.1 Pre-silicon Techniques

This technique is mostly used by the system on-chip engineers or design manufactures. Functional validation depends on functional tests of design. The formal verification approach depends on logic values verification of design, but this approach will not detect unexpected functionalities of the target design (Bugenhagen & Wiley, 2011). The code or structural analysis method is tested to identify redundancy statements in the entire design code. This technique will not detect suspicious codes or signals in the post-processing stage. The last approach is proof-carrying hardware, which uses an interactive theorem to verify security properties on soft internet protocols.

9.6.1.2 Post-silicon Techniques

This technique is again classified into two types: destructive and non-destructive methods. A reverse engineering process, such as DE packaging of integrated circuit or validation of product identity, is used in the destructive technique. The verification of integrated circuits from third-party sources or untrusted sources is explained in non-destructive techniques. The first classification of proof is the functional test. This test is performed with the help of test vectors. The next method is the side-channel analysis. The side-channel analysis method has some advantages, such as extra path delay and power and heat can be measured to detect hardware Trojans. Barker and Roginsky (2011) have demonstrated the hardware- detection Trojan based on excess path delay on 20 paths, which helps detect Trojans with greater than 80% accuracy. The next method is the Automatic Test Pattern Generation (ATPG) method, which uses the chip's digital stimulus to analyze digital output. The way (Venugopalan & Patterson, 2018) discusses the researchers designed the ATPG method to detect Trojan detection. Most commonly, a Trojan is inserted in the chip's existing logic in this method.

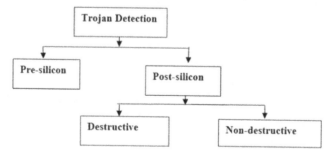

FIGURE 9.5
Hardware Trojan detection techniques.

9.6.2 Design for Trust

Conventional hardware-Trojan-detection techniques don't guarantee detection of Trojans. The procedure for trust offers a more significant potential approach for many countermeasures against hardware Trojans. It involves the prevention of hardware insertion. The following method is runtime monitoring, which monitors chip behavior to detect hardware Trojans in the chip's runtime. Runtime monitoring is classified into three types: configurable security monitors; variant-based parallel execution; and hardware-software hybrid approach. The first classification uses security monitors to monitor real-time functionality by using reconfigurable logic in the chips' system. The second classification performs continuous execution of different functions in the chips. This approach will help many computers, achieves a high level of trust for engineers and has high performance. The last classification utilises design-verification tests to identify the circuits which are not functional and marks them as suspicious.

9.6.3 Prevention of Hardware Trojan Insertion

Attackers insert a Trojan once they know the circuit's functionality. They always use reverse engineering to insert Trojans. Techniques such as obfuscation, layout filter, and camouflaging are used to prevent the insertion of hardware Trojans. The obfuscation method will modify the circuit's functionality. The chip's internal circuit is obscured, and the circuit will not work correctly. This method prevents the insertion of Trojans by attackers. The next technique which thwarts the Trojan is the layout-filtering method. Attackers insert a hardware Trojan in unused vacant spaces in the circuit. Empty spaces in the printed circuit board are filled with functional circuits using built-in self-authentication approaches. Attackers cannot replace these filter cells with Trojans. The last method is camouflaging, which is done at the layout levels. Fake connections are added in every layer to create non-functional layouts; dummy contacts will prevent attackers from obtaining a gate-level circuit of the design.

9.6.4 Split Manufacturing

The next method is to reduce risk in the integrated circuit design. This method increases the security of the integrated circuit by preventing dangerous insertion. To avoid the insertion of malicious Trojans, the system is categorised into two layers: the front end of the line and back end of the line parts for fabrication. Once these layers are designed, attackers will not insert Trojans.

9.6.5 Hardware Security Module

The upcoming generation is full of Internet-of-Things devices. These devices have some vulnerabilities, and they are prone to many attacks. The tamper-proof hardware security module protects user information such as cryptographic public and private keys, personal identification verification numbers, etc., used to provide additional security for layers of IoT devices. This physical device will be embedded into another hardware component or devices easily and act as a standalone device. It is always connected to the server. This physical device has the capability of securing private and public keys.

9.6.6 Trusted Platform Module

The trusted platform module is a secured microprocessor with many cryptographic functions. This microprocessor comes in different shapes and sizes. The trusted platform module hardware provides a trusted root and has inbuilt RSA engines. This RSA engine performs 2048-bit encryption and decryption algorithms. This secured microprocessor uses in-built hash functions to compute against hash values and prevent attacks against public and private keys.

9.6.7 Physical Unclonable Functions

The physical unclonable function is lightweight security which devises identification, key-generation algorithms, data storage, and authentication. These functions are unique because of properties in tamper evidence. They are similar to one-way functions. The silicon-based physical unclonable parts are alternatives for IoT security devices. Key generation by physical unclonable processes is tamper-resistant; ring- oscillator-based functions are excellent choices for authentication.

9.6.8 Device Identity

Device identity is compulsory for organizations around the world. Primarily the organizations depend on securing digital communications with digital certificates. These certificates implement different security controls such as authentication, secure communication and host-to-host communication using transport-layer security protocols. The manufacturer of IoT devices must consider the lifespan of digital certificates and key authentications.

Other countermeasure frameworks proposed by researchers are given below.

9.6.8.1 EPIC: Framework to Protect Smart Homes in IoT Environments

This design is proposed by Shin, Meneely, Williams, and Osborne (2010) to defend against an adversary's traffic-analysis attack. It ensures the user's data confidentiality by using secure multi-hop routing protocols.

9.6.8.2 Static Random Access Memory-Physical Unclonable Function

Authentication protocols proposed by Liu, Zhang, Zeng, Peng, and Chen (2012) check the authentication of edge nodes of IoT using unclonability device identifications. This method reduces attacks such as spoofing and man-in-the-middle.

9.6.8.3 SRPL [Secure Routing Protocol]

Hash functions provide more security in routing algorithms; hash-chain authentication prevents malicious activity in edge nodes. Attacks such as sinkhole and selective forwarding can be prevented easily with these routing protocols (Miessler, 2015).

9.6.8.4 INTI [Intrusion Detection System]

Detection of sinkhole attacks in edge nodes is proposed by Suo, Wan, Zou, and Liu (2012). It depends on 6LoWPAN protocols, analyses node behavior, and defends edge nodes against malicious attacks.

9.6.8.5 ML-IDS [Machine Learning based Intrusion Detection System]

Detection of wormhole attacks in the edge nodes is proposed by Jara, Ladid, and Gómez-Skarmeta (2013). This technique combines k- means clustering and decision trees to detect wormhole attacks.

9.6.8.6 SecTrust

Detection of Sybil attacks in edge nodes is proposed by Atlam and Wills (2020). This technique will make routing decisions in nodes, detect malicious attacks and quickly identify virus-infected nodes.

9.6.8.7 SMQTT [Secure Extension of MQTT (Message Queue Telemetry Transport)]

This protocol is proposed by Singh, Rajan, Shivraj, and Balamuralidhar (2015), and it ensures edge-node-to-edge-node security. It uses an attribute-based encryption algorithm to protect IoT networks against attacks such as man-in-the-middle and phishing.

9.6.8.8 DDoS

This protocol detects DDoS and DoS attacks in edge nodes, and is proposed by Adat and Gupta (2017). It combines analyzing the node's behavior and monitoring incoming traffic signals.

9.6.8.9 Software Defined-IoT

This framework is proposed by Yin, Zhang, and Yang (2018), uses a software-defined anything paradigm and has a technique called "cosine similarity of vectors" for detection and protection against DDoS attacks.

9.6.8.10 Lightweight Algorithm

This framework detects side-channel attacks, and is proposed by Choi and Kim (2016). It is based on lightweight-encryption algorithms and has mask-type protection of secret keys of cryptographic algorithms.

9.6.8.11 Defense Against Gray Hole Attacks in Edge Nodes

This can be classified into two approaches: the first approach has defected the nodes that are selectively forwarding based on the acknowledgments (Yu & Xiao, 2006); the second approach is to develop new routing schemes which are more resilient. Hai and Huh (2008) propose a new lightweight-detection algorithm that detects forwarding attacks in wireless sensor nodes. It is proved that this new algorithm can defend against a high probability of data collisions in IoT nodes. Brown and Du (2008) describe an efficient approach for reporting packet data drops during transmission. Wang, Zhang, Cao, and Porta (2003) develop a new framework for identifying selective forwarding attacks in the network. The sensor nodes will work as promiscuous modes in this algorithm to overhear transmission data from neighbouring nodes.

9.6.8.12 *Defence Against Sinkhole and Rushing Attacks in Edge Nodes*

Hu, Perrig, and Johnson (2003) provide a prevention protocol that defends against rushing attacks. Ngai, Liu, & Lyu (2006) presented an algorithm that first finds the suspected nodes then identifies intruders quickly using a network flow graph. Shafiei, Khonsari, Derakhshi, and Mousavi (2014) proposed another two schemes for sinkhole attack detection. The idea here is if the sinkhole is attacked, the energy in the node is depleted rapidly concerning other sensor nodes. The first scheme is based on base-station-estimator results. The base station commands all the nodes to evade the attack region in the process of routing. The second approach monitors energy levels in the nodes. Zhang et al. (2014) proposed that, based on the redundancy mechanism in wireless sensor networks (WSNs), to detect the sinkholes attacks, the messages are sent to suspicious nodes with different multipaths.

9.7 Conclusion

In any IoT networks, privacy is crucial because the user is always connected to the Internet. Some of the issues, challenges, and requirements are identified and discussed in this chapter. Other than privacy issues in the IoT network, heterogeneous attacks in the edge nodes are discussed. Furthermore, faulty nodes in the IoT networks can experience many attacks such as the denial of service, man-in-the-middle attacks, physical attacks and rushing attacks, which are launched by multiple advisories or third-party sources. Fault tolerance in the Internet of Things can be monitored inside the network and identify faulty nodes in the internetwork. In conclusion, IoT devices cannot be fully secured until a strong foundation is created with hardware components. The efforts and cost of manufacturing the devices could lead to waste if the concerned device has a hardware Trojan inserted. This hardware Trojan can be triggered anytime in the network and can quickly destroy or damage the device. Embedded hardware security in edge nodes plays an essential role in protecting the IoT device's identity against tampering attacks and preserving the user's privacy and security. Many researchers around the world are working on security issues of IoT nodes. The IoT can be explained in three or four layers, depending on the application. The four layers of IoT and their security loopholes are exploited in this chapter. Countermeasures for each layer and security solutions are also broadly explained in this chapter. Countermeasures are adopted to defend and secure the edge nodes from different attacks. The improvements are also suggested in this chapter in the edge node network to make it more secure and prevent deployment issues. The IoT will be part of daily life in upcoming generations. Extraordinary steps or procedures need to be taken by device manufactures, advisers or third-party sources to ensure user information privacy.

In this chapter, we have explored the security challenges of overall IoT systems, focusing on vulnerabilities of IoT edge nodes or devices. Security issues are broadly explained with vulnerabilities and existing problems and solutions. Discussion about the need for new and better hardware security and some open cases are extensively defined with neat diagrams. They describe hardware issues in IoT devices and hardware attacks in detail. Also, countermeasures are discussed that include detection techniques, split manufacturing for trust, design for faith, and other methods such as Hardware Security Module (HSM) and Trusted Platform Module (TPM) to improve hardware security.

References

Adat, V., & Gupta, B. B. (2017, Apr). A DDoS attack mitigation framework for internet of things. In *2017 International Conference on Communication and Signal Processing (ICCSP)* (pp. 2036–2041). Chennai: IEEE.

Alaba, F. A., Othman, M., Hashem, I. A. T., & Alotaibi, F. (2017). Internet of Things security: A survey. *Journal of Network and Computer Applications*, *88*, 10–28.

Alessio, B., De Donato, W., Persico, V., & Pescapé, A. (2014, August). On the integration of cloud computing and Internet of things. In *Proceedings of the Future Internet of Things and Cloud (FiCloud)* (pp. 23–30).

Alison, D. R. (2018). As IoT attacks increase 600% in one year, businesses need to up their security - TechRepublic. https://www.techrepublic.com/article/as-iot-attacks-increase-600-in-one-year-businesses-need-to-up-their-security/. Retrieved: September 26, 2018.

American College of Cardiology (2018). Can your cardiac device be hacked? https://www.acc.org/about-acc/press-releases/2018/02/20/13/57/canyour- cardiac-device-be-hacked

Atlam, H. F., & Wills, G. B. (2020). IoT security, privacy, safety, and ethics. In *Digital twin technologies and smart cities* (pp. 123–149). Cham: Springer.

Barker, E., & Roginsky, A. (2011). Transitions: Recommendation for transitioning the use of cryptographic algorithms and key lengths. *NIST Special Publication*, *800*, 131A.

Bhattasali, T., Chaki, R., & Sanyal, S. (2012). Sleep deprivation attack detection in wireless sensor network. *International Journal of Computer Applications*, *40*(15), 19–25.

Bhayani, M., Patel, M., & Bhatt, C. (2016). Internet of Things (IoT): In the way of the smart world. In *Proceedings of the International Congress on Information and Communication Technology* (pp. 343–350). Singapore: Springer.

Bhunia, S., Hsiao, M. S., Banga, M., & Narasimhan, S. (2014). Hardware Trojan attacks: Threat analysis and countermeasures. *Proceedings of the IEEE*, *102*(8), 1229–1247.

Botta, A., De Donato, W., Persico, V., & Pescapé, A. (2016). Integration of cloud computing and Internet of things: A survey. *Future Generation Computer Systems*, *56*, 684–700.

Brakerski, Z., & Vaikuntanathan, V. (2014). Efficient fully homomorphic encryption from (standard) LWE. *SIAM Journal on Computing*, *43*(2), 831–871.

Brown, J., & Du, X. (2008). Detection of selective forwarding attacks in heterogeneous sensor networks. In *IEEE International Conference on Communications, 2008. ICC'08* (pp. 1583–1587). IEEE.

Bugenhagen, M. K. & Wiley, W. L. (2011). Pin-hole firewall for communicating data packets on a packet network. US Patent 8,015,294, issued September 6, 2011.

Burgner, D. E., & Wahsheh, L. A. (2011, April). Security of wireless sensor networks. In *2011 Eighth International Conference on Information Technology: New Generations* (pp. 315–320). IEEE.

Capkun, S., Buttyán, L., & Hubaux, J. P. (2003). Self-organized public-key management for mobile ad hoc networks. *IEEE Transactions on Mobile Computing*, *2*(1), 52–64.

Chakraborty, C., Gupta, B., & Ghosh, S.K. (2014). Mobile metadata assisted community database of chronic wound. *Elsevier: International Journal of Wound Medicine*, *6*, 34–42. ISSN: 2213-9095, 10.1 016/j.wndm.2014.09.002

Chakraborty, C., Gupta, B., & Ghosh, S. K. (2015). Identification of chronic wound status under tele-wound network through smartphone. *International Journal of Rough Sets and Data Analysis, Special issue on: Medical Image Mining for Computer-Aided Diagnosis*, *2*(2), 56–75, 10.4018/ IJRSDA.2015070104.

Chandramouli, R., & Mell, P. (2010). State of security readiness. *XRDS: Crossroads, The ACM Magazine for Students*, *16*(3), 23–25.

Chatterjee, U., Govindan, V., Sadhukhan, R., Mukhopadhyay, D., Chakraborty, R. S., Mahata, D., & Prabhu, M. M. (2018). Building PUF based authentication and key exchange protocol for IoT without explicit CRPs in the verifier database. *IEEE Transactions on Dependable and Secure Computing*, *16*(3), 424–437.

Chen, W., Guha, R. K., Kwon, T. J., Lee, J., & Hsu, Y. Y. (2011). A survey and challenges in routing and data dissemination in vehicular ad hoc networks. *Wireless Communications and Mobile Computing, 11*(7), 787–795.

Choi, J. & Kim, Y. (2016). An improved LEA block encryption algorithm to prevent side-channel attack in the IoT system. In *2016 Asia-Pacific Signal and Information Processing Association Annual Summit and Conference (APSIPA)* (pp. 1–4).

Choi, J., Li, S., Wang, X., & Ha, J. (2012, June). A general distributed consensus algorithm for wireless sensor networks. In *2012 Wireless Advanced (WiAd)* (pp. 16–21). IEEE.

Chou, S. Y., Chen, Y. S., Chang, J. H., Chih, Y. D., & Chang, T. Y. J. (2017, February). 11.3 A 10 nm 32 Kb low-voltage logic-compatible anti-fuse one-time-programmable memory with an anti-tampering sensing scheme. In *2017 IEEE International Solid-State Circuits Conference (ISSCC)* (pp. 200–201). IEEE.

Dahbur, K., Mohammad, B., & Tarakji, A. B. (2011, April). A survey of risks, threats, and vulner-abilities in cloud computing. In *Proceedings of the 2011 International Conference on Intelligent Semantic Web-services and Applications* (pp. 1–6).

Dyn cyberattack (2018). Dyn cyberattack. https://en.wikipedia.org/wiki/2016_Dyn_cyberattack

Galleso, M. (2016). *Samsung Gear S3 Classic and Frontier: An Easy Guide to Best Features*. Lulu Press, Inc.

Gassend, B., Clarke, D., Van Dijk, M., & Devadas, S. (2002, November). Silicon physical random functions. In *Proceedings of the 9th ACM conference on Computer and communications security* (pp. 148–160).

Gaur, A., Scotney, B., Parr, G., & McClean, S. (2015). Smart city architecture and its applications based on IoT. *Procedia Computer Science, 52*, 1089–1094.

Grover, A., & Berghel, H. (2011). A survey of RFID deployment and security issues. *Journal of Information Processing Systems, 7*(4), 561–580.

Gupta, A., Chakraborty, C., & Gupta, B. (2019). Medical information processing using smartphone under IoT framework. *Springer: Energy Conservation for IoT Devices, Studies in Systems, Decision and Control, 206*, 283–308, ISBN 978-981-13-7398-5, https://doi.org/10.1007/978-981-13-7399-2_12.

Hai, T. H. & Huh, E.-N. (2008). Detecting selective forwarding attacks in wireless sensor networks using two-hops neighbor knowledge. In *Seventh IEEE International Symposium on Network Computing and Applications, 2008. NCA'08* (pp. 325–331). IEEE.

Halim, T., & Islam, M. R. (2012). A study on the security issues in WSN. *International Journal of Computer Applications, 53*(1), 26–32.

Han, J., Ha, M., & Kim, D. (2015, October). Practical security analysis for the constrained node networks: Focusing on the dtls protocol. In *2015 5th International Conference on the Internet of Things (IoT)* (pp. 22–29). IEEE.

Hashem, I. A. T., Chang, V., Anuar, N. B., Adewole, K., Yaqoob, I., Gani, A., & Chiroma, H. (2016). The role of big data in the smart city. *International Journal of Information Management, 36*(5), 748–758.

Hashizume, K., Rosado, D. G., Fernández-Medina, E., & Fernandez, E. B. (2013). An analysis of security issues for cloud computing. *Journal of Internet Services and Applications, 4*(1), 5.

Heer, T., Garcia-Morchon, O., Hummen, R., Keoh, S. L., Kumar, S. S., & Wehrle, K. (2011). Security challenges in the IP-based Internet of Things. *Wireless Personal Communications, 61*(3), 527–542.

Hu, Y.-C., Perrig, A., & Johnson, D. B. (2003). Rushing attacks and defense in wireless ad hoc network routing protocols. In *Proceedings of the 2nd ACM Workshop on Wireless Security* (pp. 30–40). ACM.

Jara, A. J., Ladid, L., & Gómez-Skarmeta, A. F. (2013). The Internet of everything through IPv6: An analysis of challenges, solutions, and opportunities. *Journal of Wireless Mobile Networks, Ubiquitous Computing, and Dependable Applications, 4*(3), 97–118.

Joye, M. (2008). On white-box cryptography. *Security of Information and Networks, 1*, 7–12.

Kandasamy, N., Telagam, N., Seshagiri Rao, V. R., & Arulananth, T. S. (2017). Simulation of analog modulation and demodulation techniques in virtual instrumentation and remote lab. *International Journal of Online and Biomedical Engineering (iJOE)*, *13*(10), 140–147.

Kandasamy, N., Ahmad, F., Reddy, S., Telagam, N., & Utlapalli, S. (2018). Performance evolution of 4-b bit MAC unit using hybrid GDI and transmission gate based adder and multiplier circuits in 180 and 90 nm technology. *Microprocessors and Microsystems*, *59*, 15–28.

Kinney, S. L. (2006). *Trusted platform module basics: Using TPM in embedded systems*. Elsevier.

Koo, D., Hur, J., & Yoon, H. (2013). Secure and efficient data retrieval over encrypted data using attribute-based encryption in cloud storage. *Computers & Electrical Engineering*, *39*(1), 34–46.

Kraijak, S., & Tuwanut, P. (2015, October). A survey on the Internet of things architecture, protocols, possible applications, security, privacy, real-world implementation, and future trends. In *2015 IEEE 16th International Conference on Communication Technology (ICCT)* (pp. 26–31). IEEE.

Kumar, S., Singh, S. P., Singh, A. K., & Ali, J. (2013). Virtualization, the great thing, and issues in cloud computing. *International Journal of Current Engineering and Technology*, *3*(2), 338–341.

Li, L. (2012, May). Study on the security architecture of the Internet of Things. In *Proceedings of 2012 International Conference on Measurement, Information, and Control* (Vol. 1, pp. 374–377). IEEE.

Li, M., Koutsopoulos, I., & Poovendran, R. (2010). Optimal jamming attack strategies and network defense policies in wireless sensor networks. *IEEE Transactions on Mobile Computing*, *9*(8), 1119–1133.

Lin, J., Yu, W., Zhang, N., Yang, X., Zhang, H., & Zhao, W. (2017). A survey on Internet of things: Architecture, enabling technologies, security and privacy, and applications. *IEEE Internet of Things Journal*, *4*(5), 1125–1142.

Liu, C., Zhang, Y., Zeng, J., Peng, L., & Chen, R. (2012, May). Research on dynamical security risk assessment for the Internet of Things inspired by immunology. In *2012 8th International Conference on Natural Computation* (pp. 874–878). IEEE.

Maple, C. (2017). Security and privacy in the Internet of things. *Journal of Cyber Policy*, *2*(2), 155–184.

Medaglia, C. M., & Serbanati, A. (2010). An overview of privacy and security issues in the Internet of things. In *The Internet of things* (pp. 389–395). New York, NY: Springer.

Mendez, D. M., Papapanagiotou, I., & Yang, B. (2017). Internet of things: Survey on security and privacy. *arXiv preprint arXiv:1707.01879*.

Merry Jr, D. E., & Hajeck, M. J. (2009). *US Patent No. 7,502,256*. Washington, DC: US Patent and Trademark Office.

Miessler, D. (2015). Securing the Internet of things: Mapping attack surface areas using the OWASP IoT top 10. In *RSA Conference*.

Nagaraju, K., & Sridaran, R. (2012). A survey on security threats for cloud computing. *International Journal of Engineering Research & Technology (IJERT)*, *1*(7), 1–10.

Nagarjuna, T., Lakshmi, S., & Nehru, K. (2019). USRP 2901-based SISO-GFDM transceiver design experiment in virtual and remote laboratory. *The International Journal of Electrical Engineering & Education*, 0020720919857620.

Nick Heath (2017). Petya ransomware: Where it comes from and how to protect yourself - TechRepublic. https://www.techrepublic.com/article/petyaransomware-where-it-comes-from-and-how-to-protect-yourself/. Retrieved: September 26, 2018.

Ngai, E. C., Liu, J., & Lyu, M. R. (2006). On the intruder detection for sinkhole attack in wireless sensor networks. In *IEEE International Conference on Communications, 2006. ICC'06* (vol. 8, pp. 3383–3389). IEEE.

O'Gorman, G., & McDonald, G. (2012). *Ransomware: A growing menace*. Symantec Corporation.

Ongtang, M., McLaughlin, S., Enck, W., & McDaniel, P. (2012). Semantically rich application-centric security in Android. *Security and Communication Networks*, *5*(6), 658–673.

Padhy, R. P., Patra, M. R., & Satapathy, S. C. (2011). Cloud computing: Security issues and research challenges. *International Journal of Computer Science and Information Technology & Security (IJCSITS)*, *1*(2), 136–146.

Patil, D. H., Bhavsar, R. R., & Thorve, A. S. (2012). Data security over the cloud. *International Journal of Computer Applications*, *5*, 11–14.

Pongle, P., & Chavan, G. (2015, January). A survey: Attacks on RPL and 6LoWPAN in IoT. In *2015 International conference on pervasive computing (ICPC)* (pp. 1–6). IEEE.

Rostami, M., Koushanfar, F., & Karri, R. (2014). A primer on hardware security: Models, methods, and metrics. *Proceedings of the IEEE, 102*(8), 1283–1295.

Shafiei, H., Khonsari, A., Derakhshi, H., & Mousavi, P. (2014). Detection and mitigation of sinkhole attacks in wireless sensor networks. *Journal of Computer and System Sciences, 80*(3), 644–653.

Shamsoshoara, A., Korenda, A., Afghah, F., & Zeadally, S. (2019). A survey on hardware-based security mechanisms for the Internet of things. *A Xiv preprint arXiv:1907.12525.*

Shin, Y., Meneely, A., Williams, L., & Osborne, J. A. (2010). Evaluating complexity, code churn, and developer activity metrics as indicators of software vulnerabilities. *IEEE Transactions on Software Engineering, 37*(6), 772–787.

Sidhu, S., Mohd, B. J., & Hayajneh, T. (2019). Hardware security in IoT devices with emphasis on hardware Trojans. *Journal of Sensor and Actuator Networks, 8*(3), 42.

Singh, M., Rajan, M., Shivraj, V., & Balamuralidhar, P. (2015, Apr). Secure MQTT for Internet of Things (IoT). In *2015 Fifth International Conference on Communication Systems and Network Technologies*, Gwalior, India (pp. 746–751). IEEE.

Soni, V., Modi, P., & Chaudhri, V. (2013). Detecting Sinkhole attack in the wireless sensor network. *International Journal of Application or Innovation in Engineering & Management, 2*(2), 29–32.

Suo, H., Wan, J., Zou, C., & Liu, J. (2012, March). Security in the Internet of things: a review. In *2012 International Conference on Computer Science and Electronics Engineering* (Vol. 3, pp. 648–651). IEEE.

Tao, H., Bhuiyan, M. Z. A., Abdalla, A. N., Hassan, M. M., Zain, J. M., & Hayajneh, T. (2018). Secured data collection with hardware-based ciphers for IoT-based healthcare. *IEEE Internet of Things Journal, 6*(1), 410–420.

Telagam, N., Kandasamy, N., & Nanjundan, M. (2017). Smart sensor network based high quality air pollution monitoring system using Labview. *International Journal of Online and Biomedical Engineering (iJOE), 13*(08), 79–87.

Telagam, N., Kandasamy, N., Nanjundan, M., & Arulanandth, T. S. (2017). Smart sensor network based industrial parameters monitoring in IOT environment using virtual instrumentation server. *International Journal of Online and Biomedical Engineering (iJOE), 13*(11), 111–119.

Tewari, A., Jain, A. K., & Gupta, B. B. (2016). Recent survey of various defense mechanisms against phishing attacks. *Journal of Information Privacy and Security, 12*(1), 3–13.

Thakur, B. S., & Chaudhary, S. (2013). Content sniffing attack detection in client and server-side: A survey. *International Journal of Advanced Computer Research, 3*(2), 7.

Uttarkar, R., & Kulkarni, R. (2014). Internet of things: architecture and security. *International Journal of Computer Applications, 3*(4), 12–19.

Valdivieso Caraguay, Á. L., Benito Peral, A., Barona Lopez, L. I., & Garcia Villalba, L. J. (2014). SDN: evolution and opportunities in the development of IoT applications. *International Journal of Distributed Sensor Networks, 10*(5), 735142.

Venugopalan, V., & Patterson, C. D. (2018). Surveying the hardware Trojan threat landscape for the internet-of-things. *Journal of Hardware and Systems Security, 2*(2), 131–141.

Vučinić, M., Tourancheau, B., Rousseau, F., Duda, A., Damon, L., & Guizzetti, R. (2015). OSCAR: Object security architecture for the Internet of Things. *Ad Hoc Networks, 32*, 3–16.

Wang, X. (2014). *Hardware Trojan attacks: Threat analysis and low-cost countermeasures through golden-free detection and secure design* (Doctoral dissertation, Case Western Reserve University).

Wang, X., Tehranipoor, M., & Plusquellic, J. (2008, June). Detecting malicious inclusions in secure hardware: Challenges and solutions. In *2008 IEEE International Workshop on Hardware-Oriented Security and Trust* (pp. 15–19). IEEE.

Wang, G., Zhang, W., Cao, G., & Porta, T. La (2003). On supporting distributed collaboration in sensor networks. In *Military Communications Conference, 2003. MILCOM'03* (vol. 2, pp. 752–757). IEEE.

Wang, X., Salmani, H., Tehranipoor, M., & Plusquellic, J. (2008, October). Hardware Trojan detection and isolation using current integration and localized current analysis. In *2008 IEEE International Symposium on Defect and Fault Tolerance of VLSI Systems* (pp. 87–95). IEEE.

Yang, K., Hicks, M., Dong, Q., Austin, T., & Sylvester, D. (2016, May). A2: Analog malicious hardware. In *2016 IEEE Symposium on Security and Privacy (SP)* (pp. 18–37). IEEE.

Yin, D., Zhang, L., & Yang, K. (2018). A DDoS attack detection and mitigation with software-defined Internet of Things framework. *IEEE Access, 6*, 24694–24705.

Yu, B., & Xiao, B. (2006). Detecting selective forwarding attacks in wireless sensor networks. In *20th International Parallel and Distributed Processing Symposium, 2006. IPDPS 2006* (p. 8). IEEE.

Zhang, F.-J., Zhai, L.-D., Yang, J.-C., & Cui, X. (2014). Sinkhole attack detection based on redundancy mechanism in wireless sensor networks. *Procedia Computer Science, 31*, 711–720.

Zhao, K., & Ge, L. (2013, December). A survey on the Internet of things security. In *2013 Ninth International Conference on Computational Intelligence and Security* (pp. 663–667). IEEE.

10

Study of Hardware Attacks on Smart System Design Lab

N. Ramasubramanian and J. Kokila

*National Institute of Technology, Tiruchirappalli,
and Indian Institute of Information Technology,
Allahabad*

10.1 Introduction

This chapter illustrates elaborately the use of technology to design a hardware platform to handle IoT edge-computing requirements to secure against physical attacks, along with countermeasures and case studies. Based on hardware requirements, there are some features that are to be comprised by the state of art through IoT devices. The power of consumption at all levels marks the major concern for specific applications. Due to upgradation of millions of heterogeneous devices at the next level of its lifetime, and the accommodation of new technology such as 5G in deep learning with an extensive amount of energy (Capra, Peloso, Masera, Roch, & Martina, 2019).

Basic hardware components involved in IoT are presented in the first subsection. IoT architecture is described with a focus on physical and data link layers. In an IoT application, edge computing is performed at the end poles which need to be customised and deployed in the platform; hence, IoT platforms with edge-computing components have also been discussed. Major components required for smart homes, industry, and agriculture are discussed with their benefits for IoT.

Hardware attacks are unpredictable and multiplying day by day. Citing recent studies, some major hardware attacks are described in detail with characteristics, implementation details, results, and countermeasures. The next subsection deals with IoT-based attacks, which are classified into physical, side-channel, cryptanalysis, software and network attacks. Some of these attacks are physical-structure-based, such as tampering, DOS, and battery draining. The side-channel based attacks are timing, power, fault, electromagnetic and environmental analysis. The hardware Trojan, botnet, ransomware, data integrity theft, and IP hijacking are crypto-based attacks. The new attacks caused by the increase in smart devices are also projected in this chapter. The last session of this chapter will suggest countermeasures for the above attacks. A recent IoT simulator is used to simulate the smart system design lab and the threat analysis is also performed for the hardware attacks.

The contributions of the chapter can be consolidated as follows:

1. The various architectures of IoT, applications involved in IoT devices, and their hardware components, such as platforms and edge-computing, are described.
2. The importance of hardware security, classification of attacks based on the edge computing, characteristics and countermeasures are studied.
3. A case study is proposed on system design lab which is simulated using TinkerCAD; its threat analysis is reported with a focus on hardware attacks.

10.1.1 Basics of IoT Devices

IoT is used in daily life to make everything smarter, including the home, transport, cities, agriculture, education and industry. It's difficult to separate an application from IoT nowadays. Something as simple as an electric bulb or as complex as a smart city with essential sensors and actuators is connected to the internet across all the physical devices in the city.

According to the user, everything should happen efficiently without the intervention of humans which is automated by an IoT technology. Billions of things around the world are based on IoT architecture. To study these things, we need a simple computation-based model for input, processing, and output. Input is used to collect external data, like temperature, light, audio, video, and any sort of analog or digital data. Processing handles the raw data at two levels: the edge level and the network level. Input may be command or mechanical means which is given at the edge level or at the network level . The machine may showcase output according to the specific application.

In Figure 10.1, an example of a simple IoT operation is illustrated. In a room, a light is automatically switched off or on based on sensed input. The sensor detects whether a person is present or absent; that sensed information is sent to the cloud center for processing. The cloud center sends a command to the actuator to push the switch off or on. The above example seems simple, but if millions of heterogeneous things like electric bulbs, AC, cameras, televisions, among others, are automated in an advanced network in various domains, this may create a challenge for the researchers.

The sensors and actuators are globally accepted as primary components in most IoT applications needed to accomplish important tasks. Such tasks are recognizing, triggering, and scheming at each level of IoT architecture. Hence, depending on the device, it is unsatisfactory to perform smart operations in most of the applications. Platform-based edge computing is required in edge devices to perform efficiently offline and provide support to make decisions. IoT edge computing is to service the multilevel servers with growing computational abilities to provide heterogeneous processing offline with the help of low power IoT, system on chip (SoC), field programmable gate arrays (FPGAs), and mobiles. Edge computing has advantages like low power and cost, sufficient bandwidth, security, and more efficiency than cloud computing. Hence, cloud computing is

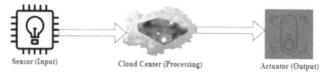

Sensor (Input)   Cloud Center (Processing)   Actuator (Output)

FIGURE 10.1
Simple components involved in IoT devices.

replaced by edge computing in most IoT applications. Edge devices will increase from 20 billion in 2020 to 100 billion by 2025. Hence, IoT edge computing is also considered a component of IoT devices.

10.2 The IoT Architecture

IoT architecture refers to the purpose, association, and execution of a set of rules to transfer data among all levels of the IoT system, since during processing, data are scattered all over the IoT network and edge devices. We must understand deeply data and proximity. If information has to be delivered faster, processing of the data should be near end devices. A clear understanding about the processing steps and the level of data transfers across the network is required . Hence, insight to IoT architecture is essential to reduce processing and enhance performance. IoT architecture is difficult to define because of billions of raw data and the heterogeneity of processing and devices available in the real world.

IoT architecture is dynamically changing based on fine tuning in each decade. Various researchers propose a multi-layered architecture featuring third, fourth, fifth and seventh layers, described in Table 10.1, with merits and demerits. The layered approach of different IoT architecture is shown in Figure 10.2.

Requirements of the future IoT architecture:

An IoT architecture is necessary for the following reasons (Yaqoob et al., 2017):

- Data must be parallelly collected from heterogeneous sensors or actuators. The support for analysing and controlling such a large amount of data is needed.
- Data must be handled efficiently to be understood; useful data must be separated from raw data.
- Efficient network communication must be set up for each IoT device with flexible, secure, and lightweight protocols.

TABLE 10.1

Different Layered Architecture for IoT with Merits and Demerits

IoT Architecture	Proposed by	Merits	Demerits
Third Layer	Wu, Lu, Ling, Sun, and Du (2010)	Defines the basic idea of IoT	Fine tuning is needed
Fourth Layer	Darwish (2015)	The Specific device information is also mentioned.	User intervention is missing at the last stage
Fifth Layer	Wu et al. (2010)	The main tasks here, are visualised and managed.	The latest technology has to be adopted such as machine learning, data science, etc..
Seventh Layer	Santos, Ameyed, Petrillo, Jaafar, and Jaafar (2020)	Each layer is fine-tuned to explain the concept with recent technology.	Security has to be maintained at each level.

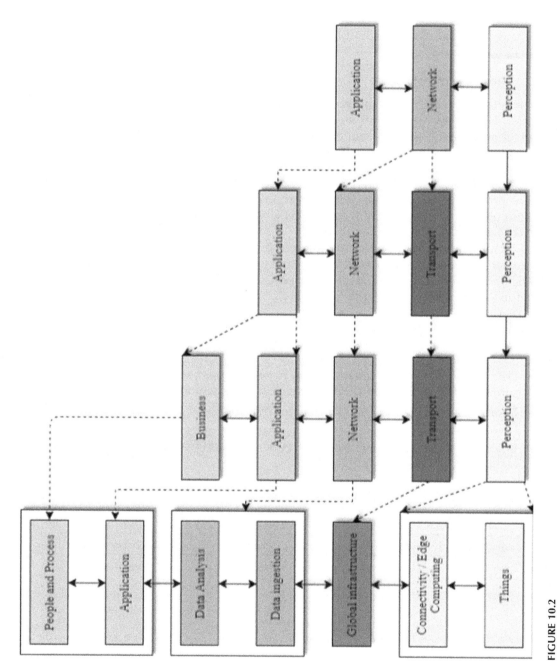

FIGURE 10.2
Three -to-seven layer architecture of IoT.

- IoT architecture should be measurable in order to add and remove the number of devices anywhere and at any time.
- Each software device in the network platform must be made more secure by monitoring and encryption mechanisms.
- IoT solutions must be accessible and reliable.
- Devices and platforms must be independent, as must the use of network and software.
- The layers of IoT architecture must support open standards and be interoperable.

10.2.1 Components of the IoT Architecture

An IoT solution can be framed based on the selected hardware; this responsibility lies in a research proposal and project execution of an engineer. Hardware choice can decide the cost, time, user-friendliness and performance of an application. The four major components of IoT architecture are edge devices, network, platform, and software; devices and platforms being the hardware components (Chanda, Das, Banerjee, & Chakraborty, 2021). Edge devices are connected using network components and applications that are developed on the platform.

10.2.2 An IoT Platform

An IoT platform is a multi-layered technology used to systemise and control the IoT device which is connected to the network. It can also be defined as a facility to bring things online for deploying specific applications. An IoT platform incorporates all other things in the IoT layer to simplify data and resource management, and provide security at different levels across the IoT. The IoT platform offers the following features: device management; customization for deployment; dynamic updating; interoperability; efficient communication; and individual service. Platforms must offer user-friendly software with the goal of solving all problems of an IoT ecosystem.

The IoT platform is mainly used to decrease complications in the organization and implementation of IoT. It also provides intellectual features for the designers, developers, and service providers through the console, software, APIs, data engine, APPs, IPs, and a modern set of algorithms. Due to the rise of smart devices, now in the trillions, the IoT network comes with challenges that must be handled by IoT platforms. The IoT application is the major cause of the challenges and the following aspects must be considered: device identification and management; updating and personalization; flexibility and data organization; security service; providers; connectivity; and abstraction in programming. Using the IoT platform simplifies the design and development process. It provides security aspects. It allows updating, adding n-devices and minimising cost and latency.

10.2.2.1 Types of IoT Platform

The roots of IoT are devices, edge components, and connectivity; therefore, the IoT platform is very modest as an APP installation that permits things to be connected to servers. Nowadays, it is difficult to evaluate the number of connected devices and things which are specifically designed for handling difficult work in support of IoT facilities. Hence, IoT platforms are in demand, with their extended properties like security and

TABLE 10.2

Types of IoT Platform Based on Leading Companies (Tzafestas, 2018)

Company's Name	Amazon	Cisco	Google	IBM	Microsoft
Platform's Name	AWS IoT Core	Cisco IoT cloud Connect	Google cloud	IBM Watson IoT	Microsoft Azure IoT suite
Device management	Yes	Yes	Yes	Yes	Yes
AI and ML capability	Yes	Yes	Yes	Yes	Yes
Flexiblity	Yes	Yes	Embedded Operating system	Yes	Yes
Device Security	Device Defender	Yes		Yes	Yes
Device Protocol	MQTT, HTTP, Web Sockets	MQTT, HTTP	MQTT, HTTP	MQTT, HTTP	AMQP, MQTT, HTTP

privacy, AI, and ML, which suit real-world scenarios. The types of IoT platform offered by leading companies (IoT platforms, 2020) are listed in Table 10.2:

10.2.3 IoT Edge Computing

The computation was performed over the end device with significant improvement in the context. Edge is not a computation layer, but it has progressed as an updated version of cloud computing in various ecosystems such as IoT (Gonzalez, Hunt, Thomas, Anderson, & Mangla, 2020). Four issues related to the progression of new edge computing are listed below:

10.2.3.1 Cloud Computing

The cloud provides computation, storage, and services related to communication for many applications, such as industries, academics and healthcare. Some examples of sources for storage are Google Cloud, Amazon and Azure with other extended capabilities, such as content delivery networks (CDN) (Banerjee, Chakraborty, & Paul, 2019).

10.2.3.2 The IoT Gateway

The IoT gateway is a device used to enable communication between local devices and cloud servers. A new protocol called "Message Queuing Telemetry" Protocol (MQTT) is required to collect data for many devices and send it to the infrastructure. The IoT gateway uses the MQTT for processing; it also filters before transferring the information to the cloud.

10.2.3.3 Artificial Intelligence

AI is also referred to as Machine Intelligence; it performs programmed or simulated actions similar to humans. Deep learning can be implemented in IoT to achieve interpretation in the real-world. Fault tolerance must be realised by considering high performance of industry equipment and devices.

10.2.3.4 5G Networks

The 5G network is the descendant of the 4G network; it was deployed in 2019. This network provides higher bandwidth and downloading speed of approximately 10 Gbit/sec (Chakraborty and Rodrigues Joel, 2020).

Hence, this network was replaced with a new device to enable new technological transfer in IoT and machine learning. Telecom has adopted the 5G network to adjust the gap between the cloud and edge computing.

10.2.3.5 Types of Platform for IoT Edge Computing

Basically, the IoT platform is divided into five types based on the commercial services delivered to the consumers at the end-level. They are end-to-end; connectivity-based; cloud-storage-based; application enabling; device management; and analytics. These platforms are used in each level of the architecture separately or collectively to deploy and implement the applications in industrial IoTs.

10.2.3.6 The architecture of IoT Edge Computing

Core IoT edge architecture consists of penta components to explain the different facilities and has the ability to support IoT systems such as data management, offline support, updated applications, AI support, and complex processing. Figure 10.3 shows the architecture of IoT Edge computing.

10.2.3.7 IoT Edge Devices for Now and the Future

Edge computing has become a feasible technology for applications such as smart houses, smart cities, smart industries, and agriculture. These applications need to collect billions of heterogeneous data for processing and analyzing it at the edge, which is a challenging task. Hence, we must study the computation and evaluation of edge devices in terms of four applications. Security threats are also increasing at the edge level of the real-world hence we need to take care of the hardware. Hardware-based attacks for IoT edge computing are listed in this chapter with examples.

10.3 Hardware and Software Components of IoT Applications

Automation can be implanted in the home, industry, agriculture, etc., with the involvement of IoT technology, making all those applications smarter. It also involves both hardware and software at each level of the system design.

10.3.1 Smart Home

To make a home smarter, the following hardware components are needed (Perry, 2018):

- An IoT computer is essential to control devices that are included in the smart house.

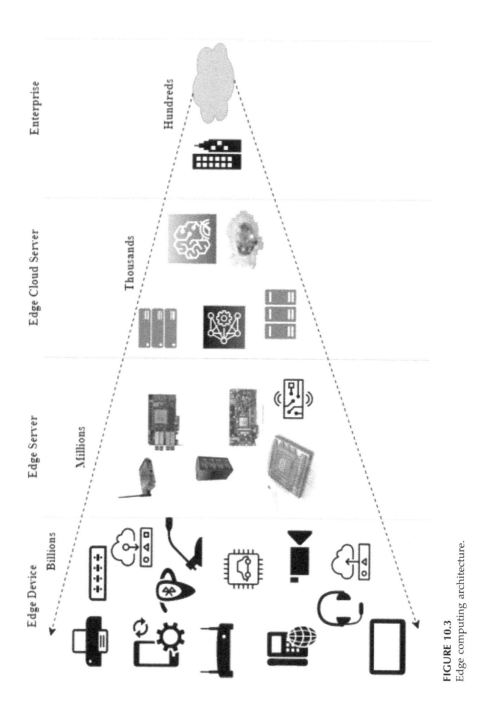

FIGURE 10.3
Edge computing architecture.

- Signals from various devices must be received and transmitted to the IoT computer using the receiver and transmitter IP.
- An opening, like a wireless electrical outlet, is needed to listen and respond to on/off signals respectively.
- An alarm system is also needed at windows and doors to sense the signals and control the status.

The requirements of software and hardware components are listed below:

- An operating system is needed which can run on the IoT computer with the latest version.
- The RF receiver and transmitter have logical units that are used to access input and output pins of the IoT computer. These logical units must be downloaded from an online source and support C or C++ library to control the specific hardware that runs on heterogeneous platforms.
- Github code for home automation is available which can be used for reference; additional features can also be added to verify the results.
- Various simulators or emulators have been developed to learn how to work IoT applications without real hardware.

Figure 10.4 shows the simple illustration of a smart home. The smart house may also need a gateway for routing the sensed data to the server; an individual device is not enough, due to lack of battery storage and network protocols (Tzafestas, 2018). The following factors are required to choose the impeccable gateway for smart home applications: basic building blocks; maintainable network protocol; secure configuration; real-time support; and MQTT, COAP, HTTP provision. In the initial stage of marketing, an IoT gateway depends on the basic building blocks and supported hybrid protocols.

IoT gateways can be created from scratch using hardware such as Intel edition, Rasberri Pi etc., or existing gateways can be used which are available in the market. These existing gateways are customizable and easily communicate to the cloud and devices with robust prototype support. The smart house can be designed using open-source IoT platforms such as Home Assistant, Calais, Domoticz, Openhab, Openmotics, LinuxMCE, iPhone etc.

10.3.2 Smart Industry

Industrial IoT includes a device which is connected across all industries such as business, healthcare, transportation, marketing, or utilities. Smart industries allow managers to gather and evaluate data automatically in order to make intelligent decisions and improve product quality.

Industries are using IoT to become smarter, acquire efficiencies and save money, but there is a practical challenge to implementing industrial IoT. Many industries initially got loans for equipment and technology and they are connected to their intranets. There are many standards which must be adopted to connect the machines to the internet and to make the existing software system manage data from different industries. The real challenge lies in figuring out the local instrument to connect the internet and respond to the cloud service for

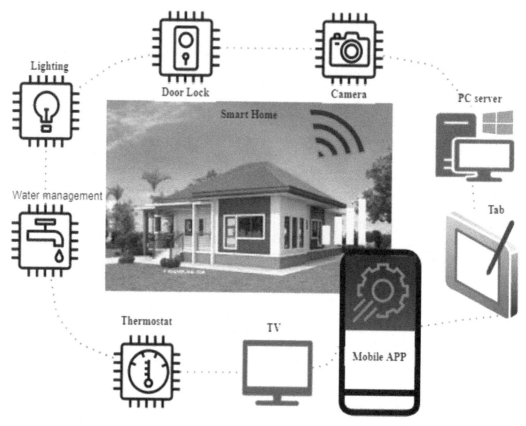

FIGURE 10.4
Features of a smart home.

evaluation. Smart industry is a network of interconnected computing devices embedded in many ways to collect and swap data, which are then useful in a business context to create operational savings and efficiencies (Saqlain, Piao, Shim, & Lee, 2019).

The benefits of the smart industry (Infinite Uptime, 2018) are listed as follows:

10.3.2.1 Improving Efficiency

Industries save money and time by controlling connected devices. Machine-level optimisation is required for examining efficiency. Many industries may also waste energy by not switching off unused machines, light, fans, etc. These issues will be taken care of by IoTs to save electricity.

10.3.2.2 Increase Uptime

Firms with inadequate funds and older machines frequently discover that they are facing a challenge. The lifetime machine has to be improved without replacing internal components. The IoT has a facility such as analytical maintenance and machine monitoring with remarkable results to manage the above challenge.

10.3.2.3 Improve Safety

IoT is also used to safeguard employees in the industries. The continuous monitoring and statistical analysis of each machine in an industry will find faulty machines. Hence, work-related injuries will be decreased to a minimum and the employees' safety will be ensured by adopting the IoT.

The Components of an IoT system are listed below:

10.3.2.4 Edge Device at the Front End

The Edge device of the IoT is the sensor and control device that focuses on the constant gathering of information and reacting in such a way that the accuracy, reliability, and consistency are maintained in a closed-loop. The gathered information is sited and controlled inside the company by existing technology. We must push this information to the new networks outside by finding different ways to reduce the complexity; there are some industries where such technology is not adopted and in that case, hardware exploration has to be initiated from the beginning. A lot of companies are available on the market that help us to automate industries with ready-to-use components. Such leading companies are Monnit and Libelium, which offer an easy way to use and simplify deployment methods.

10.3.2.5 Connected Technology

The main aim of IoT is to send and receive data between equipment connected to different industries. Hence, connectivity is essential within and outside industries, which may be wired or wireless based on the requirement. The wired option is usual, less expensive, and more reliable. Wireless technology is adopted in the majority of industries, with options such as WiFi, Bluetooth, Mesh Network, cellular Network, and LPWAN technology.

10.3.2.6 IoT Platform for Data Analytics

An IoT needs software support to analyze the collected, sent and received data. The software has to make an optimised decision for the information collected at the edge. The IoT platform has reliable software to connect the edge hardware and access node across the local network to different edge computing which is located outside the industries. There are leading companies to provide such platform-based IoT with two options: an end-to-end to solution as a whole package or a customizable building kit.

10.3.3 Smart Agriculture

Smart agriculture is defined as agriculture and food producing systems that are driven by new technology such as IoT, data analytics, and data science (Saiz-Rubio & Rovira-Más, 2020). Nowadays, IoT is mostly used in smart agriculture by means of sensors used to monitor soil, plants, climate and cattle; this directly affects the productions. These directly help farmers, actuators, smart tractors, independent robots, and drones. The fields are maintained using smart greenhouses and connected spaces. Collected data are managed, visualised, and analysed.

Using IoT in agriculture has the following benefits:

- Efficiency is enhanced by growing more crops in soil that is deficient, or despoiled by pesticides or bad weather. Automation in agriculture predicts problems such conditions cause and optimised decisions are made accordingly.
- The population in India is increasing day by day; this may lead to the scarcity of food.
- Smart closed-loop agriculture will help us to grow plants in many places, such as a rooftops, home gardens, balconies, or supermarkets.
- Many IoT applications concentrate on optimising usage of resources such as water, energy, and land. Data gathered from heterogeneous sensors will help IoT edge computing to allocate enough resources for each crop.
- The use of pesticides and fertilizer is taken care by IoT devices. Hence, organic products are grown through these methods.
- The agility of agriculture is maintained by IoT solutions using a continuous monitoring and prediction system. New features of smart agriculture will also help farmers to save their crops by quickly responding to bad weather, humidity, air quality, or health of crops.
- Crops will grow faster and product quality will be higher by using data-driven agriculture.
- The best situation can be recreated using IoT based on connected systems; nutrients can be preserved in food.

10.3.3.1 Components of Smart agriculture

The major components (Aleksandrove, 2018) used in smart agriculture are as follows:

Agriculture is support by IoT technology in many of its functions. The most common usages are: monitoring climate conditions; livestock; managing crops and end-to-end farming; precision farming; predictive analysis; and use of drones in smart farming. There are several techniques adopted by smart devices that are rooted in agriculture to improve productivity and quality. The components are: hardware; use of AI; device maintenance; and flexibility.

10.3.3.2 Hardware

Devices must be created or customised using specific sensors that are chosen for the agricultural process. Sensors are chosen based on the type of data needed and the determination to solve problems in the agriculture domain. Sensors measure features such as climate, lighting, temperature, soil condition and humidity to collect data accurately and improve productivity.

10.3.3.3 The Uses of AI

Collected data should be more accurate and sensible so that AI can be used for decision making. The data analysis should be done by using machine learning and powerful algorithms to attain a clear understanding of the gathered data.

10.3.3.4 Device Maintenance

Devices used in agricultural fields can be damaged easily by the elements. Hence, the maintenance of such devices is challenging. Hardware devices chosen for smart agriculture must be robust to maintain sensors which may need to be replaced often, increasing costs and decreasing performance and quality.

10.3.3.5 Flexibility

Applications developed for smart agriculture should be personalised to practice in the fields. The owner of the field must be able to access easily applications and gather information from anywhere through his PC or smartphone. Each connected device should be independently and flawlessly connected to the internet to sense information correctly, then send and receive data to and from servers.

10.4 Hardware Security in IoT Edge Computing

In the early 2010's, designers were not concerned about securing devices, because things and their related apps were not connected to the internet. But in 2020, simple things such as bulbs, cameras, speakers and microphones, among others, are connected to the internet. This creates major threats in all such simple things and security issues should be addressed from the beginning stages.

In 2020, securing IoT devices from recent threats to safeguard data is a challenging task for designers; hence, security must be included in the device at both hardware and software levels. A platform-based way to deal with security offers different layers of safeguards by exploiting the most recent security progresses in both hardware and software to execute inside-out, far-reaching assurances. This chapter deals with hardware attacks on IoT platforms which are intended for edge computing.

To enable hardware security in IoT edge computing (Xiao, Jia, Liu, Chang, Yu, & Lv, 2019), the following are required:

- Key management must guarantee that keys are not vulnerable in a secure transmission. The things must be able to create and store keys safely, including private keys, to authorize devices.
- Encryption is recognised by hardware, hashing, and random number generation; it quickens cryptographic procedures on the gadget. This hardware saves both time and effort.
- Memory has to be secured to protect explicit areas of RAM and Flash memory from unapproved access. Sensitive code and data in memory are separated from non-sensitive to provide security, while write-once code in memory protects them from reprogramming.
- The hackers' access to attack vectors such as debugging and programming interfaces must be minimised by protecting the attack vectors.

Edge computing offers a practical processing mechanism for IoT applications and cyber security. Its increased use has introduced many security attacks in the real-world that can be put forth by four viewpoints (Ansari, Alsamhi, Qiao, Ye, & Lee, 2020):

- Limitations of edge server: In Edge computing, billions of edge devices are connected to edge servers; most of the processing is carried out in the edge server instead of the cloud server. Since the edge devices are simple and fragile, many physical attacks can damage them. The computation of cloud servers is more powerful than that of edge servers; hence, edge servers have recently become vulnerable to attacks.
- Unknown status of the Edge device: The edge device is as small as an LED and availability is in terms of billions. They do not have an identity to be remembered in the server; hence, the status of the edge device is always unknown.
- Heterogeneous OS and Protocols: Edge servers and devices can use any real-time OSes and protocols. There is no standardization for the same. Hence it is difficult to design a protection mechanism for edge computing.
- Access control mechanisms: To enable many applications, edge computing is becoming more complex in order to handle fine-grained access control; therefore, coarse-grained access control is used by IoT edge computing.

Devices at the edge level are vulnerable to attacks in numerous ways because most of the processing is done at the edge device or servers rather than data centers (Shapsough, Aloul, & Zualkernan, 2018). Some of the ways are:

- Edge devices in the network must send data to various servers for processing. Cloud servers must trust that the data sent by the device is authentic, so each device in the IoT network must be protected from false identity.
- Attacks may change the configuration of the device by malicious software or hackers; hence, the integrity of the device is lost by these attacks.
- The right level of security has to be provided in each stage of the device from edge to user. This may cause hackers to mount malware on the device and cause serious damage.
- It frequently occurs through the illegitimate usage of trade websites. Examples of such attacks are man-in-the-middle and eavesdropping.

10.5 Hardware Attacks

Hardware attacks try to obtain, modify, terminate, remove, embed, or expose information from the device through remote access or by direct physical access without strong authorization. In this section, the hardware-based IoT attacks are classified based on perception layers and how each attack affects edge computing; their special characteristics are also explained. Physical-layer-based attacks have been separately considered in regard to edge computing. Hardware attacks on IoT devices on the edge level are classified as invasive, semi-invasive, and non-invasive.

We will start with a review of the security attacks edge computing faces. These attacks are caused during the design, synthesis, and implementation phases of IoT edge computing. Protection mechanisms are the next step in distinguishing normal from harmful operation. Root causes of attacks were studied finally with some practical protection mechanisms. The taxonomy of hardware attacks is illustrated in Figure 10.5.

10.5.1 Invasive Attacks

Invasive attacks can acquire direct access to the core components of the IoT device and physically damage the device or its components. This attack requires significant knowledge, expensive equipment, and time to damage the device physically (Ibrahim, Sadeghi, & Tsudik, 2018).

10.5.1.1 Physical Attacks

IoT devices are deployed in various domains which are usually hostile; such things are susceptible to physical access. The adversary will attack the hardware or software of the things physically by deriving the secret key, altering the software, and damaging the circuits. This may lead to the destruction of hardware or software.

Physical attacks are used repeatedly to exploit recent IoT weaknesses. The attacker initially buys a physical device form the market and tries to get physical access to the IoT device to attack it. They implement a test attack on the device using reverse engineering, de-packaging of the chip, layout reconstruction, micro-probing or particle-beam techniques to find what type of outputs are obtained through the operation.

These attacks depict the vulnerabilities of the system. Examples of physical attacks (Abdul-Ghani, Konstantas, & Mahyoub, 2018) are as follows:

10.5.1.2 Tampering

Edge devices are deployed in different areas; this provides hackers with opportunities to tamper. This may damage the whole device or data can be obtained from it. Data tampering may also lead to wrong computing at the edge server.

10.5.1.3 Micro-probing

During device analysis by micro-probing, the analyst employs the same thought processes as when troubleshooting a full-size device. Micro-probing is only a tool for the analyst to access IoT nodes on the equipment while analyzing the behavior of its parts. The process of electrically pinpointing a failure site is known as failure isolation, which requires the analyst to identify abnormal voltages and/or currents. Voltage and current dimensions are achieved by electrical-dimension instruments attached to a probe-like device with a needle through the micromanipulator.

Common tools attached to the probe station are voltmeters, curve tracers and oscilloscopes, among others. Device excitation from voltage supplies, waveform generators and the like may be supplied to the device in the same manner.

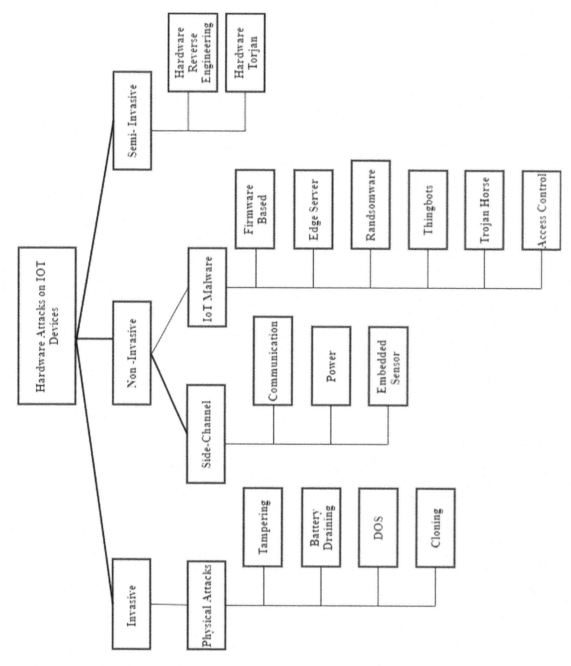

FIGURE 10.5

10.5.1.4 Battery Draining

Most edge devices have a battery to power them up. Frequent physical access of the device may drain battery power, which may slow down the processing of the whole system.

10.5.1.5 DOS Attacks

Edge servers are mainly used to send and receive data from edge devices. They also perform important computation and analysis of edge device output and send it to the next level. Physical access to an edge device can allow the hacker to replicate more devices. These replicated devices can flood the communication which will slow down or stop computation. Physical access to edge servers is also possible due to limited security. This will cause damage to edge devices and cloud servers.

10.5.1.6 Cloning Attack

IoT devices can be accessed physically by adversaries who can then clone them. These devices can become entry points by capturing edge devices of network infrastructures.

10.5.2 Non-Invasive

Such attacks extract secret information without accessing core components by data manipulation (Park, Rahman, Vassilev, Forte, & Tehranipoor, 2019). Signals, power, time delay, and input and output of embedded sensors are the data exploited by these attacks. The tools needed for these attacks are quite small and low-cost; hence, they can exploit IoT devices by installing and transferring software or information. These attacks can be further classified as side-channel and IoT malware attacks, each of which is explained below.

10.5.2.1 Side-channel Attacks

A hacker continuously gains side-channel data from the targeted system and uses recent algorithms and machine-learning techniques to find sensitive information like secret keys, accounts, and login details. Edge-computing-based side-channel attacks are classified as communication signal, power consumption, and embedded sensors.

10.5.2.2 Communication-Signal-Based Attacks

Communication attacks occur in the edge computing environment; the attacker need not be on an edge device or edge server. The attack can be launched from an outside node, which watches signals and net traces to extract secrets from the environment. Two types of communication signals are packet and wave signals. Edge computing uses a small network, with or without wires. A packet is a formatted unit used among such networks for effective data transfer. A sequence of a packet transferred between the edge device and edge serve will allow the attackers to infer sensitive information such as TCP packet number, daily activities, and user presence.

Electromagnetic signals are the next element in the edge-computing environment for leaking secret information. Recent sources of signals are modern televisions,

IoT sensor devices, fingerprint movement, WIFI and EEG. These signals are monitored at the edge device and server level, and allow attackers to infer secret information such as video content, input or output of edge sensors, and passwords, among others.

10.5.2.3 Power-based Attacks

Power is a measure of the electric energy utilization of a device. Every device has its own profile to intimate power consumption at its operating and rest states. Hence, power consumption of each device at each computation varies based on its hardware and software components, which relate secret data to it. Two types of power-based side channels are the gathering of power consumption using a meter, and the oscilloscope.

A smart meter is used to measure correctly power consumption at the office, home and industries. Attackers can monitor, measure, and analyze the power trace of simple edge devices such as TVs, washing machines, laptops, smartphones or air conditioning to infer secret information such as household activity, webpage details, and to inject malicious code.

An oscilloscope is an electronic device used to measure the voltage and current of hardware. Cryptosystems hardcode secrets into modern embedded systems. These secrets can be obtained from power traces of the hardware by continuous monitoring and analysis of the traces. Researchers have reported that all cryptosystems are vulnerable to power-analysis attack. Cryptosystems such as ECC in FPGA, and firmware in smart lights leak secrets using SPA, DSA, and CPA.

10.5.2.4 Embedded Sensor-based Attacks

Several embedded sensors are available on the market which are integrated into smartphones and tablets. Though the use of such sensors is more attractive, security remains a great challenge. Some examples are: passcodes that are leaked by the sound emitted from keystrokes of early smartphones; screen patterns that are leaked by the reflected signal of the fingertips over microphones; keystrokes that are inferred by the user's eye movement from recorded video of smartphone cameras.

10.5.2.5 IoT Malware Attack

This attack is caused by the installation of malicious software, in the form of firmware, into IoT edge devices. The major threats to hardware security are malware attacks (Kumar et al., 2019). In the case of edge computing, engaging powerful firewalls and protection mechanisms is not possible, due to the lightweight and low- power-consumption nature of edge devices. Hence, malware attacks are more predictable on IoT devices. Malware attacks are based on the infectious nature of IoT edge-computing architecture. Table 10.3 lists some firmware-based IoT malware attacks.

10.5.2.6 Edge Server Attacks

Edge server attacks are further classified into five types: SQL injection; cross-site scripting; and Request Forgery; Server-Side Request Forgery (SSRF); and Extensible Markup Language (XML) signature wrapping, which is explained in the following reference (Martin & Lam, 2008).

TABLE 10.3

IoT Malware Attacks Based on Firmware

Name	Infected	Through	Exploiting
IoT Reaper	Millions of IoT devices	Internet protocol and Wi-fi	30 RCE vulnerabilities existing in 9 different IoT devices ranging from the network router to IP camera
Remote firmware update	LaserJet printers	Online firmware updates	Remote firmware update adopted by printer
Firmware update	Smart Nest Thermostat	USB connection	Arbitrary firmware
Firmware update	Logitech G600 mouse	Networking or USB	Firmware of Mouse
Firmware modification	IoT devices	Zigbee light link protocol	Remotely and contactlessly inject malicious firmware
Passively inject malicious UIs	Mobile Devices	Benign apps	Android Task Structure
Backdoor Code Injection	Mobile Devices	Found that 6.84% of official Android apps from the Google Play Store and 2.94% of iOS apps from the Apple App Store	Libraries such as PhaLibs
Android WebView	Android devices	Malicious website	Remotely inject malicious apps

10.5.2.7 Ransomware

IoT needs to address a few digital difficulties. One of the most significant is Ransomware attacks. Ransomware is a harmful sort of programming that confines permission to essential data and request installment for gaining permission to this data. There are two types of ransomware: crypto and locker based (Humayun, Jhanjhi, Alsayat, & Ponnusamy, 2020).

10.5.2.8 Thingbots

Thingbots contain numerous connected devices such as computers, laptops, smartphones and tablets, that transfer vast amounts of data indirectly to the internet (Mannilthodi & Kannimoola, 2017). These attacks can send thousands of message requests to a target; the platform may crash when handling such a huge amount of data.

10.5.2.9 Trojan Horse

IoT generates a new ecosystem where malware can be used to produce powerful botnets. Mirai, a new Trojan virus for Linux, is tough to identify and which would have occurred in the remote systems. The risk is an advanced variant of Gafgyt, (aka BASHLITE, aka Torlus) malware, which has been castoff by DDoS service providers.

It realizes as an ELF Linux executable and drives mainly on DVRs, routers, web IP cameras, Linux servers, and other devices running Busybox, a mutual tool for devices.

Access control for devices and servers in edge computing is achieved by the authentication and authorization processes. These processes are vulnerable to attacks. Four categories of access control attacks are listed below (Ni, Lin, & Shen, 2018):

- Attacks target weaknesses in the authentication process by employing loopholes in WPA/WPA2 security protocols.
- Attacks target logical weaknesses or design flaws that may present in authorization protocols used by the edge computing systems.
- Dictionary attacks use a password dictionary to get earlier credentials of authentication systems.
- In over-privileged attacks, the attacker tricks the victim system into conveying higher (than required) access rights to an app or device, which can then be used to perform malicious activities inside the network.

10.5.3 Semi-Invasive

For semi-invasive attacks, an adversary needs to contact the surface during the design, synthesis, implementation, or fabrication phase of the device to add a faulty circuit as part of the device; however, the attack will not damage the outer layer of the device. These attacks are more effective than invasive, and have lower costs and are less time-consuming than non-invasive (Skorobogatov, 2005).

Hardware Reverse Engineering: adversaries access the main part of the device to understand how it works. A competitor may buy a new technology processor in the market, disassemble it, and produce a similar new processor (Gordon, Kilgore, Wylds, & Nowatkowski, 2019). The steps involved in reverse engineering are:

- IoT devices are physically reviewed.
- Device firmware images and file systems are abstracted by evading boot-time security of firmware and improving the information with out-of-band means.
- Firmware images are analyzed and its secrets are retrieved.

Hardware Trojan (HT): a simple malicious circuit can be added to damage the device after two or three uses or by triggering it, which may not be known to the user. An adversary may access the device and inject hard code locally or remotely to damage the device. These attacks can fully damage the IoT edge device and cause a denial of service to edge servers. They block security measures remaining in the device, causing the infected device to function incorrectly. The HT can lead to information leakage, evade the security of the system, or cause damage to the whole device.

HTs are a direct threat to already vulnerable IoT. The behavior of an HT cannot be changed once it is inserted. Hardware Trojans are different from Trojan horses because they cannot be removed just by a firmware update. They are very harmful and more challenging to eliminate. HTs may cause much damage to the IoT ecosystem, such as information leakage, denial of service (DoS) attacks, service degradation, and failure of the device. The HT is classified based on physical characteristics of the attacks that are referred to in (Sidhu, Mohd, & Hayajneh, 2019).

10.6 Countermeasures

A countermeasure is a technique that reduces an attack by eliminating it, by reducing the harm it can cause or reporting it so that accurate action can be taken. Countermeasures for such attacks are listed below:

The process for counteracting physical attacks:

- Ensure that the devices and edge servers have no unprotected ports or connectors.
- Implement locks or other ways to ensure that only authorised people can get access to your product.
- Identify the things in the ecosystem to be installed externally from common reach
- The physical unclonable function (PUF) is considered the solution for physical attacks (Babaei & Schiele, 2019) since they would harm the internal construction of the PUF if we try to access it when it is in operation. The PUF is a dynamic secret generator; hence, we cannot access the key randomly.

Preventing side-channel attacks is very challenging due to their passive nature, but it can be achieved if sensitive data are not accessed or leaked by the side channel. Side-channel-based obfuscation and Trust-Zone hardware are used to prevent these attacks (Strackx & Piessens, 2017). Data perturbation and differential privacy are the general defense mechanisms recommended by recent researchers.

A promising technique is detect-and-filter to prevent server-side injection attacks. The fine-grain contact regulator and fixed scrutiny of malicious programs are the major mechanisms used for injection attacks (Halfond, Viegas, & Orso, 2006). Research on creating means to mitigate firmware modification is also accepted for defending against such attacks.

Two-factor authentication is a well-known robust mechanism against dictionary attacks. To enhance the security of the authentication protocol, cryptographic implementation should be strong and new secure protocols with enhanced features should be used, such as OAuth 2.0. The operating system on edge devices must use a resilient authorization model to resist over-privileged attacks (Wang & Chen, 2014).

There are many studies taking advantage of well-known dynamic or static analysis for detecting IoT malware; however, static-based methods are more effective when addressing the multi-architecture issue. A thorough survey of static IoT malware detection is specified in (Ngo, Nguyen, Nguyen, & Le, 2020).

To protect IoT devices from ransomware, resetting the device is not alone sufficient, and also some data files have been installed periodically. Mitigation should be enabled in IoT networks to cope with the heterogeneous property of edge networks (Humayun et al., 2020).

Hardware Trojans are prevented in the insertion level using obfuscation in the functional and structural modules of the design.

Another obfuscation technique, termed camouflaging, adds dummy contacts which keep invaders from attaining the correct netlist from the layout level. Invaders insert malicious code in idle empty spaces in the circuit to create HTs. If these spaces are occupied by functional cells, hackers cannot populate them with Trojans because of the functionality.

10.6.1 Security Measures for IoT Devices

- A quick search for IoT device track records for security is essential.
- The software used in the device should be up to date to avoid recent attacks.
- IoT devices which cannot be repaired should be used minimally to reduce risk and prevent the execution of malicious code.
- Use a lightweight protocol to secure devices from entry-point attacks.
- Since the devices are simple, use a complex password and avoid the use of default.
- Advanced configuration must be used to set up IoT security.
- Secure connection to IoT devices is ensured by the latest security protocols and strong passwords.
- Limit direct access of IoT devices; this may lead to device tampering.
- Deactivate universal plug-and-play provisions whenever possible.
- Common malicious codes are stored in volatile memory; they can be removed by periodically switching off and restarting the devices.

10.7 Case Study on Smart Lab

The article (Knight, Kanza, Cruickshank, Brocklesby, & Frey, 2020) presents a smart lab with a case study on a prototype system, which uses voice, text, and visual dashboards. This system combines sensors and raspberry pi's to enable communication inside a lab. The above article inspired me to write a case study on smart labs. Automation of a hardware laboratory for an engineering student is a serious issue because of the following:

- Physical access to the device is not permitted.
- Software is node-locked for the device and there is no trial version available outside the lab.
- Learning how to connect each component in the breadboard and know to handle each device with proper care.
- Code can be implemented for the specific application and embedded in heterogenous kits to view the outputs.
- Output can be viewed physically, which creates an interest for the student to learn and explore more in hardware labs.
- Logical and physical errors can be analyzed and reported.

It is better to safeguard students and faculty by providing them a platform through optimization of the hardware lab using IoT platforms, such as Orange kit and MPSoC.

In this case study, we have considered a lab titled "Computer-system Design lab". This laboratory uses hardware and software listed in Table 10.4.

TABLE 10.4

Hardware and Software Specification of System Design Lab

Development Kits	Specification	Software tools
Mbed NXP LPC11U24	32-bit ARM Cortex-M0 microcontroller	Free online IDE and C++ compiler
Mbed NXP LPC1768	ARM Cortex-M3 based microcontroller	High level C/C++ SDK
Raspberry Pi	7000MHz Broadcom BCM2835 CPU with 512MB RAM	Installation package version of logi.RTS
Scientech 6205 IoT Builder	Processor: 64bit ARMv7 Quad Core Processor 1.2GHz	C Programming and Python
Zybo Zynq-7000	Features the Xilinx Zynq^{TM} Z-7010 AP SoC 512MB DDR3/128Mb Serial FLASH	Xilinx Vivado Design Suite. Xilinx SDSoC v2016.3 (25 User Licenses from Xilinx)
Zedboard Zynq-7000 APSoC	Zynq-7000 AP SoC XC7Z020-CLG484Dual-core ARM Cortex A9	Xilinx Vivado Design Suite. Xilinx SDSoC v2016.3 (25 User Licenses from Xilinx)

10.7.1 Automation of Smart Lab

The steps to create a smart system design labs are as follows:

1. Select an academic platform for the smart lab.
2. Select the required hardware and software for automation.
3. Consider progress in terms of scalability.
4. Choose applications that are extremely fast.
5. Discuss hardware attacks.

10.7.1.1 Select an Academic Platform for Smart Lab

Numerous platforms can be used to build a smart lab. Supporting technologies and costs are to be considered while selecting the platform. These are the recognised platforms for smart lab:

FIWARE, Kaa, ThingWorx, Salesforce, and Zetta are various open-source tools used in real-time for end-end IoT solutions with protocols such as MQTT and CoAP.

10.7.1.2 The Hardware and Software Requirements

To atomise the lab these modules are involved: a lab monitoring and control system; a student attendance system; a lab manual system; and kits which monitor and evaluate.

10.7.1.3 Lab Monitoring and Control System

To check the status of the system before the lab session, lights, fan, and air conditioners are switched on and off based on requirements. Embed boards are placed in front of each system and their operations are verified before the lab sessions. Sensors used are photonic sensors, temperature sensors and thermostats.

10.7.1.4 Attendance System

The students are given a token for system access in the lab based on admission details. A specific user name and password are also assigned to them to authorise use of the lab resources.

10.7.1.5 Lab Manual System

Study materials and online classes are available each hardware equipment so that students can clarify thoughts and post queries regarding hardware, software, and experiments.

10.7.1.6 Kits usage Monitoring and Evaluation

Since the hardware kits used are heterogeneous, the processes of monitoring and evaluation are challenging. The kits use different communication modes such as local networks, Wi-Fi, and individual systems to synthesis and execute code. Classification has to be made to separate the kit's usage and evaluation software, developed for each student individually.

Based on institute requirements, practical assessment labs, end semester labs can be conducted and the node lock can be enabled at the time of the exam. Since heterogeneous kits are used, priority based on device location is enabled for easy evaluation. Assessment may be calculated on regular practice, model lab, and end-of-semester performance.

10.7.1.7 Consider Progress in terms of Scalability

Scalability is achieved by using smaller and more particular processes. The most important characteristic of IoT is scalability, which means that at each level the device, edge computing, and cloud computing will increase in terms of billions, thousands, and hundreds respectively. These include features such as service in terms of hardware, software, and network. The scalable things of smart labs are number of students, IoT devices, development boards, and usability of the software.

All these things must support the scalability concept in platforms used in the IoT ecosystem.

10.7.1.8 The Operation of the Application should be Extremely Fast

The physical deployment of things using hardware or software while connected to a faster and more intelligent network makes smart labs extremely fast. In order to offer minimal device access as a package of speed up the steps are listed as follows.

- When devices are powered on and being used by the student, a message is sent to the software management system of the lab.
- This software will enroll the unique device ID, student ID and network usage.
- Continuous monitoring can be done on the device characteristics so that no manual entry is needed and error-free operation can be carried out.

Figure 10.6 shows the automated lab system design, for which components used for each module at the edge level are specified. Edge devices are connected to the Rasberry pi, Zynq MPSoC, and IoT builder for performing intelligent actions by collecting sensor information. These levels are termed edge servers because all collected information from edge devices is efficiently manipulated. The next higher level is the edge cloud, in which high-performance networks play an important role. Monitoring at this level will be done by the institute's computer support group.

10.7.2 Simulation of System Design Lab

TinkerCAD and Circuits TinkerCAD provide a browser-based application for designing, simulating electronic circuits, and creating PCB boards. An autodesk circuit simulator can simulate Arduino-based projects for testing designs and programs before creating them in real life. The system design lab consists of four modules in which lab monitoring and control systems and student attendance systems are simulated using TinkerCAD.

10.7.2.1 Lab Monitoring and Control System

A simulation is developed in TinkerCAD for keeping track of the number of students in a lab. When students enter the lab, PIR sensors at the door detect them, and the count of students in the lab is incremented. Along with that, appliances such as lights, fan, and air conditioning are switched on automatically. The appliances stay on as long as even one student is in the lab. When a student exits the lab, it is also detected by the PIR sensors at the door. This changes the count accordingly. When the count variable is zero, the appliances are turned off automatically.

Components used are Arduino, Breadboard, 16 x 2lcd, two Pir sensors, two leds (represent two lights), two dc motors (represent two fans), resistors and wires.

10.7.2.2 Attendance System

For each lab session, the student will receive a passcode to their registered mail ID according to the specific equipment they are going to use. This passcode will be unique so that the full lab session is monitored for each student; credentials are stored with login. If the passcode is correct, the student's attendance will be recorded, otherwise he is considered absent for the session. The student enters the password on a keypad. The key entered is checked by the Arduino code. If the key matches that in the code, specific hardware access is allowed for the related login account. After three hours, access to the hardware and its accounts are restored automatically.

Components used are Arduino, Breadboard, 16 × 2 lcd, 4 × 4 keypad, a servo motor (which represents the access control), resistors, and wires. Simulation results for both modules are shown in Figure 10.7.

10.7.3 Security Threat Analysis

A security threat analysis typically consists of finding the specific IoT device to be protected, as well as classifying and estimating potential threats. The IoT devices used in the smart lab are based on the four modules specified in section 10.7.1.

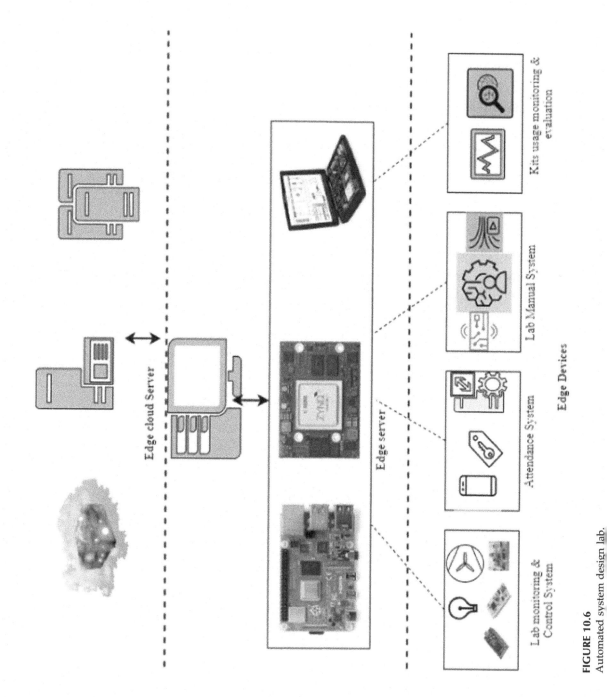

FIGURE 10.6

Automated system design lab.

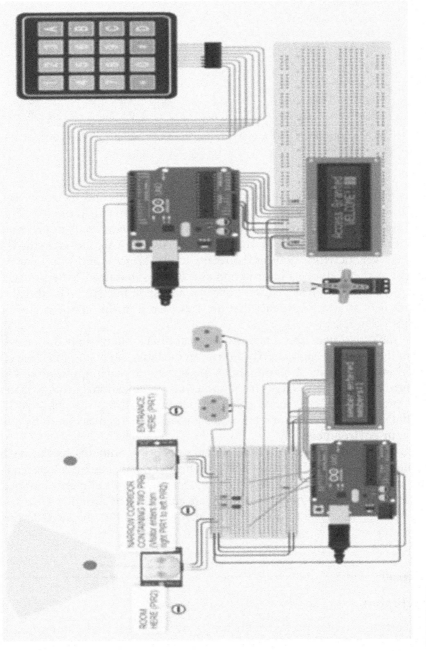

FIGURE 10.7
Simulations report for lab monitoring and control system and student attendance system.

TABLE 10.5

Security Threat Analysis Report for System Design Lab

Outcome /Possibility		IoT Hardware	Edge Server	Network	Data
Invasive	Tampering	C / C	R / C	D / C	N / D
	Battery Draining	R / A	R / B	R / B	N / D
	DoS	R / B	R / B	R / B	N / D
	Cloning	R / C	N / D	N / B	R / C
Non-Invasive	Side-channel	C / B	R / B	D / B	C / B
	IoT Malware	C / C	R / C	D / C	C / C
Semi-Invasive	Reverse Engineering	C / C	N / C	N / C	R / C
	Hardware Trojan	R / B	R / B	R / B	N / D

Outcomes: C—critical, R—restricting, D—disruptive, N—no impact Possibility: A—Certain (1) B—likely (0.25) C—unlikely (0.75) D—impossible (0).

A record is maintained to collect information on components and their security threats for the specific ecosystem. An example of such a record is shown in Table 10.5. The results displayed in this record were confirmed during the requirements analysis process and are specific to a certain lab. The results of the threat analysis may vary based on the hardware components, server, network devices, software used, services, and data. For example, security threat analysis involves information and devices that need to be protected from hardware attacks such as invasive, non-invasive, and semi-invasive. The threat analysis is united with lists of attacks, threats that are considered in this chapter, and potential vulnerabilities.

The above table describes the outcomes of each attack along with the possibility of occurrence in the system design lab. The entries in the table can be modified based on the specific application. Security threat analyses are specific to each application. One technique to reduce bias is to have agents from numerous organizational groups contribute to the analysis process. This gets many different perspectives into the analysis. It is also suggested that you review your threat analysis periodically, such as annually, to identify changes in your environment.

Smart labs are vulnerable to physical attacks, since the students are using the kits physically and with less care. The plug-and-play firmware must be updated frequently to avoid malware attacks. A rigorous security protocol is needed to detect unauthorised or unintended actions. The components in the lab are controlled by an aging system which is too specialised and expensive to update frequently and integrate the latest technologies.

10.8 Conclusion

IoT applications are tightly packed with hardware, software, and communication. We have tried our best to give detailed analysis of hardware components along with their potential attacks in specific cases. On the other hand, software and communication-based literature still need to be explored keenly. Simulation results are also included with threat analysis for the system design lab. This may help us to implement an automation lab with prevention techniques for hardware attacks.

References

Abdul-Ghani, H. A., Konstantas, D., & Mahyoub, M. (2018). A comprehensive IoT attacks survey based on a building-blocked reference model. *IJACSA) International Journal of Advanced Computer Science and Applications*, 9(3), 355–373.

Aleksandrove, M. (2018). IoT in agriculture: Five technology uses for smart farming and challenges to consider. https://dzone.com/articles/iot-in-agriculture-five-technology-uses-for-smart. Retrieved: September, 2020.

Ansari, M. S., Alsamhi, S. H., Qiao, Y., Ye, Y., Lee, B. (2020). Security of distributed intelligence in edge computing: Threats and countermeasures. In *The cloud-to-thing continuum* (pp. 95–122). Cham: Palgrave Macmillan.

Babaei, A., & Schiele, G. (2019). Physical unclonable functions in the Internet of Things: State of the art and open challenges, *Sensors, 19*(14), 3208.

Banerjee, S., Chakraborty, C., & Paul, S. (2019). *Programming paradigm and Internet of Things* (pp. 148–164). CRC: A Handbook of Internet of Things & Big Data. ISBN 9781138584204.

Capra, M., Peloso, R., Masera, G., Roch, M. R., & Martina, M. (2019). Edge computing: A survey on the hardware requirements in the internet of things world. *Future Internet, 11* (4), 100.

Chakraborty, C., & Rodrigues Joel, J. P. C. (2020). A comprehensive review on device-to-device communication paradigm: Trends, challenges and applications. *Springer: International Journal of Wireless Personal Communications, 114*, 185–207. doi: 10.1007/s11277-020-07358-3.

Chanda, P.B., Das, S., Banerjee, S., & Chakraborty, C. (2021). *Study on edge computing using machine learning approaches in IoT framework* (1st ed., Ch. 9, pp. 159–182). CRC: Green Computing and Predictive Analytics for Healthcare.

Darwish, D. (2015). Improved layered architecture for Internet of Things. *International Journal of Computing Academic Research (IJCAR), 4* (4), 214–223.

Gonzalez, J., Hunt, J., Thomas, M., Anderson, R., & Mangla, U. (2020). Edge computing architecture and use cases. *IBM*. https://www.lfedge.org/2020/03/05/edge-computing-architecture-and-use-cases/. Retrieved: August, 2020.

Gordon, T., Kilgore, E., Wylds, N., & Nowatkowski, M. (2019). Hardware reverse engineering tools and techniques. In *SoutheastCon* (pp. 1–6). IEEE.

Halfond, W. G., Viegas, J., & Orso, A. (2006). A classification of SQL-injection attacks and countermeasures. In *IEEE International Symposium on Secure Software Engineering* (vol. 1, pp. 13–15), IEEE.

Humayun, M., Jhanjhi, N. Z., Alsayat, A., & Ponnusamy, V. (2020). Internet of things and ransomware: Evolution, mitigation and prevention. *Egyptian Informatics Journal*, 22(1), 105–117.

Ibrahim, A., Sadeghi, A. R., & Tsudik, G. (2018). Us-aid: Unattended scalable attestation of iot devices. In *37th Symposium on Reliable Distributed Systems (SRDS)* (pp. 21–30), IEEE.

Infinite Uptime (2018). Smart factory and its benefits on manufacturing industry. https://infinite-uptime.com/blog/smart-factory-benefits-manufacturing/. Retrieved: September, 2020.

IoT platforms – IoT platform definitions, capabilities, selection advice and market. https://www.i-scoop.eu/internet-of-things-guide/iot-platform-market-2017–2025/. Retrieved: August, 2020.

Knight, N. J., Kanza, S., Cruickshank, D., Brocklesby, W. S., & Frey, J. G. (2020). Talk2lab: The smart lab of the future. *IEEE Internet of Things Journal*, 7(9), 8631–8640.

Kumar, R., Zhang, X., Wang, W., Khan, R. U., Kumar, J., & Sharif, A. (2019). A multimodal malware detection technique for Android IoT devices using various features. *IEEE Access*, 7, 64411–64430.

Mannilthodi, N. & Kannimoola, J. M. (2017). Secure IoT: An improbable reality, *IoTBD*, vol. 1 - 978-989-758-245-5, pp. 338–343.

Martin, M. C. & Lam, M. S. (2008). Automatic generation of XSS and SQL injection attacks with goal-directed model checking. *USENIX Security Symposium*, 31–44.

Ngo, Q. D., Nguyen, H. T., Nguyen, L. C., & Le, V. H. (2020). A survey of IoT malware and detection methods based on static features. *ICT Express*, 6(4), 280–286.

Ni, J., Lin, X., & Shen, X. S. (2018). Efficient and secure service-oriented authentication supporting network slicing for 5G-enabled IoT, *IEEE Journal on Selected Areas in Communications, 36*(3), 644–657.

Park, J., Rahman, F., Vassilev, A., Forte, D., & Tehranipoor, M. (2019). Leveraging side-channel information for disassembly and security. *ACM Journal on Emerging Technologies in Computing Systems (JETC), 16*(1), 1–21.

Perry, J. S. (2018). Setting up the hardware for home automation system. *IBM Developer.* https://developer.ibm.com/technologies/iot/tutorials/iot-smart-home-01/. Retrieved: August, 2020.

Saiz-Rubio, V., & Rovira-Más, F. (2020). From smart farming towards agriculture 5.0: a review on crop data management, *Agronomy, 10*(2), 207.

Santos, M. G. D., Ameyed, D., Petrillo, F., Jaafar, F., & Jaafar, M. (2020). Internet of Things architectures: A comparative study, *arXiv preprint arXiv:2004.12936.*

Saqlain, M., Piao, M., Shim, Y., & Lee, J. Y. (2019). Framework of an IoT-based industrial data management for smart manufacturing. *Journal of Sensor and Actuator Networks, 8*(2), 25.

Shapsough, S., Aloul, F., & Zualkernan, I. A. (2018). Securing low-resource edge devices for IoT systems. In *International Symposium in Sensing and Instrumentation in IoT Era (ISSI)* (pp. 1–4), IEEE.

Sidhu, S., Mohd, B. J., & Hayajneh, T. (2019). Hardware security in IoT devices with emphasis on hardware Trojans. *Journal of Sensor and Actuator Networks, 8*(3), 42.

Skorobogatov, S. P.(2005). Semi-invasive attacks: A new approach to hardware security analysis, University of Cambridge. Computer Laboratory: Technical Report (04 2005).

Strackx, R. & Piessens, F. (2017). The Heisenberg defense: Proactively defending SGX enclaves against page-table-based side-channel attack, *arXiv preprint arXiv:1712.08519.*

Tzafestas, S. G. (2018). The internet of things: A conceptual guided tour. *European Journal of Advances in Engineering and Technology, 5*(10), 745–767.

Wang, J., & Chen, Q. (2014). ASPG: Generating android semantic permissions. In *IEEE 17th International Conference on Computational Science and Engineering* (pp. 591–598). IEEE.

Wu, M., Lu, T. -J., Ling, F. -Y., Sun, J., & Du, H. -Y. (2010). Research on the architecture of Internet of Things. In *3rd International Conference on Advanced Computer Theory and Engineering (ICACTE)* (vol. 5, pp. V5–484). IEEE.

Xiao, Y., Jia, Y., Liu, C., Chang, X., Yu, J., & Lv, W. (2019). Edge computing security: State of the art and challenges. *Proceedings of the IEEE, 107*(8), 1608–1631.

Yaqoob, I., Ahmed, E., Hashem, I. A. T., Gani, A., Ahmad, A. I. A., Imran, M., & Guizani, M. (2017). Internet of things architecture: Recent advances, taxonomy, requirements, and open challenges. *IEEE Wireless Communications, 24*(3), 10–16.

11

A Novel Threat Modeling and Attack Analysis for IoT Applications

C. K. Uday Karthick and R. Manimegalai
PSG Institute of Technology and Applied Research

11.1 Introduction

Technologies take giant leaps as years pass. One such remarkable leap was the introduction of the concept of "Internet of Things", also called IoT. An IoT device is just like any other device used to complete a particular task, but unlike ordinary devices, IoT devices have an in-built feature that allows them to interact with the internet, send or receive messages, or perform tasks remotely on their own or with the help of the internet. IoT improves communication between machines, reduces potential latency due to human intervention, reduces overall operational and functional costs, and saves a lot of time. IoT devices can perform a wide variety of tasks, including data collection, compilation and processing, automated decision making, self-regulation in accordance with the environment, content and context-aware judgments, and remote control. As mentioned above, the scope of IoT device use is growing: it is expected soon to be an inevitable part of daily life. This proposed approach for threat modeling in IoT devices can be used to increase the number of threats detected in IoT devices; therefore, overall security will be improved.

11.1.1 Security in IoT Devices

As these devices become a part of daily life, manufacturers and users pay little attention to their security. Manufacturers often treat security of devices as a non-functional requirement and are always ready to sacrifice security. Users also lack awareness of security, which leads to major security breaches and violation of their privacy. Threats to a device must be properly identified, classified, and analyzed, then appropriate measures must be taken for their mitigation. The more complex a system gets, the more security it needs. Adding a layer of security on these devices may turn out to be very difficult as IoT devices have limited memory and processing power. So, adding security will consume considerable resources, which is why proper decisions must be made to understand which risks need to be addressed. It must also be made clear that if one device on a network can be compromised, the whole network can be compromised easily, which makes IoT devices easy and attractive targets. Threats to each device must be understood to the fullest to implement cost-effective controls to reduce the risk to an acceptable level.

11.1.2 Organization of the Chapter

The organization of the chapter is as follows: Section 11.1 is the introduction. Section 11.2 presents the literature survey. It presents details of existing studies which have contributed to similar topics and identifies gaps in existing methods. Section 11.3 describes some basic terminology and the proposed threat-modeling approach for IoT devices. This section also provides a comparison between traditional approaches for threat modeling and the proposed approach, with the help of the experimental results achieved. Section 11.4 adapts the proposed IoT-TMA to various applications to demonstrate the effectiveness of the approach. Section 11.5 discusses various mitigation techniques that can minimise or eliminate threats to IoT devices. Section 11.6 concludes the chapter.

11.2 Literature Survey

Seeam, Ogbeh, Bellekens, and Guness (2019) have made a study about the threat-modeling process and security issues involved in IoT, which provides a brief idea on how threat modeling can be used to better identify security issues and possible attacks on a particular system. Aufner (2019) has made his point in a work which looks at the gaps between threat modeling and its implementation. The work details the different types of threat-modeling frameworks. Three different types of IoT layer models are also explained to provide a better understanding about IoT devices and how to secure them. Heartfield et al. (2018) have come up with a study which outlines the potential impact of cyber threats in smart home devices. Saini, Duan, and Paruchuri (2008) have expressed the effective usage of attack trees to perform threat modeling. The study discusses a practical usage of attack trees and the different types of attack trees that could be used to execute improvised threat modeling and enhance the output of the process. Omotosho, Haruna, and Olaniyi (2019) have proposed their work on a threat-modeling procedure for IoT health devices. The STRIDE and DREAD models were used to identify threats and find their impacts.

Mena, Papapanagiotou, and Yang (2018) have conducted a survey of the security posture of IoT devices. Core information security concepts were identified as Confidentiality, Integrity and Availability. Maple's (2017) study outlines various privacy and security considerations that are spotted with IoT devices. The traditional CIA triad was used to identify issues in various fields where IoT devices are deployed. Security operations like IAM, implementation, updating, etc., were discussed. Reddy and Chandrashekara (2018) have proposed their observations on various threats for IoT devices where privacy-related threats were a notable point of focus. Rafferty, Iqbal, and Hun (2017) have presented their analysis on security for a smart home network. The study focuses on dynamic agents that may be connected to the network, such as smart toys. Ahlawat, Sangwan, and Sindhu (2020) have made their study on the architecture of IoT devices, their various levels, and security threats posed at each level were listed. Ecclesie Agazzi (2020) accomplished a study that discussed possible ways in which a smart home system could be exploited. Bou-Harb and Neshenko (2020) evaluated

TABLE 11.1

References used in the Work

Ref. No	Year	Title	Author	Findings
1	2019	Threat Modeling and Security Issues for the Internet of Things.	Amar et al.	Better Usage of Threat Modeling to identify threats in IoT devices.
2	2019	The IoT security gap: a look down into the valley between threat models and their implementation.	Peter Aufner	Identifies Gaps between threat modeling and implementation.
3	2018	A taxonomy of cyber-physical threats and impact in the smart home.	Ryan et al.	Potential impact of cyber threats in smart home devices.
4	2008	Threat Modeling Using Attack Trees.	Saini et al.	Usage of attack trees for threat modeling
5	2019	Threat Modeling of Internet of Things Health Devices	Omotosho et al.	Threat modeling for IoT healthcare devices.
6	2018	Internet of things: survey on security	Mena et al.	Survey of the current security posture in IoT devices.
7	2017	Security and privacy in the internet of things	Maple	Privacy and Security concerns in IoT devices.
8	2018	A study on Internet of Things (IoT) threats.	Vikas Reddy and Chandrashekara	Privacy-related threats based on IoT devices
9	2017	A Security Threat Analysis of smart home Network with Vulnerable Dynamic Agents.	Rafferty et al.	Security for smart home network.
10	2020	IOT System Model, Challenges and Threats.	Ahlwat et al.	Various levels of security threats to IoT devices.
11	2020	smart home, security concerns of IoT.	Alessandro Ecclesie Agazzi	Different ways in which smart homes could be exploited.
12	2020	Cyber Threat Intelligence for the Internet of Things.	Bou-Harb et al.	Evaluation of multidimensional attack vectors for IoT devices.

multidimensional attack vectors for IoT devices. Their study also made notable points on exploiting vulnerabilities and the process of remediating the vulnerabilities. From the literature review, it was observed that there is no fixed process for threat modeling that is universally applied. It is also seen that traditional threat-modeling methodologies can aid in software threat-modeling processes. These methods are used for IoT devices, making the number of threats discovered significantly less. Table 11.1 below shows the references used for literature survey.

11.3 Terminology used in Proposed Threat Modeling for IoT Devices

11.3.1 Basic Terminology

To perform threat modeling, one must clearly understand basic security concepts. In this section, definitions for basic terms are given.

11.3.1.1 CIA Trait

This is also called the AIC Trait, which stands for Confidentiality, Integrity and Availability (CIA). Other traits which must be given serious attention are non-repudiation, authentication, authorization and accountability. Manufacturing companies must decide which traits which they need to pay the most attention to in order to introduce perfect security controls for their devices.

11.3.1.2 Vulnerabilities

Vulnerabilities can be explained as absence of security control or presence of weakness in an asset. Vulnerability may be exploited by a threat actor who may harm the asset or the organization.

11.3.1.3 Threat

Threat is any event or occurrence that may harm the asset. Threats are performed by threat actors. Threat actors may be classified into various types like hacktivist, advanced persistent threats, script kiddies, organized crime, etc. Each type of threat actor has a different goal. The goal may be money, fame, disruption of other businesses, etc.

11.3.1.4 Risk

Risk is the probability or possibility of the asset being harmed or attacked. Risks can be quantitative, which can be measured in terms of monetary value, or qualitative, which cannot be assigned a definite value and depend on the opinion of individuals. Risks can be calculated by multiplying the probability of the event occurring and the impact which the event brings (Chakraborty & Kumari, 2016).

11.3.1.5 Threat Modeling

In simple terms, threat modeling may be defined as the identification, classification and analysis of possible threats to any asset or system in order to prioritise the threats and introduce cost effective controls, which trim the attack landscape and reduce threats to an acceptable level. Performing threat modeling of an IoT device involves nearly the same steps as any other threat modeling, but with a few minor changes that tailor it specifically to IoT devices. Threat modeling plays an important role in securing IoT systems, as there may be many security loop holes when building a system with limited memory and processing resources. Threat modeling also helps manufacturers understand the entire threat landscape and allocate resources according to risks posed by. This helps to achieve a monetary balance between cost of countermeasures and impact of threats. Threat modeling can be performed proactively, where threats are mapped during the design of the device, or retroactively after the device is built. It is well-known that not all threats can be figured out proactively, but identifying threats during the initial stage of system or application development can help build better security and make it cost effective, as opposed to adding a layer of security after development.

As examined in the literature review, threat-modeling frameworks and models currently used for IoT are originally made to identify software threats; traditional models miss a lot of threats which may be specific to IoT devices. It is also made clear that there is no universally- accepted formal method for threat modeling of IoT devices. The above observation may help create a threat-modeling process that considers problems that were overlooked in the traditional threat-modeling process. The paper also introduces a model called "PSM", which stands for Privacy threats, Safety threats and Malfunction threats, that need to be paid attention to in IoT devices. This proposed methodology can be used specifically for the purpose of threat modeling of IoT devices. Threat modeling for a smart home environment has been performed to demonstrate the effectiveness of the methodology. Finally, mitigation techniques for threats in IoT system have been proposed which ensure the hardening of IoT devices.

11.3.2 Steps Involved in Threat Modeling

The proposed IoT-TMA has a seven-step process. Each step is discussed in detail in the upcoming sections. All of the following steps must be performed to effectively identify the maximum number of existing threats.

Step 1: Determine the scope
Step 2: Identify and prioritise assets
Step 3: Perform decomposition analysis
Step 4: Realise existing controls
Step 5: Identify and categorise threats and vulnerabilities
Step 6: Analyze the hardware situation
Step 7: Prioritise to respond

11.3.3 Determine the Scope

The first step is to determine the scope of the project and understand the scope of the threat modeling to be performed on the assets. Clear boundaries must be drawn which indicate the parts of the asset that are to be evaluated and the parts that are not. If threat modeling is performed for a single unit or device, the entire unit may be involved in the scope. If threat modeling is performed for an environment which consists of multiple devices, those devices must be mentioned clearly. Clear differentiation and understanding of tangible and intangible assets must be made. Scope must also be defined in order to avoid confusion and misconceptions about the threat modeling to be performed. The scope must be documented in a clear and understandable manner.

11.3.4 Identify and Prioritise Assets

An asset is anything that has a value or can attract the attacker. Assets can be tangible, like sensors, processors and physical devices, or intangible, such as customer data that might be stored in the device. Not all assets are of same value; the organization must be able to distinguish among and prioritise them so that resources can be allocated accordingly. The understandings must be properly documented for future reference. It is to be noted that the asset list may not include every single one of the assets. Only those with the greatest value may be included, as mitigation of threats consumes a considerable amount of resources.

11.3.5 Perform Decomposition Analysis

Decomposition analysis is also called reduction analysis, where the entire device is broken down into individual components to gain a better and granular understanding of the device. Such analysis also helps one to better understand the logic of the device. Having a clear understanding of the device or system subject to threat modeling is cardinal during the process. It also helps one to better understand the essential parts of the device. The greater the understanding the individual has of the device logic, the greater the chance of identifying further threats. The input interfaces of the device, the output interfaces of the device and the data flow paths must also be identified.

11.3.6 Realise Existing Controls

This step is often overlooked in performing threat modeling. Before identifying any threats, existing controls must be taken into account and threats to them mitigated. It must also be ensured that the controls introduced in the future should not interfere with the existing set of controls, which may cause a downgrade in the security of the device. Understanding the existing controls also helps one to design new controls effectively, so there will not be any redundancy in functions performed by the controls. Threats which have already been addressed may reappear and consume resources.

11.3.7 Identify, Classify and Prioritise Threats

Once a clear understanding of the asset and existing security controls has been established, new vulnerabilities and threats can be discovered. Discovery of new threats is the first step towards implementing controls and defenses in order to protect the asset. Threats may be in many forms which may include Denial of Service, spoofing of identity, physical destruction of the asset, unauthorised access, authentication bypass, and sensitive data disclosure, among others. There may be more threats than the ones that are realised, such as STRIDE, CORAS and LINDDUN. Threat-modeling frameworks can be used to classify and identify threats. Threats with most impact and highest probability should take priority. Diagrammatic representation of the attacks can also be made with the help of attack trees. Threats discovered must be put into standard classes. Some of these are: threats based on threat actor types; threats based on software; or threats based on environment; among others. All discovered threats must be properly documented.

11.3.8 Analyze the Hardware Situation

As discussed earlier, the problem with well-known software threat-modeling techniques are that they do not consider some threats which are specific to IoT devices. The three main types of threats to consider for IoT devices which directly affect the users are "PSM": Privacy, Safety and Malfunction. The most overlooked feature of IoT devices during threat modeling is the physical or hardware feature. IoT devices can also cause serious safety issues not used appropriately. Manufacturer may be found liable in case the device has a neglected issue which compromises the user's safety. Since IoT devices may also have physical parts, malfunction of any part may also be considered a serious threat, even if there is no well-defined threat actor behind the action. So controls and safety measures might be put in place in case if any device malfunctions. IoT devices are equipped with input devices like cameras, microphones, etc., which may compromise the privacy of users. It may also lead to spying on people in real time. Since devices may be placed anywhere, even all around the house, serious privacy issues arise other than webcams in laptops. Respective privacy issues for any particular IoT device must be addressed. Identified threats must be documented for future use.

11.3.9 Prioritise to Respond

Even when all possible threats have been identified, not all can be mitigated. The ones with the highest possibility of occurrence or highest impact should be mitigated first, as per the organization's discretion. The DREAD system can achieve a better ranking of threats. The DREAD consists of Damage potential of the attack, Reproducibility of the attack, Exploitability, Affected users and Discoverability of weaknesses. High, medium, or low values can be assigned to each threat, or points can be awarded according to damage potential of each threat. Since no organization has infinite resources, proper prioritization is critical for efficient use of them. Threats that are accepted also must be formally documented and provided with a reason for acceptance, so as to avoid any future confusion.

11.3.10 Experimental Results

Table 11.2 below shows a comparison between STRIDE, and STRIDE coupled with PSM. As part of the work, we have implemented the proposed IoT-TMA for three IoT devices. A total of 18 most common threats for IoT devices were identified. Out of the 18, a traditional threat-modeling approach was able to identify 9, or 50% of all threats being identified, whereas the proposed threat-modeling approach was able to identify 14, or 77.77% of all threats being identified. There was a significant improvement of 27.77% of threats being discovered while using the proposed seven -step approach. The increase in the total number of threats being identified shows that the proposed seven-step approach has a better chance of finding threats which may be missed if traditional approaches were used. Hence, using the proposed approach over the traditional approach is recommended.

11.4 Adopting the Proposed IoT-TMA for Various Applications

11.4.1 Smart Home Environment

The proposed IoT-TMA has been implemented in smart home environments which consists of multiple individual devices. An environment can be made up of multiple devices. A smart lock example is used for demonstration; similar procedures can be followed for other devices present in the environment. Some of the devices present in a smart home environment may be smart televisions, smart air conditioners, smart locks, smart cameras, smart fire extinguishing system and smart temperature control, among others.

11.4.1.1 Determine the Scope

The scope of the system is the devices that are involved; below are mentioned devices of the entire system. These devices include smart televisions, smart air conditioners, smart

TABLE 11.2

Comparison between STRIDE and STRIDE + PSM

THREAT EVENTS	STRIDE	STRIDE + PSM
Device Malfunction	Not Addressed	Addressed by Malfunction Threats
Physical Attack on Users	Not Addressed	Addressed by Safety Threats
Physical Attack to Environment	Not Addressed	Addressed by Safety Threats
Home Invasion	Not Addressed	Addressed by Privacy, Safety Threats
Threats due to Poor Build	Not Addressed	Addressed by Malfunction Threats
Data Theft	Addressed by Information Disclosure	Addressed by Information Disclosure, Privacy
Identity Theft	Addressed by Information Disclosure	Addressed by Information Disclosure, Privacy
Sinkhole Attacks	Addressed by Denial of Service	Addressed by Denial of Service
Destruction of device	Addressed by Denial of Service	Addressed by Denial of Service
Remote Vehicular Hijack	Addressed by Spoofing, Tampering	Addressed by Spoofing, Tampering
Denial of Service	Addressed by Denial of Service	Addressed by Denial of Service
Man-in-the-Middle	Addressed by Spoofing, Tampering	Addressed by Spoofing, Tampering
Ransomware Attacks	Addressed by Denial of Service	Addressed by Denial of Service
Using Botnets	Addressed by Spoofing, Denial of Service	Addressed by Spoofing, Denial of Service
Social Engineering	Not Addressed	Not Addressed
Lack of User Awareness	Not Addressed	Not Addressed
Zero Day Attacks	Not Addressed	Not Addressed
Inadequate Patch Management	Not Addressed	Not Addressed

cameras, smart locks, smart fire systems and smart temperature controls. The above devices are included as a part of the threat-modeling process; possible threats to the devices will be identified.

11.4.1.2 Identify and Prioritise Assets

As mentioned earlier, an asset is any product with a specific value assigned to it. Smart lock systems consist of assets like physical locks, user applications, databases used for remote authentication and data storage, and sensors.

11.4.1.3 Perform Decomposition Analysis

Decomposition analysis is illustrated for smart locks below to give a better understanding of the working of the system under study. Figure 11.1 depicts decomposition analysis.

11.4.1.4 Realise Existing Controls

The goal of this step is to realise if any existing security controls are present so functions of newly-implemented controls do not overlap with them. In the above smart lock system, an authentication system exists in order to validate users. There is no need to add another main authentication system, but a two-factor system may be added to increase strength without disturbing the existing authentication system. Existing controls must be identified for all devices present in the scope.

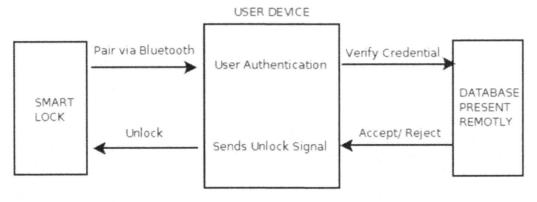

FIGURE 11.1
Decomposition of a smart lock system.

11.4.1.5 Identify, Classify and Prioritise Threats

Threats and vulnerabilities for every single device present in the system must be identified. Different models such as STRIDE, CORAS, LINDDUN, etc., exist to identify different types of threats. The STRIDE model was introduced by Microsoft, and is one of the most popular models used in the industry. The STRIDE model is used for the smart lock system below to ensure a detailed understanding. Attack trees could also be used in order to better understand threats to a particular system. An attack tree is a diagrammatical representation of how a particular device, system, asset or process can be attacked or compromised by a malicious actor. Attack trees follow a top-down approach. Table 11.3 below demonstrates STRIDE and Figure 11.2 displays an attack tree.

11.4.1.6 Analyze the Hardware Situation

The hardware situation for every asset present with the scope is evaluated. For the example being used, the hardware situation of a smart lock is quite simple. Table 11.4 below depicts the hardware situation using the PSM model.

TABLE 11.3

STRIDE for Smart Home Environment

S	The attacker could spoof Bluetooth signals as if the signals appear from a legitimate end user.
T	The attacker could tamper with communication between the lock and the mobile by placing appropriate tools in the middle of two devices.
R	Any user who has the access to the UI and knows the credentials can enter the house and perform any task with no accountability such as unique identification.
I	Unencrypted communication between the user device and the remote DB can be obtained using proper tools, leading to information disclosure.
D	The attacker could use a jamming device to prevent communication between the user device and lock in order to deny entry.
E	With no proper authorization, any user could access any room in the building, which is an escalation of privilege.

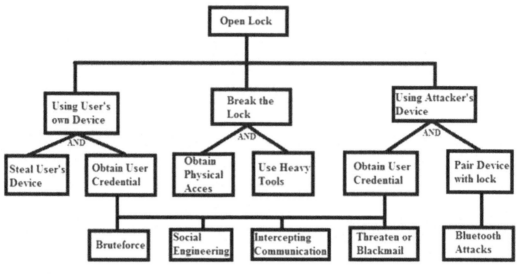

FIGURE 11.2
Attack tree for smart lock system.

TABLE 11.4

PSM for Smart Home Environment

P	No direct privacy concerns.
S	An attacker may try to alter the battery of the lock in an attempt to make it explode which leads to user safety concerns.
M	An attacker may pour water on the lock to make it malfunction and may try to open it.

11.4.1.7 Prioritise to Respond

Now that possible threats and attacks for each asset are identified, they are prioritised. Attacks with more probability and higher impact must be addressed first. As mentioned earlier, no organization has infinite resources, so existing resources must be spent wisely to reduce risk to the bare minimum and to an acceptable level. In the above smart lock example, the chance of an attacker breaking a lock is less, as it requires physical access to the place and draws too much attention when trying to break a lock. So, it can be dealt with after addressing threats like unencrypted communication or using weak credentials.

11.4.2 IoT-based Garment Unit

The smart garment unit is an IoT enable device that is used to observe the performance of the production of a garment and can estimate the time required for the completion of a particular task. It operates autonomously and increases overall efficiency of the process.

11.4.2.1 Determine the Scope

Here the scope is the all the individual devices that enable the smart garment unit. Since the garment unit needs all the devices to function effectively, the entire unit is included in the scope in order to effectively perform threat modeling.

11.4.2.2 Identify and Prioritise Assets

The assets include the USB HD camera, sensors used to collect information from the sewing machine and pedal motors, sensors used to collect texture-based information, the master system, and the smart phone application the user interacts with. Now, prioritization must be done based on a particular metric. In this case, smart phone applications are a more likely target than sensors that collect information. So, potential threats for the application are given priority.

11.4.2.3 Perform Decomposition Analysis

Here, we assume the asset selected is the smart phone app. Its architecture is analyzed by breaking it down in order to perform threat modeling effectively and find the maximum number of threats possible. Figure 11.3 displays the decomposed layout of a smart garment unit.

11.4.2.4 Realise Existing Controls

Analysis is conducted on the asset to find any existing control, if present. The application contains an interface that displays the observations made. The observations are received from the cloud system. No existing controls exist in the asset, so there is no possibility for control overlap or existing control malfunction.

11.4.2.5 Identify, Classify and Prioritise Threats

As mentioned earlier, the STRIDE model is an excellent threat model which can be used to identify potential threats. The STRIDE model is illustrated in Table 11.5 below.

11.4.2.6 Analyze Hardware Situation

The main goal of this step is to figure out PSM threats in the asset. Table 11.6 below briefly describes the threats.

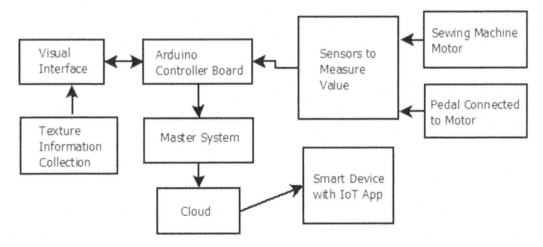

FIGURE 11.3
Decomposition analysis of a smart garment unit.

TABLE 11.5

Applying 'STRIDE' Model for Smart Garment Unit

S	An attacker can spoof his credentials to pose as a legitimate user in order to access data from cloud.
T	The values in the cloud environment could be spoofed to provide incorrect data.
R	Any user who has credentials can modify data and deny it.
I	Information could be disclosed if the communication between the Unit and the cloud is compromised.
D	Legitimate users may be denied access to data stored in cloud by flooding the cloud with fake requests.
E	The application can be made to run in privileged mode to perform unauthorised activities.

TABLE 11.6

Applying 'PSM' Model for Smart Garment Unit

P	The camera could be used to know about the environment and working nature.
S	No direct safety issues.
M	The sensors can be made to malfunction, thus not reporting accurate data.

11.4.2.7 Prioritise to Respond

Now that the potential threats have been identified, they are mitigated according to their priority. As mentioned earlier, the smartphone app is an attractive target for an attacker and a considerable amount of resources may be spent in order to mitigate the risks in the application.

11.4.3 IoT-based Water Quality Monitoring System

The IoT-based water quality monitoring system (Water-QMS) is used to monitor the quality of water and make observations that decide the composition and quality of water. It performs real-time water monitoring with the help of IoT. The system uses cloud and deep learning in order to perform monitoring and reporting effectively.

11.4.3.1 Determine the Scope

Since the (Water-QMS) is a single unit, the scope of threat modeling includes the entire system. The entire system is included to consider all possible threats and determine every possible attack surface.

11.4.3.2 Identify and Prioritise Assets

Water-QMS consists of many individual assets, such as wireless sensor nodes, buoy systems, cloud data storage systems and the communication interface. The assets are prioritised based on probability of attack. The probability of attacking a cloud database is much higher than attacking a sensor node. Prioritization also may be done based on other metrics like monetary value.

11.4.3.3 Perform Decomposition Analysis

Decomposing the Water-QMS allows us to have a granular view of the application. It also helps in better understanding the workflow of the overall process. Cecomposition analysis is illustrated in Figure 11.4 for Water-QMS and potential threat analysis.

11.4.3.4 Realise Existing Controls

Analyzing the current structure, there is an authentication system which is used to grant access to collected and analyzed data to legitimate users. So, the authentication system should not be disturbed by newly-implemented controls and no overlaps must exist. There is no other control in place to be considered.

11.4.3.5 Identify, Categorise and Prioritise Threats

Now threats can be identified with the help of STRIDE model and classified based on a particular metric. Table 11.7 below identifies threats based on STRIDE.

11.4.3.6 Analyze the Hardware Situation

The threats to Water-QMS based on PSM should be understood. Table 11.8 below shows threats based on PSM.

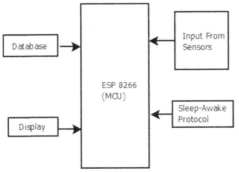

FIGURE 11.4
Decomposition analysis of water-QMS.

TABLE 11.7

Applying 'STRIDE' Model for Water-QMS

S	The false data sent by attacker to the cloud can be spoofed as if it arrives from a legitimate source.
T	The values stored in the database could be tampered with to provide false data.
R	With no proper logging, any user can deny an action that was performed by the user.
I	The communication between the buoy device and database could be rigged to disclose information.
D	An attacker may physically break the device which may result in Denial-of-Service.
E	With no proper authorization, any user may escalate the level of privilege in order to perform unauthorised activities.

TABLE 11.8

Applying 'PSM' Model for Water-QMS

P	No direct **privacy** issues.
S	By tampering with the device to provide false readings may not lead to the detection of any potential harm in water which may harm the health consumers.
M	A **malfunction** which is not noted may also cause serious issues to the consumers.

11.4.3.7 Prioritise to Respond

Now that threats have been identified and classified, they must be prioritised in the order that the user wants to eliminate them. Easy targets for an attacker are network nodes and cloud database. Attackers always looks for the weakest link in the chain to exploit it.

11.5 Mitigation Techniques for Threats in IoT Devices

11.5.1 Network Segmentation

As mentioned earlier, if a single device is compromised in a network, the whole network can be compromised in no time. So, to avoid such problems, it is better to isolate IoT devices to a separate network using network segmentation techniques such as VLAN (Virtual Local Area Network) or physical network segmentation, giving IoT devices their own separate network. Network segmentation not only provides an improved security posture, but also provides an option for QoS (Quality of Service) (Chakraborty, Gupta, & Ghosh, 2013). So, even if one of the IoT devices is compromised, it would not affect or harm the regular network. Even if other IoT devices are compromised if one device is compromised, the overall risk is lower in the former over the latter.

Many studies have proved that users do not change default credentials provided by the vendor. It is effortless for an attacker to try out a vendor- supplied default credential available on the internet. Users must equip themselves with adequate knowledge to change default credentials and regularly change their credentials, as the longer they stay the same, the more vulnerable they become. Manufacturers must have features built into applications that force users to change credentials during their first login session, along with frequent reminders to do so regularly. The devices must have features that force users to have alphanumeric passwords of adequate length.

11.5.2 Effective Encryption

Because IoT devices have finite computing resources, manufacturers skip encryption, which consumes significant processing power. It is evident that encryption can prevent eavesdropping and information disclosure. Manufacturers must design products with the thought of adding encryption up front, which costs less, rather than adding it after the design or device is built. Data-in-motion and data-at-rest could be encrypted in order to protect the confidentiality of data and provide adequate privacy to users. Encryption algorithms like AES (Advanced Encryption Standard) and Twofish are popular. Storing the password as their hashes also increases security if a collision free hashing algorithm is chosen. SHA-256 is a widely-used hashing algorithm which can be used with Salt to further increase the security.

11.5.3 Effective Patch Management

Users often forget that IoT devices also may receive regular security patches but they go unnoticed. Users must be educated to look for security patches frequently. The manufacturer also should take appropriate steps to ensure that release of patches is well publicized to users. Automatic approval and application patches may be a good option for domestically-used IoT devices but not so good for devices used in industry, as it may violate change management policies, if any. Specific roles must exist in an organization or industry to check for patches regularly and apply them properly, according to organizational policy.

11.5.4 Disabling Unnecessary Features

Users must be advised about the basic configuration of the device being used. By doing so, users can change the configuration according to their needs. The default configuration in which the device is released is not suited for every user. By disabling unwanted features, the overall attack surface is reduced. Having the device at its best configuration improves the security of the device, helps the user to achieve a better understanding of it and gives the use granular control over it. For industrial devices, proper configuration management policies must be maintained to avoid unwanted security loop holes.

11.5.5 Proper Physical Security

No matter how secure the device is, if the attacker gains physical access to it, then the attacker has a higher probability of infecting or harming the system. At worst, the attacker could physically damage the device or even try to steal it. IoT devices must be placed in a spot where there is limited access. If the device must be placed in a place with easy physical access, then it must be properly hardened and appropriate safety measures must be put in place to reduce the effectiveness of an attack. Care must also be taken to assure that only the appropriate amount of resources is spent for physical security. Implementing excessive system controls may reduce the usability of the device or may lead to an increase in production costs.

11.6 Conclusion

The proposed novel methodology can be adopted for the purpose of threat modeling for IoT devices. The seven-step methodology addresses various threats in IoT devices that directly affect users' Privacy, Safety and threats if the device Malfunctions. Coupling

PSM, the proposed new methodology, and existing threat models and approaches leads to improved threat-modeling results. Improved threat-modeling results help to identify and mitigate more threats and hence lead to a safer device or environment. The mitigation techniques which were mentioned toward the end of this chapter can be used to increase the security posture of devices by making simple changes to the devices and their functions.

11.6.1 Future Work

The proposed novel methodology focuses on eliminating threats to IoT systems which affect users' Privacy, Safety and threats due to Malfunction. IoT devices are loaded with newer features like Bluetooth Low Energy (BLE) and potential features like NFC. These proposed protocols have their own way of functioning and threats exist that are unique to those protocols. Soon, IoT devices will be equipped with these protocols as standard issue. An extension to this proposed methodology would be to include steps in threat modeling that pay special attention to newly-incorporated communication protocols in IoT devices. The inclusion of the steps could further increase the total number of threats found and thereby increase the overall security posture of the device. The current trends in IoT must always be monitored and specific changes must be made in the threat-modeling approach in order to identify newer threats to the devices.

References

Ahlawat, B., Sangwan, A., & Sindhu, V. (2020). IOT system model, challenges and threats. *International Journal of Scientific and Technology Research, 9*(3), 6771–6776.
Aufner, P. (2019). The IoT security gap: A look down into the valley between threat models and their implementation. *International Journal of Information Security, 19*, 3–14. doi:10.1007/s10207-019-00445-y.
Bou-Harb, E. & Neshenko, N. (2020). *Cyber threat intelligence for the Internet of Things*. Springer. doi: 10.1007/978-3-030-45858-4.
Chakraborty, C. & Kumari, S. (2016). Bio-metric identification using automated iris detection technique. In *3rd International Conference on Microelectronics, Circuits, and Systems* (pp. 113–117). IEEE.
Chakraborty, C., Gupta, B., & Ghosh, S. K. (2013). A review on telemedicine-based WBAN framework for patient monitoring. *Mary Ann Libert Inc.: International Journal of Telemedicine and e-Health, 19*(8), 619–626. ISSN: 1530-5627, 10.1089/tmj.2012.0215
Ecclesie Agazzi, A. (2020). Smart home, security concerns of IoT.
Heartfield, R., Loukas, G., Budimir, S., Bezemskij, A. , Fontaine, J., Filippoupolitis, A., & Roesch, E. (2018). A taxonomy of cyber-physical threats and impact in the smart home. *Computers and Security, 78*, 398–428. doi: 10.1016/j.cose.2018.07.011
Maple, C. (2017). Security and privacy in the internet of things. *Journal of Cyber Policy, 2*(2), 155–184. doi: 10.1080/23738871.2017.1366536.
Mena, D. M., Papapanagiotou, I., & Yang, B. (2018). Internet of Things: Survey on security. *Information Security Journal: A Global Perspective, 27*(3), 162–182. doi: 10.1080/19393555.2018.1458258.

Omotosho, A., Haruna, B. A., & Olaniyi, O. M. (2019). Threat modeling of Internet of Things health devices. *Journal of Applied Security Research*, 14(1), 106–121. doi:10.1080/19361610.2019.1545278.

Rafferty, L., Iqbal, F., & Hun, P. C. K. (2017). A security threat analysis of smart home network with vulnerable dynamic agents. In J. K. T. Tang & P. C. K. Hung (Eds.), *Computing in smart toys*. Cham: Springer International Publishing. doi:10.1007/978-3-319-62072-5_8.

Reddy, S. & Chandrashekara, S. N. (2018). A study on Internet of Things (IoT) threats. *International Journal of Engineering and Technology*, 7, 109–111.

Saini, V., Duan, Q., & Paruchuri, V. (2008). Threat modeling using attack trees. *Journal of Computing Sciences in Colleges*, 23, 124–131.

Seeam, A., Ogbeh, O., Bellekens, X., & Guness, S. (2019). Threat modeling and security issues for the Internet of Things. In *IEEE Nextcomp 2019*. Mauritius: IEEE. doi: 10.1109/NEXTCOMP.2019.8883642.

12

Trust Management in Internet-of-Things Devices

Sandeep Mahato, Subrata Dutta, and Kailash Chandra Mishra

NIT Jamshedpur, India

12.1 Introduction

The IoT is a dream area in which different tools and intelligent objects interact separately. As a core component of the Internet of Things (IoT) or Cyber Physical System (CPS), a large-scale Wireless Sensor Network (WSN) is to be fully incorporated into the internet. It is important to consider various security concerns that come with IoT / CPS, such as the detection of malicious attacks (Chen, Chang, Sun, Li, Jia, & Wang, 2011). The knowledge these systems provide is the real benefit of these intelligent IoT applications. The IoT view is designed to facilitate the transmission of vital data through different sensors, including human body parameters and environmental measurements (Azzedin & Ghaleb, 2019). Each node within the IoT network is both an information service provider and the applicant or the referrer. When questions are disseminated across the network to determine what is most suitable, the variety of obtained knowledge opens up a challenge. Moreover, despite their ability to defend them from external adversaries, the internal adversaries of a benign node, which become hostile and disrupt transmission, cannot be handled by conventional protection mechanisms. Such deceptive activity can only be identified by trust models by businesses with IoT devices to distinguish honest nodes. In order to ensure authentication, confidentiality and protection of connected items, such a trust model is needed to take into account the complex and heterogeneous existence of interconnected devices in the IoT (Mohammadi, Rahmani, Darwesh, & Sahafi, 2019). The management of behavioral issues is difficult due to site independence, data source heterogeneity, distributed participation and varying clients. Confidence in behavior is about a broader concept of the credibility of a unit. Therefore, to achieve the maximum advantages of intelligent applications, it is necessary to build trustworthy IoT environments. Building trustworthy IoT environments also reduces irretrievable and unpredictable damage so that intelligent sensor-driven systems are safe, reliable, and scalable (Azzedin & Ghaleb, 2019). Initially, trust management systems were developed to resolve problems with access control and to unify authentication and authorization in distributed systems.

It is not possible to create trust between sensor nodes by simply using traditional trust mechanisms. In social and commercial ties, trust is one of the most complex, fuzzy and challenging concepts. Several scholars have suggested a variety of meanings for faith.

Confidence is the subjective likelihood, according to Gambetta D, whereby a person, A expects another person, B, to take a certain action, on which depends his welfare. Depending on the trust rating, the node or data operation is rated as good or bad (Chen et al., 2011). In a confidence relationship, there are two people involved, a trustor and a trustee. For mutual gain, they depend on each other. The trust relationship is comprised of the atmosphere of trust, the intent of trust and the risk of trust. In Thirukkumaran and Kannan (2019) trust, direct trust, indirect trust and recommended trust were classified. The confidence in all trusting institutions is focused on the experiences or findings, i.e., the direct experiences between the confidence node and trustee node. Indirect confidence in the past does not encourage the trustor or the trustee to communicate or experience together. In this case, the view and advice of the other nodes create trust. Indirect trust can be formed if the target node's contact behavior cannot be controlled directly. Trust checks should include filtering guidelines, as guidelines of third parties are not always reliable.

A trust and reputation model is recognised as an essential approach to defending large distributed sensor networks against malicious node attacks in IoT/CPS. Trust establishment mechanisms can promote cooperation between distributed computing and communication entities, encourage the identification of untrustworthy entities, and help multiple protocol decision-making processes (Chen et al., 2011; Chakraborty & Rodrigues, 2020).

This chapter contributes to fundamental concepts of trust models like reputation, honesty, accuracy, trust assessment, trust calculation in IoT, goals, objectives, major challenges and issues of trust in the IoT environment and some recently-proposed trust models. This chapter can help researchers better understand trust concepts and models in IoT environment.

The organization of this chapter is such that section 12.2 presents fundamental concepts of trust model, section 12.3 discusses trust management goals, section 12.4 concludes objectives of trust management in different layer of IoT, section 12.5 includes trust models in WSN, section 12.6 presents transport systems trust management, section 12.7 discusses trust management in P2P networks, section 12.8 presents trust management in Social IoT, section 12.9 explains some recently proposed trust management techniques in IoT, section 12.10 focuses on issues and challenges in trust and section 12.11 discusses trust applications.

12.2 Fundamentals of the Trust Model Concept

Azzedin and Ghaleb (2019) defined some basic terms for the trust model principle:

Behavior Trust—Behavior trust, as used by Azzedin (2014),means that the system is trustworthy if its capacity to function as expected is transparent. Such conviction isn' t correlated with a fixed value of the computer, but subject to the device's behavior, and only applies within that device's reach.

Reputation—Devices are needed to handle risks during interactions in a complex atmosphere with other computers. Because of uncertainties and incomplete or distorted awareness of each other in a complex world, devices are susceptible to dangers. Confidence is built by integrity as one solution to this dilemma. In trust models, the concept of credibility is already used. The principle of the legitimacy of a body

(Azzedin, 2014) is to recognise its behavior on the basis of judgments made by others or a common perception of past conduct of an individual in a specific sense.

Honesty—The concept of honesty used by (Azzedin, 2014) maintains that a suggestion is said to be truthful if the information that is obtained from an individual in a particular sense obtained the same details as the majority believes is at a particular time.

Accuracy—The definition of precision used by (Azzedin, 2014) is that a recommendation is considered accurate if the difference between the knowledge derived from it and the truthfulness of the system at that time lies within a specific framework; the entity's real trustworthiness lies within the level of accuracy.

Trust assessment: Trust evaluation is a technical approach to representing trust relationships in the communication system framework for assessing trust-influenced properties.

Trust Calculation: Karthik and Dhulipala (2011) proposed an algorithm for trust calculation. The first stage in confidence calculation involves measuring the trust value of the node through past interactions and the feedback of the networking neighbors. This calculation of trust values is called the node's indirect trust value. The trust's relative value is considered the trust's original value. When the node's initial trust value is communicable, the node will communicate. If not, the direct value of trust is calculated.

As part of the trust system, direct confidence assessments include three models: the safety node model; the node model of mobility; and the stability node model. For every model, the node determines the value of trust. Contact starts when confidence in the security model is adequate. Otherwise, trust will be calculated for the mobility model. Where there is not adequate performance in the mobility model to interact, the reliability model of the node will be assessed. If the node trust model is unsatisfactory, a node shall add to the calculation of total trust an indirect trust and a direct trust. If the overall confidentiality of the node for safe interaction is insufficient, the node will reject the contact request of the node.

12.3 Esteem Assets and Trust Management Goals

According to Yan, Zhang, and Vasilakos (2014), trust has a tangible and non-quantifiable effect and is regarded as very complex. It is highly security-related, as maintaining security of systems and user safety is a necessary means of gaining confidence. Trust is more than protection, however. It involves not only security, but many other things, such as goodness, power, resilience, availability, flexibility or other business features. The definition of trust extends beyond protection, making it harder and more complicated to build, protect and preserve faith in the short term than protection alone. The author proposes another significant principle relating to the trust: confidentiality, which is an organization's capacity to decide whether, when and for whom it is to reveal or disclose information. Five classes of assets affect trust:

- Objective properties of the trustee, such as security and efficiency of the trustee. In particular, credibility is a public evaluation of the trustee's past conduct and performance.

- Subjective assets of the trustee, such as fiduciary integrity, benevolence and fairness.
- Subjective assets of the trustee, such as confidence or disposition, and ability to trust.
- Objective properties of the trustee, such as conditions or policies defined by the trust or decided by the trustee.
- Background of trust relationships in, say, trust purpose, trust environment (e.g., time, place, behavior, equipment used, way of operating, etc.) and the danger of trust. Trust relationships are present. It describes any information that can be used to identify the circumstances of the persons concerned.

Context is an essential aspect that influences trust. It shows the state of confidence. IoT trust management issues were studied for different reasons, part or all of the trust assets mentioned above. In IoT confidence assessment, the proposal of Bao, Chen, and Guo (2013) was based on three characteristics of confidence: honesty; collaboration; and group interest. The reliable trust property assumes that not every node is truthful. Cooperative trust property assesses whether the trustee is a socially-cooperative trustee or a trust for the public interest, or the trustee belongs to or has similar skills as the same social or other groups (e.g., co-locations or partnerships).

12.4 Objectives of Trust Management in Different Layers of IoT (Yan et al., 2014)

1. Faith and decision-making: faith management is an efficient way of assessing IOT organizations' trust relationships and allowing them to make wise decisions to communicate and cooperate. Trust relationship evaluation is a big part of intelligent and autonomous trust management of all IoT organizations at all levels.

2. Trust in data perception: IoT can provide accurate data sensing and collection. In this connection, we pay attention to the properties of trustworthiness, such as sensitivity, precision, protection, reliability, durability, and collection performance, i.e., the empirical qualities of the client in an IoT layer. Trust in data collection equals trust in IoT data when the physical perception layer is collected and preprocessed.

3. Data security: user privacy should be protected flexibly in accordance with the policies and preferences of IoT users, including user data and personal information. This objective concerns the intent properties in general of the IoT system.

4. Data fusion and mining trust (DFMT): a significant quantity of IoT data should be processed and analyzed reliably with regard to the collection, confidentiality and accuracy of data processing. This is also a matter of confident social computing to satisfy users' needs because of their social experiences, social

discovery and study. DFMT relates to the IoT network layer data processor's objective characteristics (Sucharitha, Chakraborty, Srinivasa, & Reddy, 2021).

5. Data communication and trust: the data in the IoT system should be securely shared and exchanged. In communications and data transmission, unauthorised computer entities cannot access the private data of others. This goal focuses on the privacy and security of the IoT system, which includes light security, trust and information security solutions. Confidential IoT network routing and management are two important issues to resolve to achieve this goal (Liu & Wang, 2010).

6. Quality services of IoT: quality of IoT services should be guaranteed. 'Here only, only me, only now' services mean IoT services can, at the appropriate time and location, be tailored and delivered to the right individual reliably. This concerns mainly the management of trust in the IoT application layer but must be assisted in other layers. In addition to the objective characteristics of IoT services (trustee), IoT quality services goals are also applicable to objective and subjective quality attributes (trustee) and context (Chakraborty, Gupta, & Ghosh, 2014).

7. System Safety and Robustness: IoT trust management can resolve system attacks efficiently to allow users in the IoT system to acquire confidence. This goal involves all levels of the system focusing on the protection of the system and its efficiency (including trustworthiness and accessibility).

8. Generality: Generality is superior to generically and generally applied for the objective characteristics of the system, and trust management for various IoT systems and facilities.

9. Human-computer trust interaction: trust management ensures user acceptance, and guarantees safe human-computer interaction. This criterion focuses on the trustor's subjective characteristics on the application layer (i.e., the IoT users).

10. Identity Trust: IoT-device identities for secure IoT purposes are well managed. IoT requires flexible and efficient identity management. This refers to the need to support all layers. It deals with the objective characteristics of an IoT device (i.e., privacy of identity) and quality features of IoT individuals (i.e., user hope).

Esteem management supporting the above points is the key to achieving trustworthy IoT in general. Trust management should cover all IoT levels, not just boost security, confidentiality and confidence on-layer. At all levels of IoT, reputation management and effective co-operation between confidence management strategies is required.

Confidence in WSN is cooperative, non-transitive, temporal, dynamic and not monotonous, according to Karthik and Dhulipala (2011). The importance of WSN trust depends on the protective characteristics, mobility and reliability of the node.

Dynamic WSNs offer more coverage and are more complex than static sensor networks or static WSNs, and can lead to increased reliability. For this reason, Tian, Wang, Xiong, and Ma (2020) proposed a blockchain-based (BC-EKM) secure key management system in DWSNs. The blockchain networks employed were H sensor nodes, which replaced most BS work in the entire system to achieve secrecy. In the security evaluation, our device was robust against a pseudo-BS attack, node compromise and others. In overhead storage and energy usage, simulation testing showed positive results.

Khan et al. (2019) maintain that, because of their low performance in cooperation, teamwork and overhead memory needs, current WSN confidence management frameworks are not efficient. Their proposal is to construct a new systematic trust estimate method of large WSN clusters to establish cooperation, confidence and safety through detection of malicious (faulty) sense nodes with reduced resource consumption. The proposed scheme, Lightweight Time Synchronization (LTS), has unique features, including rigorous confidence measurement, an aggressive approach, and a simple cluster-heading confidence category. Data and communication trust play an important role in handling malicious nodes. LTS works on two levels: IT, and an IT-distributed approach to ensure accurate and efficient decisions with minimum overheads on sensor nodes. LTS is an autonomous network and is not influenced by any particular routing scheme. Without taking into account the weight and scale of wrongdoing, LTS is scarcely able to minimise on-off attacks and collusion attacks. Another weakness seems to be that LTS is only suitable for homogeneous WSN.

She, Liu, Tian, Sen Chen, Wang, and Liu (2019) conclude that malicious node detection in WSNs appears to be one-time centralised decision-making, to avoid the traceability of the original data. It is difficult to reproduce and validate the detection process or prevent error and false positive problems. A trust model, Balanced Transmission Mechanism (BTM), of the wireless sensor network for malicious node detection was proposed. This initially gives the entire confidence model structure. After that, it creates a blockchain data system that can be used to identify malicious nodes. Finally, security flaws are detected in 3D space by using a consensus mechanism and WSN quadrilateral-calculation location; voting results are recorded in the distributed blockchain. The model can efficiently classify malicious nodes in WSNs on the basis of the results obtained and ensure traceability of the detection process.

Kim et al. (2019) suggest a trust management blockchain model to improve the trust relationship between lightning nodes and exclude malicious nodes from the Wireless Sensor Network. This composite confidence evaluation involves conduct trust and data-based confidence. The behavioral trust of beacon nodes is measured with different metrics such as closeness, credibility, familiarity and frequency of interaction. The trust evaluation phase for the decentralised WSN blockchain output is performed successfully by this method. Data-based trust evaluation is the direct and indirect confidence between participating beacon nodes, which is a core aspect of data confidence. To build a decentralised blockchain trust management model, the composite Trust Wert of each beacon node is forwarded to BS. Esteem management systems are important methods to detect malpractice from malicious nodes. The trust and credibility assessment framework is the most powerful security measure to protect WSNs from internal attacks.

In order to test WSN nodes' trust and reputation, Zhao, Huang, and Xiong (2019) implemented an Exponential Faith and Reputation Evaluation Process. It is used to track behavior of nodes and characterise their confidence distribution exponentially. The node's trust is used to check for reliable nodes to send data and prevent malicious attacks within wireless sensor networks. This strategy would create an efficient trust protection mechanism without taking other states into account. The element of confidence is redefined and the number of successful interactions can be calculated. The element of confidence may lead to a rapid decline in direct trust that can undermine the effect of malicious nodes. Simulation results show that node trust can be reasonably and internally calculated; in particular, on-off attacks can be overcome. For the data aggregation method, the trust of the nodes was used.

Arifeen et al. (2019) introduce a confidentiality model for Underwater WSN privacy improvements. A test of sensor node viability with the Markov Decision Method was carried out using an Adaptive Neuro Fuzzy Inferring System. In each state of the Markov Decision Process(MDP), the sensor node defines the remittance node trust behavior by following the rules of FIS and selects a trustworthiness node, preventing any malicious or affected node from violating its privacy. Esteem findings suggest that a minimum value of confidence is 91 %. This approach combined a trust management with a privacy model to improve the privacy of the location and eradicate malicious nodes. Wu (2019) proposes a Beta and LQI trust Model (BLTM) for WSNs for beta and link quality indicators (LQI). In this model, touch trust, energy trust and data confidence have been considered for the first time in the calculation of direct trust. Thereafter, weight calculations were made for touch confidence, energy confidence and data trust. Finally, a LQI analysis framework was proposed to ensure continuity and reliability in trust values of ordinary nodes within a low-quality network. For future work, BLTM can be combined with cryptographing devices such as SERP, SIA, and Spins etc. It can also provide a complete way of assessing malicious node confidence values, coexisting with low-network connections, and selecting the correct weight and threshold value for highly integrated sensor networks.

12.5 Transport Systems Trust Management

Several confidence models have been suggested to determine vehicle reliability on the Vehicle Network Internet. Cheng, Liu, Yang, Member, Sun, and Approach (2019) suggest a three-layer distributed model to detect malicious vehicles in the vehicle network, allowing them to be objective and subjective. Faith evaluation of vehicles is done in a distributed manner. Trust establishment in internet of vehicle (IoV) is a core security issue that is continually constrained by scalability challenges. Javaid, Aman, and Sikdar (2020) propose a blockchain-based IoV protocol using the Consensus Algorithm Smart Contracts, Physical Unclones, Physical Unclonable Function (PUFs), certificates, and Dynamic proof of work (dPoW). In combination with contracts, the blockchain in this protocol ensures that trusted vehicles are registered and malicious objects are blocked. PUFs are used to allocate each vehicle from which a unique faith identity can be generated. Certificates issued by roadside units preserve vehicle privacy, whereas the dPoW consensus allows the protocol to be scaled in line with incoming vehicle traffic. Cinque, Esposito, Russo, and Tamburis (2020) suggeste effective confidence protection for IoT and IoV by leveraging the potential continuity and security assurances of blockchain. This mechanism has a good balance between robustness and efficiency. Johannes, Meuser, Steinmetz, and Buchholz (2019) has suggested a subjective logic-based process, that amends knowledge based on the reliability of data exchanged in the multi-agent framework (MAS). If several officers investigate the same incident, they may combine their details. The system should identify and separate wrongdoing so that high trust can be sustained. An attacker model was defined, which includes both defective and malicious agents. The process has been applied to Intelligent Transportation Systems (ITS). In simulation, the method was well scaled to the size of the MAS and was capable of detecting and isolating misbehaving agents efficiently.

12.6 Trust Management in P2P Networks

Currently, peer-to-peer file-sharing networks, used to exchange and distribute information, are attracting much attention. Kamvar, Schlosser, and Garcia-molina (2003) have established a pair-to-peer file sharing network algorithm that assigns each pair a unique global value based on their peer uploading history, to minimise the number of non-authentic file downloads. They suggested a distributed and secure approach to calculate global trust value based on power shift. The use of these confidence values for P2P simulation bias downloads reduces the volume of non-authentic files on the network in many threat circumstances. It enables peers to use these global confidence values to pick only pairs from which they have downloaded. Meng (2019) suggests TrueTrust, a confidence model for the introduction of P2P, with retributive justice that enables peers to provide genuine services and input. First, they define some principles and transaction processes of the TrueTrust model, and then divide the feedback into two groups, positive and negative, on the basis of which they extend the model. TrueTrust's reputation is closely linked to applicant reviews on previous peer services. Awasthi, Singh, and Member (2020) present a local trust-aggregation algorithm for the P2P network. Without normalization, this aggregation was a true representation of past behavior of peers in the network. Global trust calculation was done using an optimization approach, converging to a particular value. This new algorithm rates peers around the network as well as provide an absolute representation of their credibility. As a consequence, positive peers or malicious peers can be detected. By selecting the required values of the parameters, this algorithm will converge much faster with any global consensus.

12.7 Trust Management in Social IoT

The Social Internet of Things (SIoT) is a network standard in which IOT devices interact, and a social network is formed between devices and users. The strength of social ties between users is focused on users' services and IoT devices. In this context, it is critical that the protection of IoT devices and users is assessed so that malicious actors are unable to spread malicious content or disrupt the network.

A new privacy-preserving mechanism within the SIoT ecosystem has been proposed. Azad, Bag, Hao, and Shalaginov (2020) advocate for measuring the reliability of IoT devices and users at the same time. The proposed system acquires a homomorphism encryption system with decentralization characteristics, self-enforcement and privacy security. In malicious parties and conspirators, the proposed framework ensures proper computing, privacy and user security. Self-reinforcing computing is the unique feature of the proposed method, because without involving any trusted third party, the entire computing process is carried out. This allows authentication of score integrity in an autonomous manner. Talbi and Bouabdallah (2019) suggest an interest-based trust management system for SIoT. The scheme focuses on affinities between objects to determine trust and promote cooperation in an interconnected world. In the proposed trust scheme, when the two parties share history, the trustee is judged subjectively per the interest desires of the trustee. Otherwise, the trustee depends on suggestions made by a third party (i.e., the recommendation). In order to promote the position of the necessary facilities, these recommendations consider both confidence and similarity in terms of the interest expectations perceived by the trustee. (Xia, Xiao, Zhang, Hu, & Cheng, 2019) focus on exploring an effective context-aware confidence inference method to build trustworthy links autonomously between SIoT artifacts. The proposal divides confidence into two types: trust in familiarity; and faith in similarity, with the sociological and psychological principles of generating trust between human beings. Trust in familiarity is calculated on the basis of direct trust; faith in similarity is calculated on the basis of trust of external and internal similarities. In this schema, unique methods are implemented for estimating the various elements of trust and a kernel-based, nonlinear, grievous prediction model has been developed to predict the direct trust of a specific individual. This is a key element in the system as a whole. By considering the fuzziness and complexity of the idea of trust, this scheme has implemented a fuzzy logic method for the synthesis of these trust elements.

12.8 Trust Management Techniques in IoT

A variety of researchers have suggested different trust management strategies. Pourghebleh, Wakil, and Navimipour (2019) categorise selected trust management strategies into four groups. Recommended techniques are based on estimation, prediction-based, policy and repudiation (Table 12.1).

TABLE 12.1

Summary of IoT Trust Techniques

Technique	Contribution	Limitation
Monir, Abdelaziz, Abdelhamid, and Ei-Horbaty (2016) Recommended based	Considered situation awareness that informs IoTartifacts of the situation and uses it to deal with IoT military security threats.	Heterogeneity of the nodes was not considered.
Shabut, Dahal, Bista, and Awan (2015) RpR Model	Included reputations, recommendations and information.	Due to difficulty of assessing subjective properties numerically, knowledge TM was not considered to calculate the final trust scores.
Chen, Bao, and Guo (2016) Recommended based	Recommendations for the estimation of the importance of confidence, the background of encounters and direct observations have been used. Privacy has also been retained.	Low scalability
Ben Abderrahim, Elhdhili, and Saidane (2017) Prediction based	Kalman filtering method was applied to determine the value of trust and avoid potential attacks.	Unable to manage a heterogeneous environment
Jayasinghe, Otebolaku, Um, and Lee (2017) Prediction based	A collective filtering method was employed to predict trust values among users and specific information sources with some criteria.	Low scalability
Ben Abderrahim et al. (2017) Prediction based	Versatile trust model considers the dynamic aspect of IoT.	Reliability and heterogeneity issues of objects were not considered.

(Continued)

TABLE 12.1 (Continued)

Technique	Contribution	Limitation
Gu, Wang, and Sun (2014) Policy based	Layer trust management architecture encourages privacy protection and identity authentication.	Accessibility, reliability and integrity, were not assessed.
Al-Hamadi and Chen (2017) Policy based	Properly assesses data and source reliability for correct decision making by considering the quality, seriousness, and possibility of loss of health.	Social IoT characteristics for peer-to-peer trust assessment were not considered.
Chen (2018) Policy based	Fusing MRC and SC methods with TM enhanced efficiency and solved the security problems of the application layer.	The trade-off between trust and QoS degree was not considered.
Li, Song, and Zeng (2018) Policy based	Using contextual knowledge and anomalous IoT data, data reliability and IoT node attributes were analyzed for trust management.	New devices or new normal findings may be treated as a malicious node.
Chen et al. (2011) Reputation-based	Fuzzy credibility was considered for trust management.	Global trust update and changes in local trust were not considered.
Nitti, Girau, Atzori, Iera, and Morabito (2012) Reputation-based	The variety and variability of IoT objects and network structures were considered for trust assessment.	Availability and reliability of the system was ignored.
Xiao, Sidhu, and Christianson (2015) Reputation-based	To preserve trust and detect malicious nodes, credit and credibility were considered.	Data handling problems for reliability and efficiency were not considered.
Varghese, Chithralekha, and Kharkongor (2016) Reputation-based	Meta-information and self organization were considered for trust and energy-efficient clustering.	Credibility and consistency of the data are not recognised.
[51TWGA] Reputation-based	Enhanced Trust results when domains share device ID, signatures and public keys to establish a trust-ID between home device and service provider.	Intruder may inject false data which leads to repudiation attack.
Premarathne (2018) MAG-SIoT Relation based	This is suitable to establish relationship based on node affinity.	On increasing number of attributes, the affinity matrix expands, giving inappropriate results.
Son, Kang, Gwak, and Lee (2017) ATES	Provides Ideal trust value on first time interaction with device.	Due to fewer situational characteristics, prediction of accuracy is difficult.
Rehiman and Veni (2017) TMSMD	Protects data using public key cryptography	Consumes maximum power for using public key cryptography with integer factorization
Caminha, Perkusich, and Perkusich (2018) SMA	Extracts text and numerical data from IoT devices to enhance trust	Increases computational overhead
Wang, Wang, Zhang, and Cao (2017) ABAC	Provides secure authorization, augments scalability in explicit processes of decision making	Difficult to predict trust value when one node interacts with several at a time
Hellaoui, Bouabdallah, and Koudil (2016) TAS-IoT	Receiver legitimates messages from the sender with no further need for authentication.	If trust value is less than the threshold then the messages are manually authenticated
Guo, Chen, and Tsai (2017) IoT-HiTrust cloud based	Considers attacks robustly to achieve appropriate trust properties in large IoT systems.	Intruder detection was not considered

Recommended based techniques: Trustworthiness of these techniques is measured by the knowledge and advice of the trusted parties (Monir et al., 2016). These techniques discover incorrect nodes to make a more informed decision on the choice of the route, although the nodes in the path do not communicate directly (Shabut et al., 2015). Trust management strategies based on the recommendations will define credible recommendations and include information of the trust measurement process (Chahal, Kumar, & Batra, 2020).

Glowacka, Krygier, and Amanowicz (2015) suggest a trust-based approach to situation awareness that informs IoT artifacts of the situation and uses it to deal with IoT military security threats. Confidence is specified in this process as the degree of trust that can be given to the object on the basis of suggestions and direct comments. The heterogeneity of the nodes is not considered by the writers. Jayasinghe, Truong, and Lee (2016) suggest a new method of trust computing based on three trust criteria: reputation, recommendations and information. They isolate and define the definitions of suggestions and reputations and the significance of those criteria. These definitions are then used to assess the trust of the objects. Simulation investigates the essential properties of accuracy, convergence and deception resistance. The system provides a comprehensive way of quantifying trust in a few iterations for thousands of objects. The downside is that implementations are unique to low-reliability service providers. Chen et al. (2016) suggest an adaptive trust management approach for complex and social IoT systems where the key concept was to allocate trust values between IoT objects. Here, every system saves the value of trust between users and devices. Recommendations for estimation of the importance of confidence, the background of encounters and direct observations are used. Various trust standards are taken into account, such as honesty, quality of service (QoS) and social cooperation among IoT objects. Privacy is retained in this form, though scalability remains low.

Prediction-Based Techniques: In this type of technique, each object assesses the reliability of other artifacts, and can not only recognise malicious artifacts, but also improve the efficiency of network protection and robustness. These approaches are used when evaluating the reliability of a new object with limited knowledge of the properties of the object (Somu, Gauthama, Kalpana, Kirthivasan, & Shankar, 2018).

Ben Abderrahim et al. (2017) offer an IoT-clustering trust management architecture. Each cluster includes artifacts of the same interest that form an interest group headed by a trusted administrator. To determine the value of trust and avoid potential attacks, the Kalman filtering method is applied. Each node keeps the confidence values of the cluster nodes for a predefined period in the trust table for the scalability goals, and these tables are cleaned. The downside is that they cannot manage the heterogeneous IoT environment. Jayasinghe et al. (2017) propose a data-centric confidence assessment and prediction model that measures data confidence and entity trust. The proposed approach consists of many different stages, including data confidence estimation, data confidence aggregation, evaluation and prediction. After trust values are collected, a filtering method is employed to forecast trust values between users and specific information sources. Such filters use different criteria, such as completeness, uniqueness, promptness, validity, accuracy and consistency. This retains the integrity, accuracy and low scalability of the results obtained. The centerpiece of trust management is applied by Ben Abderrahim et al. (2017) to restore trust value and operational value to its trusted products. The versatile confidence model offers a structure that takes account of the dynamic aspects of IoT: context and the capacity of the object. Reliability issues and heterogeneity of objects were overlooked here.

Political techniques: Policy is a restriction on the system's operation that can be represented in either natural language or mathematical notation (Qwasmi & Liscano, 2020). In accordance with predefined rules, a policy-based technician responds automatically to network conditions (Siddiqui & Ahmed, 2012). Such techniques employ a collection of confidence appraisal rules, like the establishment of a minimum confidence level to grant access and track authorization (Monir et al., 2016).

Gu et al. (2014) introduce a layered trust management architecture in IoT settings. The sensor, centre and system layers are needed for this operation. Self-organisation and a variety of programmes monitor individual trust management levels. The service requestor performs the final decision-making process in compliance with policies and confidence information obtained. The access control policy encourages privacy protection and identity authentication, but certain fundamental confidence metrics, including accessibility, reliability and integrity, are not assessed. Al-Hamadi and Chen (2017) propose a trust-based IoT policy for a healthcare plan that takes into account quality, seriousness, and possibility of loss of health. The mechanism properly assesses data and source reliability for correct decision making and modifications, reducing its confidence value to alter the behavior of nodes. It maximises opportunities to determine trust value correctly and eliminates members who submit incorrect data. The mechanism considers, only sensor nodes that cannot serve IoT service purposes consist of the IoT environment. In addition to a range of approaches, Chen (2018) implements confidence management methodologies, such as integration selection (IS) and maximum ratio confidence (MRC) schemes. Before combining them with the control information, the indicated parameters are extracted and weighed by an estimated value. In order to calculate the confidence value, data generated in the MRC is transferred to the SC. The degree of Quality of Service (QoS) depends on the trust value created in the preceding stage. One drawback is the use of a small set of nodes to evaluate the proposed process, which may not guarantee inherently scalability. Li, Song, and Wang (2018) propose a policy-based, accurate and stable sensing technique. Data reliability and IoT node attributes are analyzed using contextual knowledge and anomalous IoT data. Predefined rules are used to test the reliability of various cases. Simulation results show that the suggested approach can evaluate the reliability of the information in IoT nodes accurately and efficiently. The downside of this approach is that new devices or new normal findings can be deemed as malicious by an obsolete law.

Reputation-based techniques: Reputation is closely related to trust and can be used in a number of areas, including social science and digital sciences. It also acts as a reason for faith or distrust on the basis of good or bad past experiences, as well as historical knowledge (Chen et al., 2019). This technology is a critical trend in the decision-making process for applications from service providers. The central concept of reputation mechanisms is to allow entities to rate each other, collect feedback for each entity, centrally or distributively aggregate the information provided and to produce a reputation score (Kapoukakis and Pappas, 2013).

Chen et al. (2011) suggest a system of confidence management focused on a fuzzy credibility. Direct and indirect observations are used for each object to establish credibility. Any object identifies malicious objects and makes decisions within the wireless network with the aid of reputations. QoS consists of metric sensor nodes, like energy consumption and packet distribution ratio; this only applies if unique IoT nodes are active. Nitti et al. (2012) propose a model for an assessment of trustworthiness, where each object measures the faith of other objects, on the basis that common objects share awareness and understanding of potential service providers. Accordingly, each object is capable of selecting the service provider with the highest trustworthiness. Certain forms of malicious activity are intended to provide the need for services and information to trusted objects only. Variety and variability of IoT objects and network structures are taken into account, but availability and reliability of the system are not . Xiao et al. (2015) suggest a trust model based on credibility and a guarantor. An item is demanded by the guarantor for service and then discovered by the guarantor for obtaining the requested services at the agreed commission cost. There are two standards for preserving trust and detecting malicious nodes: credit and credibility. In this scenario, reputation can only be changed by the server, and each object's reputation is determined by its ranking. Any credit as a commission to be charged when a correct service is given by an item, otherwise some other type of credit for the objects has to be charged as a tax. The mechanism is scalable, but data handling problems for reliability and efficiency are not considered. A power-efficient method for trust modeling in IoT has been proposed by Varghese et al. (2016). The system consists of two modules, trust and energy-efficiency clustering, focused on meta-information and self-organization. Each entity maintains a trust table containing the values of its adjacent trust entity. Based on packet transmission and information about each entity, trust value is modified to create an efficient cluster. Though it has the advantages of acceptability and scalability, this scheme does not recognise the credibility and consistency of data.

In addition, Ud Din, Guizani, Kim, Hassan, and Khan (2019) discuss a range of IoT trust models and their corresponding techniques, along with summaries, which follow the comprehensive analysis of the trust management system. E-LITHE (Enhanced Light weight scheme) used to enhance security and to provide Lightweight Datagram Transport Layer Security (DTLS) for IoT which uses a hidden key that can be shared between two devices to avoid DoS attacks, thereby improving security for restricted devices and implementing a power-reducing compression strategy. However, when numerous requests from various devices are made by an attacker, battery drainage becomes critical for handling regular computations. Trusted Gateway Architecture has been established for an IoT environment by (Kim & Keum, 2017) without the intervention of a heavyweight personal security system. This guards against malicious attacks like spoofing and DS. Trustworthy gateway architecture (TWGA) consists of four parts: i) the trajectory between trust domains; ii) transmission of data to intelligent houses; iii) public or private key identification by means of an identification path; and iv) trajectory selection by an engine with an ID packet. This device offers useful security through private or public keys; however, if the intruder injects fake data it cannot prevent repudiation assaults. The author's multiplicative attribute graph for social IoT (MAG-SIoT) (Premarathne, 2018) is based on four relationships: (i) the relationship between the properties of the object, (ii) the relationship between the co-locations of the things, (iii) the relationship between parental objects and (iv) the relationship between social objects. This model tests the value of trust on the basis of trust indicators, consisting of social interactions and the setting in which the relationship is communicated. Thus, this model is acceptable for creating a relationship based on the affinity of nodes. When the number of attributes increases, an acceptable relationship of trust cannot be created. For the management of IoT computers with personal and non-personal confidence values, the Adaptive Trust Estimation Scheme (ATES) (Son et al., 2017) is proposed. There are three ways to calculate the personal confidence of the system: i) a current device scenario, (ii) a device history and (iii) a model for M5 tree retrieval. The non-personal trust value of other objects' information is used. The model is capable of generating the optimal trust value for first-time contact with other devices. However, more situational characteristics are responsible for the accuracy of the performance. Confidence is maintained for each layer of the network in the Trust Management Model for Sensor enabled Mobile Devices (TMSMD) model proposed by Rehiman and Veni (2017). Data integrity and privacy are supported by a physical layer, while confidentiality of services is protected by the application layer. This model creates confidence by reducing overheads and uses a public key to secure data against power drain due to the use of public key cryptography based on integer factorization. The Distributed Trust Management Scheme (DTMS) is built on Mendoza et al.'s distributed method to provide different IoT-setting services. Through direct observation, the trust value of each node is calculated. When every node is delivered in a timely manner, a penalty is paid for service delivered by each node unless the new nodes are found. In a confidence-managing model, this model works well to attempt selective attacks but increases the likelihood of Bad-Mouthing attacks. Caminha et al. (2018) suggest an automated method for the recognition of IoT devices, the computation of their semantic attributes and the calculation of their trust value. Based on text attributes, the Smart Middleware Architecture (SMA) measures the trust and semanticization value of IoT objects. SMA is more effective in gathering text and numerical data across the network from IoT devices. The downside is increased computing overhead for the use of textual and numerical knowledge for resource discovery and trust score computation. The Attribute-based Access Control (ABAC) model is suggested by Wang et al. (2017) to protect data from malicious nodes. The framework consists of three modules: trust assessment;, access decision and authentication. Since the level of trust varies with the actions of the node, safe authorization is given. Ud Din et al. (2019) conclude that the Adaption Trust Based Protocol (ATBP) has been established to allow for security measures between nodes of the social network. ATBP uses a trust policy which must obey all the nodes in the network. It considers integrity to be a confidence property for the management of Bad-Mouthing attacks. The model is useful to determine the best route to prevent congestion and crashes and to ensure safe and smooth driving. A variety of modules for the trust access object, trust service provider, decision and prediction agent, trust agent, data repository, trust computing and trust value calculations are included within the data-centric trust evaluation and prediction framework (DCTEPF). Trust value can be calculated by the trust-centric data evaluation and prediction framework. This would be helpful if inappropriate data were filtered. Contextual knowledge is not beneficial for confidence predictions. The Monitor, Analyze, Plan, Execute, Knowledge (MAPE-K) model helps to increase the complex level of trust management through a self-adaptation process. When data attributes cross the mark, the problem arises. The model Trust-based Development Framework for Distributed System (TDFDS) consists of four modules which identify various consumer, corporate and technological trust variables. The variable environment involves technological and monetary, cultural and religious factors. The consumer comprises the intelligence of human beings and their physical abilities. These variables for market criterion includes characteristics that affect trust. The Context-based Social Trust Model for the Internet of Things (CBSTM-IoT) has been developed for the cooperation of nodes and to restrict the interaction of suspicious nodes. In this model, confidence is higher if the value of the relationship is high. The Timely Trust Structure identifies the demand for IoT through GVT (Global Virtual Team) and demonstrates how rapid trust creation in GVTs can be influenced by different cultures. GVTs have common goals on which to operate across regional borders, and rely on technologies such as computers for communication. The Distributed Trust and Reputation Model (DTRM) is based on the distributed environment to render IoT devices capable of processing. In addition, the model proposes different protection levels appropriate in the IoT environment for sensitive devices. It could protect against Bad-Mouthing, Good-Mouthing and voting machines, but couldn't handle certain attacks, like Distributed denial of service (DDOS), Man-In-The-Middle (MIM)and wormhole. Application-Driven Network Trust (ANT) Zones split the network in confidential areas to look for new entry nodes and reconfigure current confidential areas. The proposed model puts all devices on the medium access control (MAC) layer in a separate network to mitigate malicious node results. Reconfiguration of trust zones helps limit remote communications and secure the network from a variety of attacks. The Trust-based Adaptive Security in IoT (TAS-IoT) model proposed by Hellaoui et al. (2016) divides nodes into two groups: legitimate and illegitimate. The valid node is connected to the message authenticator. Each node, based on observations, experience and suggestions are associated with the value of confidence. A function is used to decide if a message needs authentication after measurement of trust value. The Centralized Trust Management Mechanism for the Internet of Things (CTM-IoT) efficiently exchanges information among IoT nodes. In each cluster, a trustee, or master node, splits the IOT network into different clusters. The model also consists of a supernode that stores trusted data of all master and cluster nodes in the central repository. The supernode also monitors traffic and confidence management on all IoT devices. On the basis of direct and indirect tests, trust management system based on communities of interest for the Social Internet of Things (TMCo-I-SIOT) are intended to combine various confidence attributes. The proposed architecture uses the clustering principle and divides nodes into groups on an interest basis, where the network consists of a SIoT server and grouped nodes and a security management trust administrator. The Context based Trust Management System-Social Internet of Things (CTMS-SIOT) has been developed to concentrate on complex trust values and a relative context for different tasks. This model is based on the complexity of computation, where the existence of a node decreases by a decentralised architecture's cache of knowledge. The two CTMS-SIOT modules are responsible for qualitative trust and integrity. The user confidence request activates the discovery process; the user sends a confidence request to the server due to lack of knowledge in the local trust table. In CTMS-SIOT, social similarities between the requester and the chosen node are quantified. With efficient installation, this model provides a dynamic atmosphere but decreases device reliability. Trust Management Framework- Vehicular Social Networks (TMF-VSN) was proposed for VSN, which consists of three layers of confidence: Global Trust Manager (GTM); Domain Trust Manager(DTM); and Vehicular Trust Monitor (VTM). The GTM is at the top level and houses vehicle profile authentications. The DTM preserves the history, domain and relationship profiles of each individual vehicle, while the VTM preserves vehicle details. The proposed model consists of four modules: friend trust; neighbor trust; global trust; and background trust. System efficiency of the network may be enhanced by the packet delivery ratio, but the validity of the experiments can be influenced by node density. IoT-HiTrust is suggested by Guo et al. (2017) where the trustworthiness of all IoT devices has been computed in the cloud. The framework was built with three layers: a cloud layer, and a computer layer. At the level of cloud services, each IoT system has a unique identity, used to manage user data. Requests and responses from devices are exchanged only within their cloud regions, along with their stored information. If the Internet link is terminated, the cloudlet will be responsible for responding to user requests in disconnection mode.

12.9 Issues and Challenges in Trust

A system identified by Thirukkumaran and Kannan (2019) as a confidence and reputation system faces a broad range of challenges such as heterogeneity, scalability, infrastructure, identity, credibility and networks.

1. Heterogeneity must be recognised by confident and trustworthy systems because the future internet will be heterogeneous (web, digital, cyber-physical, etc.).
2. The second is scalability; trust and reputation need to be improved by increasing the number of devices in order to stay fully functional.
3. The third issue of infrastructure must be understood by the Trust and Reputation Program because information needs to be collected. Companies need to be able to connect with them and find them in the network.
4. IoT identity management offers both protection challenges and opportunities. The fundamental structure and identity of things is not identical and this is the key aspect of this challenge.
5. Preventing unauthorised alteration of hardware and software is guaranteed by the principle of credibility. Unregistered or unauthorised staff shall not change the data and shall be consistent internally and externally.
6. The last issue is the links between various items and network capabilities. This means that variations in bandwidth, availability and latency must be taken into account, particularly if certain aspects of the interactions are critical in time.

Some confidence Issues were also coined by Pourghebleh et al. (2019):

1. Privacy-Preserving Method: Most researchers have overlooked the privacy question in the trust management techniques discussed. In reality, the privacy protection approach is a vital solution for enhancing public safety and encouraging new IoT systems. It can be considered an important subject for further study, including the identification of data protection models in different IoT scenarios for the solution of the scalable and complex environments.
2. Heterogeneous IoT networks: The transmission and machine confidence of heterogeneous networks are difficult problems. Confidence management strategies used in different networks should be based on the same objectives and measures. Moreover, the use of Internet software for measuring trust, which can reduce the burden of the method, is another major issue for future study.
3. Aggregation of confidence: A confidence category is a remarkable subject for future studies. It is recommended to build a trust management scheme that aggregates or combines heterogeneous trust approaches using different types of IoT objects across various types of networks.
4. Trust security of mobile nodes: Mobile objects are included in IoT and can be modified at any time. Another significant issue that needs further research is the provision of a rapid method of confidence management, authentication and privacy security. It is important that mobility trust management strategies be stable and scalable to support accurate location updates and to address possible weaknesses in security and privacy.

5. With regard to the environment and various application scenarios, IoT networks have a highly heterogeneous environment and different application scenarios must be discussed during the decision-making phases, with regard to the facilities and applications. It cannot be concluded that the trust management process is sufficient for any one application type.

6. Energy Efficiency Aspects: Energy efficiency is a major research concern, in particular in the trust calculation field, when IoT nodes have limited computer energy. This topic has not been explored by current trust management studies. Therefore a lightweight Trust Management Framework is the critical work to be performed next.

7. Blockchain Vulnerabilities: Blockchain implementations are often insecure, although they preserve IoT ecosystems. In order to compromise blockchain accounts, private keys can be used. Further study is needed to find effective methods to avoid attacks and ensure the privacy of transactions.

12.10 Trust Applications

Key uses of this trust model include regulation of access for users or devices, protected routing, selection of cluster headers and malicious node detection (Thirukkumaran & Kannan, 2019). D. Airehrour et.al suggested the SecTrust, a lightweight, confidential routing system. Based on previous successful interactions, the trust value of the IoT sensor nodes is calculated. Assailants are removed from the network according to the efficiency of the routers. Khan et al. (2017) suggest a stable, low-power routing scheme for network routing protocol failure. The trust of each node of the network is measured by the parameters of faith, incredulity and incertitude. Ali, Abdulsalam, and AlGhemlas (2018) propose an IoT-enabled sensor network cluster head selection scheme. The head node is selected only for trusted nodes. The value of trust is calculated by successful interplay. The factor of forgetfulness is used to assign old entries a different weight. Grandison and Sloman (2000) also deal with an application focused on trust. Some trust-based applications are shown in Figure 12.1.

Medical information systems: Medicine has a wide variety of fields, each with its own confidence issues and sound ethics base. The difficulty in computerizing some aspects of medicine is in translating ethics into digital systems. The following subject addresses only the provision of medical records and health information systems. Al-Hamadi and Chen (2017) introduce a new confidence-based decision-making protocol that utilises information sharing between health IoT devices in order to build an environmentally-sustainable mutual knowledge base at a particular location and time. This knowledge requires an IoT device to operate on behalf of the user who decides to visit this site or environment for health reasons. Abou-Nassar et al. (2020) propose a Decentralized IoHT(Healthcare IoT) Health-based Blockchain Platform called DIT (Decentralized Interoperable Trust) for IoT zones that would ensure budget authentication via smart contracts and a reduction of the Semantic Lack of Confidence. It also increases TF estimation through network nodes and edges by the (Indirect Confidence Inference System) ITIS.

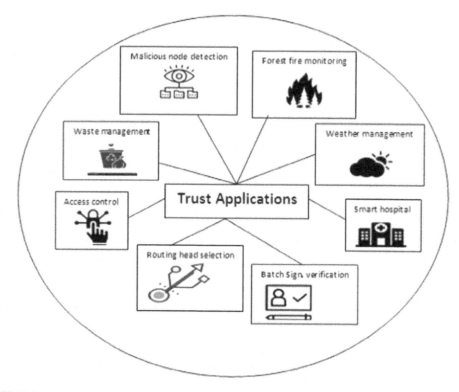

FIGURE 12.1
Trust applications.

Information Retrieving Systems: Referee and Policy Maker, combined with PICS, are the only systems which specifically respond to confidence in information recovery systems.

Mobile Code: Mobile agents relay code and data for the execution of tasks on behalf of the client. It could be used for network management, but even better, local agents can provide information on behalf of mobile users with restricted data-computing capacity and wireless communication, such as message filtration, email-to-voice translation and filtering.

Event Monitoring and Reporting: Vikash and Varma (2019) suggest an architecture based on IoT, based on TBED (trust-based event driven). This design offers a forest-monitoring solution where multiple sensor data are fused to provide true and accurate information in accordance with the trust value of the sensor nodes. Consumers with an estimated affected region record production dynamically.

A trust model for IoT has been developed by KITUR and A Pais (Kittur & Pais (2020), which helps the portal node to identify trustworthy batch verification sensor nodes. Batch check of digital signatures is an application of trust management in which a number of the Gateway node signatures are received and the signatures are checked by batch. This confidence reduces the risk of selecting unstable check nodes and the load of the device on the gateway node significantly. The results for batch verification of digital signatures when implementing the defined confidentiality model are presented.

12.11 Conclusion

In recent years, IoT has emerged as a disruptive technology in a broad range of applications like gaining unprecedented interest and trustworthiness to a large range of stakeholders from different realms. However, it is important to consider the issues and challenges that go hand in hand with such unreached domains and applications, especially those relating to security, privacy, and data protection in the sense of trust. This chapter introduces the concept of trust and trustworthiness in IoT and emphasises the importance of security, privacy and protection of nodes as the key pillars of IoT trustworthiness. In each chapter, sections provide an overview of the recent status of IoT science. The paper then presents the challenges and objectives of IoT trust's study and standardization activities to assist technical committees to set out their change roadmaps for future direction. At the end of the day, some recent requests for confidence in IoT have also been addressed here for potential reach in the future.

The challenges and objectives of IoT trust study in this chapter can be helpful for researchers for better understanding of trust in IoT and its various applications. Study of different techniques of trust management and assessment can open the possibility of various applications and implementation in various fields of IoT environment.

References

Abou-Nassar, E. M., Iliyasu, A. M., El-Kafrawy, P. M., Song, O. -Y., Bashir, A. K., & El-Latif, A. A. A. (2020). DITrust chain: Towards blockchain-based trust models for sustainable healthcare IoT systems. *IEEE Access, 8,* 111223–111238.

Al-Hamadi, H. & Chen, I. R. (2017). Trust-based decision making for health IoT Systems. *IEEE Internet Things Journal, 4* (5), 1408–1419.

Ali, B. A., Abdulsalam, H. M., & AlGhemlas, A. (2018). Trust based scheme for IoT enabled wireless sensor networks. *Wireless Personal Communication, 99* (2), 1061–1080.

Arifeen, M., Islam, A. A., Rahman, A., Taher, K. A., Islam, M., & Kaiser, M. S. (2019). ANFIS based trust management model to enhance location privacy in underwater wireless sensor networks. *International Conference on Electrical, Computer and Communication Engineering* (pp. 1–6).

Awasthi, S. K., Singh, Y. N., & Member, S. (2020). AbsoluteTrust: Algorithm for aggregation of trust in peer-to-peer networks. *IEEE Transactions on Dependable and Secure Computing, PP*(c), 1.

Azad, M. A., Bag, S., Hao, F., & Shalaginov, A. (2020). Decentralized self-enforcing trust management system for social Internet of Things. *IEEE Internet Things Journal, 7*(4), 2690–2703.

Azzedin, F. (2014). Taxonomy of reputation assessment in peer-to-peer systems and analysis of their data retrieval. *Knowledge Engineering Review, 29*(4), 463–483.

Azzedin, F.& Ghaleb, M.(2019). Internet-of-things and information fusion: Trust perspective survey, *Sensors (Switzerland), 19*(8), 19–29.

Bao, F., Chen, I. R., & Guo, J. (2013). Scalable, adaptive and survivable trust management for community of interest based internet of things systems. In *Proceedings of the 2013 11th International Symposium Autonomous Decentralized System ISADS 2013.*

Ben Abderrahim, O., Elhdhili, M. H., & Saidane, L. (2017). TMCoI-SIOT: A trust management system based on communities of interest for the social Internet of Things. In *2017 13th International Wireless Communication and Mobile Computing Conference IWCMC 2017,* pp. 747–752.

Ben Abderrahim, O., Elhedhili, M. H., & Saidane, L. (2017). CTMS-SIOT: A context-based trust management system for the social Internet of Things, *2017 13th International Wireless Communication on Mobile Computing Conference IWCMC 2017*, pp. 1903–1908.

Caminha, J., Perkusich, A., & Perkusich, M. (2018). A smart middleware to perform semantic discovery and trust evaluation for the Internet of Things. *CCNC 2018 - 2018 15th IEEE Annual Consumer Communications and Networking Conference*, 2018(January), 1–2.

Chahal, R. K., Kumar, N., & Batra, S. (2020). Trust management in social Internet of Things: A taxonomy, open issues, and challenges. *Computer and Communications*, 150(October 2019), 13–46.

Chakraborty, C. & Rodrigues, J. J. P. C. (2020). A comprehensive review on device-to-device communication paradigm: Trends, challenges and applications, *Springer: International Journal of Wireless Personal Communications*, 114, 185–207. doi: 10.1007/s11277-020-07358-3

Chakraborty, C., Gupta, B., & Ghosh, S. K. (2014). Mobile metadata assisted community database of chronic wound. *Elsevier: International Journal of Wound Medicine*, 6, 34–42. ISSN: 2213-9095, 10.1 016/j.wndm.2014.09.002

Chen, J. I. Z. (2018). Embedding the MRC and SC schemes into trust management algorithm applied to IoT security protection. *Wireless Personal Communication*, 99(1), 461–477.

Chen, I. R., Bao, F., & Guo, J. (2016). Trust-based service management for social Internet of Things systems. *IEEE Transactions on Dependable Secure Computing*, 13 (6), 684–696.

Chen, J., Tian, Z., Cui, X., Yin, L., & Wang, X. (2019). Trust architecture and reputation evaluation for internet of things. *Journal of Ambient Intelligence and Humanized Computing*, 10(8), 3099–3107.

Chen, D., Chang, G., Sun, D., Li, J., Jia, J., & Wang, X. (2011). TRM-IoT: A trust management model based on fuzzy reputation for internet of things. *Computer Science and Information Systems*, 8(4), 1207–1228.

Cheng, T., Liu, G., Yang, Q., Member, S., Sun, J., & Approach, A. P. (2019). Trust assessment in vehicular social network based on three-valued subjective logic. *IEEE Transactions on Multimedia*, 21(3), 652–663.

Cinque, M., Esposito, C., Russo, S., & Tamburis, O. (2020). Blockchain-empowered decentralised trust management for the Internet of Vehicles security. *Computer & Electrical Engineering*, 86, 106722.

Glowacka, J., Krygier, J., & Amanowicz, M. (2015). A trust-based situation awareness system for military applications of the internet of things. In *Proceedings of the IEEE World Forum Internet Things, WF-IoT 2015*, pp. 490–495.

Grandison, T.& Sloman, M. (2000). Trust in internet applications. *Communications*, 4, 2–16.

Gu, L., Wang, J., & Sun, B. (2014). Trust management mechanism for Internet of Things. *China Communication*, 11 (2), 148–156.

Guo, J., Chen, I. R., & Tsai, J. J. P. (2017). A mobile cloud hierarchical trust management protocol for IoT systems. In *Proceedings of the 5th IEEE International Conference on Mobile Cloud Computing Services and Engineering MobileCloud 2017*, pp. 128–130.

Hellaoui, H., Bouabdallah, A., & Koudil, M. (2016). TAS-IoT: Trust-based adaptive security in the IoT. In *Proceedings of the Conference Local Computer Networks, LCN*, pp. 599–602.

Javaid, U., Aman, M. N., & Sikdar, B. (2020). A scalable protocol for driving trust management in Internet of vehicles with blockchain, *IEEE Internet Things Journal*, 4662(c), 1–1.

Jayasinghe, U., Truong, N. B., & Lee, G. M. (2016). RpR: A trust computation model for social Internet of Things."

Jayasinghe, U., Otebolaku, A., Um, T. W., & Lee, G. M. (2017). Data centric trust evaluation and prediction framework for IOT. *Proceedings of the 2017 ITU Kaleidoscope Academy Conference Challenges a Data-Driven Society ITU K 2017*, 2018 (January), 1–7.

Johannes, M., Meuser, T., Steinmetz, R., & Buchholz, M. (2019). A trust management and misbehaviour detection mechanism for multi-agent systems and its application to intelligent transportation. *IEEE 15th International Conference on Control and Automation (ICCA)*, Edinburgh, Scotland, July 16–19, 2019.

Kamvar, S. D., Schlosser, M. T., & Garcia-molina, H. (2003). The eigen trust algorithm for reputation management in P2P networks. 640–651.

Kapoukakis, A. & Pappas, C. (2013). Design and assessment of a reputation-based trust framework in wireless testbeds utilizing user experience, *7960*, 1–12.

Karthik, N.& Dhulipala, V. R. S. (2011). Trust calculation in wireless sensor networks. *ICECT 2011 - 2011 3rd International Conference on Electronic Computer Technology*, 4(April), 376–380.

Khan, Z. A., Herrmann, P., Ullrich, J., & Voyiatzis, A. G. (2017). A trust-based resilient routing mechanism for the internet of things. *ACM International Conference Proceeding Series*, Part F1305, 1–6.

Khan, T., Singh, K. Son, L. H., Abdel-Basset, M., Long, H. V., Singh, S. P., & Manjul, M. (2019). A novel and comprehensive trust estimation clustering based approach for large scale wireless sensor networks, *IEEE Access*, 7, 58221–58240.

Kim, E. & Keum, C. (2017). Trustworthy gateway system providing IoT trust domain of smart home. In *International Conference on Ubiquitous Future Networks, ICUFN*, pp. 551–553.

Kim, T., Goyat, R., Rai, M. K., Kumar, G., Buchanan, W. J., Saha, R., & Thomas, R. (2019). A novel trust evaluation process for secure localization using a decentralized blockchain in wireless sensor networks. *IEEE Access*, 7, 184133–184144.

Kittur, A. S.& Pais, A. R. (2020). A trust model based batch verification of digital signatures in IoT. *Journal of Ambient Intelligence and Humanized Computing*, 11(1), 313–327.

Li, W., Song, H., & Zeng, F. (2018). Policy-based secure and trustworthy sensing for Internet of Things in smart cities. *IEEE Internet Things Journal*, 5 (2), 716–723.

Liu, Y. & Wang, K. (2010). Trust control in heterogeneous networks for Internet of Things. *ICCASM 2010 - Proceedings of the 2010 International Conference on Computer Application and System Modelling*, 1(Iccasm), 632–636.

Meng, X. (2019). TrueTrust: a feedback-based trust management model without filtering feedbacks in P2P networks. *Peer-to-Peer Networking and Applications*, (c), 9–11.

Mohammadi, V., Rahmani, A. M., Darwesh, A. M., & Sahafi, A. (2019). Trust-based recommendation systems in Internet of Things: a systematic literature review. *Human-centric Computing and Information Sciences*, 9(1).

Monir, M. B., Abdelaziz, M. H., Abdelhamid, A. A., & Ei-Horbaty, E. S. M. (2016). Trust management in cloud computing: A survey. In *2015 IEEE 7th International Conference on Intelligent Computing and Information Systems ICICIS 2015*, pp. 231–242.

Nitti, M., Girau, R., Atzori, L., Iera, A., & Morabito, G. (2012). A subjective model for trustworthiness evaluation in the social Internet of Things. In *IEEE Symposium on Personal, Indoor and Mobile Radio Communications PIMRC*, pp. 18–23.

Pourghebleh, B., Wakil, K., & Navimipour, N. J. (2019). A comprehensive study on the trust management techniques in the Internet of Things. *IEEE Internet Things Journal*, 6(6), 9326–9337.

Premarathne, U. S. (2018). MAG-SIoT: A multiplicative attributes graph model based trust computation method for social Internet of Thing. *Proceedings of the 2017 IEEE International Conference on Industrial and Information System ICIIS 2017*, 2018 (January), 1–6.

Qwasmi, N. & Liscano, R. (2020). Distributed policy based management for wireless sensor networks to support the Internet of Things environment Nidal Qwasmi University of Ontario Institute of Technology 1 Introduction 2 Motivations 3 TinyPolicy: A distributed policy framework, 24, 335–338.

Rehiman, K. A. R. & Veni, S. (2017, December). A trust management model for sensor enabled mobile devices in IoT. In *Proceedings of the International Conference on IoT Social, Mobile, Analytics and Cloud, I-SMAC 2017*, pp. 807–810.

Shabut, A. M., Dahal, K. P., Bista, S. K., & Awan, I. U. (2015). Recommendation based trust model with an effective defence scheme for MANETs. *IEEE Transactions on Mobile Computing*, 14(10), 2101–2115.

She, W., Liu, Q., Tian, Z., Sen Chen, J., Wang, B., & Liu, W. (2019). Blockchain trust model for malicious node detection in wireless sensor networks. *IEEE Access*, 7, 38947–38956.

Siddiqui, M. S. & Ahmed, S. H. (2012). Policy-based network management in a machine-to-machine (M2M) network. In *2012 15th International Multitopic Conference INMIC 2012*, pp. 387–393.

Somu, N., Gauthama, G. R., Kalpana, V., Kirthivasan, K., & Shankar, S. S. (2018). An improved robust heteroscedastic probabilistic neural network based trust prediction approach for cloud service selection. *Neural Networks, 108*, 339–354.

Son, H., Kang, N., Gwak, B., & Lee, D.(2017). An adaptive IoT trust estimation scheme combining interaction history and stereotypical reputation, In *2017 14th IEEE* Consumer *Communications & Networking Conference CCNC 2017*, pp. 349–352.

Sucharitha, M., Chakraborty, C., Srinivasa, S. R., Reddy, V. S. K. (2021). Early detection of dementia disease using data mining techniques. *Springer: Internet of Things for Healthcare Technologies. Studies in Big Data, 73*, 177–194, https://doi.org/10.1007/978-981-15-4112-4_9

Talbi, S. & Bouabdallah, A. (2019). Interest - based trust management scheme for social internet of things, (0123456789).

Thirukkumaran, R., & Kannan, P. M. (2019). Survey: Security and trust management in Internet of Things. In *Proceedings of the - 2018 IEEE Global Conference on Wireless Computing and Networking, GCWCN 2018*, pp. 131–134.

Tian, Y., Wang, Z., Xiong, J., & Ma, J. (2020). A blockchain-based secure key management scheme with trustworthiness in DWSNs. *IEEE Transactions on Industrial Informatics, 16*(9), 6193–6202.

Ud Din, I., Guizani, M., Kim, B. S., Hassan, S., & Khan, M. K. (2019). Trust management techniques for the internet of things: A survey. *IEEE Access, 7*, 29763–29787.

Varghese, R., Chithralekha, T., & Kharkongor, C. (2016, March). Self-organized cluster based energy efficient meta trust model for internet of things. In *Proceedings of the 2nd IEEE International Conference on Engineering and Technology ICETECH 2016*, pp. 382–389.

Vikash & Varma, S. (2019). Trust-based forest monitoring system using Internet of Things. *International Journal of Communication Systems* (June), 1–17.

Wang, J., Wang, H., Zhang, H., & Cao, N. (2017). Trust and attribute-based dynamic access control model for Internet of Thing. *Proceedings of the International Conference on Cyber-Enabled Distributed Computing and Knowledge Discovery CyberC 2017, 2018*(January), 342–345.

Wu, X. (2019). BLTM: Beta and LQI based trust model for wireless sensor networks. *IEEE Access, 7*, 43679–43690.

Xia, H., Xiao, F., Zhang, S. S., Hu, C. Q., & Cheng, X. Z. (2019). Trustworthiness inference framework in the social Internet of Things: A context-aware approach, *Proceedings of the IEEE INFOCOM, 2019* April (2), 838–846.

Xiao, H., Sidhu, N., & Christianson, B. (2015). Guarantor and reputation based trust model for Social Internet of Things. In *IWCMC 2015 - 11th International Wireless Communication Mobile Computing Conference*, pp. 600–605.

Yan, Z., Zhang, P., & Vasilakos, A. V. (2014). A survey on trust management for Internet of Things. *Journal of Network and Computer Applications, 42*(June), 120–134.

Zhao, J., Huang, J., & Xiong, N. (2019). An effective exponential-based trust and reputation evaluation system in wireless sensor networks. *IEEE Access, 7*, 33859–33869.

Index

A

128 Bit AES Design, 19
Access control based attacks, 22, 23
access control, 2, 9
Accountability, 7
accuracy, 15
actor, 7, 11, 12
Actuator, 8, 11, 19
Address Resolution Protocol (ARP), 24
adversary, 1–2, 4, 5, 7–5, 8, 18, 19–3, 22, 33, 37, 38, 39
AES, 26
AES one round encryption, 12
anonymization, 28
Application layer, 14
Arduino, 30, 31
artificial intelligence, 10–1, 36–2, 39–1, 40–1
assailant hub, 12
assets, 9, 10, 15, 19, 22
attack trees, 4, 11, 17
attacks, 1, 3, 5, 10, 11, 12, 13, 16, 17, 18, 19, 20, 21, 22, 24, 25, 26, 27, 30
Attacks: edge level, 3, 16, 17, 29
attacks: in general, 1–2, 2–4, 3–5, 4, 5–6, 10–4, 11–1, 17–1, 18–1, 22–4, 23–2, 32–3, 39–3, 41–1
    hardware, 5–4, 36–5, 37–5, 39–3, 43–1
    invasive, 5–5, 6–10
    security, 4–16, 7–6, 8–2, 40–1
    invasive, 5–5, 6–10
    non-invasive, 7–10, 8–5
    physical, 4–8
    semi invasive, 7–8
    sleep deprivation, 5–2
Authentication, 3, 6, 10, 18, 22, 28
Authentication, 1–4, 6, 9, 14–16, 22, 24–26, 28–36
Authentication, 7, 11, 16, 23
authentication, 7, 9, 15, 21, 22, 26
Authorization, 7
authorization, 8, 9
Autodesk circuits simulator, 30
Availability, 4

B

Battery draining, 2, 19, 41
BDR, 9
Biometric, 1, 2, 15, 24–29, 32–36

Biometrics, 1, 11, 21
biquad filter, 2–2, 3–22, 11–2, 12, 12–1, 14–2, 14–1, 15–2, 15–1, 16–1, 24–5, 25–4, 26–1–1, 26–3, 27–3, 28–3, 28–2, 30–4, 31–5, 31–1, 32–6, 33–2, 34–1, 35, 36, 42
Bit Correlation Effect, P.G. 20
BLE, 27
Blockchain, 24, 30
blockchain, 16
Bluetooth Low Energy, 2
browser, 15

C

Cameras, 12, 15
Cancelable biometric, 34
capabilityswitch, 3
Cellular Network, 24
Challenges in Security IoT nodes, 5
Channel Reciprocity, 8
channels, 10, 13, 30
CIA, 7, 27
ciphertext, 13
Circuit, 10–12, 18, 23–26, 38
cloud, 20, 21, 22, 24
cloud, 3–3, 40
cloud, 5, 8, 10, 16
CoAP, 29, 30
commands, 24
communication, 2, 3, 7, 10, 12, 13, 14, 22, 26, 27
Compiler driven, 1, 5–9, 29, 32, 37–38
compromise, 14
computing nodes, 2, 3
Conclusion, 19
Confidentiality, 3, 27–29
Confidentiality, 4, 7, 25
confidentiality, 1, 12, 21, 27
Configuration, 26
Control Data Flow Graph, 5–10, 20–21
controls, 11, 15, 16
Coprocessor, 1–2, 4, 10, 13, 19
CORAS, 1, 11, 17
Countermeasures, 1, 15
credential, 25
credentials, 13, 21, 27
credit scoring, 15
cryptography keys, 13
Cryptology, 28